T0321114

Actions of Groups

Using the unifying notion of group actions, this second course in modern algebra introduces the deeper algebraic tools needed to get into topics only hinted at in a first course, like the successful classification of finite simple groups and how groups play a role in the solutions of polynomial equations. Because groups may act as permutations of a set, as linear transformations on a vector space, or as automorphisms of a field, the deeper structure of a group may emerge from these viewpoints, two different groups can be distinguished, or a polynomial equation can be shown to be solvable by radicals. By developing the properties of these group actions, readers encounter essential algebra topics like the Sylow Theorems and their applications, Galois theory, and representation theory. Warmup chapters that review and build on the first course and active learning modules help students transition to a deeper understanding of the ideas.

JOHN MCCLEARY is Professor of Mathematics on the Elizabeth Stillman Williams Chair at Vassar College. He received his PhD in mathematics from Temple University, where he completed a thesis on algebraic topology under the direction of James Stasheff. The author of several books, including *A User's Guide to Spectral Sequences and Geometry from a Differentiable Viewpoint*, McCleary has published extensively on topology, the history of mathematics, and other topics. His current research focuses on algebraic topology – specifically, where algebra reveals what topology may conceal.

This book contains the perfect mix of all the ideas central to a second course in algebra. The focus on group actions will serve any student well as they continue their study of mathematics. The text goes at a good pace and the "Get to know" exercises sprinkled throughout the chapters provide excellent opportunities for students to pause, work with ideas presented, and master the content. This textbook is definitely unique, and I have not seen its equal.

JOSEPH KIRTLAND, Marist College

When I was young, I really enjoyed my first class in group theory, spending delightful (and at times frustrating) hours at a blackboard in an empty classroom trying to solve the homework problems. But I really didn't see why groups were all that important in the rest of mathematics. Yes, I got that the two main examples of groups (permutation groups and invertible matrices) were natural to study, and I heard my professor tell us that groups permeate modern mathematics. But I didn't see it (though I was willing to believe that my professor was probably right.) This is one of the books I wish I had had, as it lets people start down the road to understanding that groups are indeed one of the unifying concepts of modern mathematics.

THOMAS GARRITY, Williams College

This book provides a wealth of topics building on what is likely to be available to students after a first course on basic abstract algebra. A major theme is that of "groups in action," and most of the topics involve groups in a leading role (e.g., Galois theory and representation theory). The book would certainly be a useful text for a course, and also provide independent reading for a motivated student. It should also provide a good preparation or revision before deeper study at graduate student level.

ANDREW J. BAKER, University of Glasgow

Actions of Groups

A Second Course in Algebra

JOHN McCLEARY

Vassar College, New York

 CAMBRIDGE
UNIVERSITY PRESS

Shaftesbury Road, Cambridge CB2 8EA, United Kingdom

One Liberty Plaza, 20th Floor, New York, NY 10006, USA

477 Williamstown Road, Port Melbourne, VIC 3207, Australia

314–321, 3rd Floor, Plot 3, Splendor Forum, Jasola District Centre,
New Delhi – 110025, India

103 Penang Road, #05-06/07, Visioncrest Commercial, Singapore 238467

Cambridge University Press is part of Cambridge University Press & Assessment,
a department of the University of Cambridge.

We share the University's mission to contribute to society through the pursuit of
education, learning and research at the highest international levels of excellence.

www.cambridge.org
Information on this title: www.cambridge.org/highereducation/isbn/9781009158121

DOI: 10.1017/9781009158107

First published 2023

A catalogue record for this publication is available from the British Library.

*A Cataloging-in-Publication data record for this book is available from the Library
of Congress*

ISBN 978-1-009-15812-1 Hardback
ISBN 978-1-009-15811-4 Paperback

Additional resources for this publication at www.cambridge.org/mccleary

To my students

Contents

Preface

> For, all that creates the beauty and at the same time the difficulty of this theory is that one has ceaselessly to indicate the path of the analyses and to anticipate the results, without ever being able to carry them out.
>
> ÉVARISTE GALOIS, *Discours préliminaire*, IX.1830

For students of mathematics, the first course in Modern Algebra is an invitation to a new world. Visiting any place for the first time, one wants to see the grand sights, taste the local cuisine, and master a bit of the local language. All in a semester. Of course, that leaves a lot of places to come back to and deeper dives to make. This book is an answer to the question "What should I visit next?"

The standard first course in Algebra introduces groups and the fundamental concepts of subgroups, homomorphisms and isomorphisms, normal subgroups and quotient groups, and product groups, up to the Fundamental Theorem of Finite Abelian Groups. Rings are next with the analogous fundamental concepts of subrings and ideals, ring homomorphisms and isomorphisms, quotient rings, leading up to the notion of divisibility, especially in polynomial rings. If time remains (it often does not), then fields and their extensions are introduced, together with finite fields. The promise of a synthesis of these ideas in Galois theory is often made. Many second courses in Algebra focus on this beautiful topic.

Galois theory is difficult to teach (see the epigram of Galois that precedes this Introduction) and to learn. The accumulation of insights against the less sophisticated background of the first course grows into a labyrinth for students in which there are unmarked forks in the path without clear motivations for choice.

When I was a graduate student, the buzz at many colloquium talks was

ix

the number

$$808\,017\,424\,794\,512\,875\,886\,459\,904\,961\,710\,757\,005\,754\,368\,000\,000\,000$$
$$= 2^{46} \cdot 3^{20} \cdot 5^9 \cdot 7^6 \cdot 11^2 \cdot 13^3 \cdot 17 \cdot 19 \cdot 23 \cdot 29 \cdot 31 \cdot 41 \cdot 47 \cdot 59 \cdot 71,$$

which is the order of the *Monster*, a simple group whose existence would complete the project of the classification of finite simple groups. It was conjectured independently in 1973 to exist by BERND FISCHER and ROBERT GRIESS, and many things were known about it before its existence, such as its 194×194 character table [15]. How could such numbers like the order be found? Why is the monster the largest and last sporadic simple group? What was this character table, and what did it have to do with the existence of the group? The announcement in 1981 of a construction of the Monster by Griess [36] excitedly brought the classification of finite simple groups to completion. How can that achievement be understood?

After a conversation with a colleague (from Williams College), I decided to veer off the usual path and teach the second course in Algebra at Vassar College on representation theory. The beauty of these ideas rivaled that of Galois theory, and I began to get s small idea of how representations could be used in the study of finite groups. The representation theory provides a bridge from the use of group actions on sets via representations on vector spaces to groups acting on field extensions. This point of view is the basis for this book.

The story comes in three parts: Groups act on sets, groups act on vector spaces, and groups act on field extensions. In the first part, I provide a warmup chapter that focuses on the Jordan–Hölder Theorem, which associates to a finite group a list of simple groups that can be assembled into a given group. The notion of a solvable group is introduced, and the family of alternating groups A_n for $n \geq 5$ are shown to be simple groups. Some of the this material may have been taught in the first course and may be reviewed as a warmup. The second chapter treats groups acting on sets, from the familiar appearance of the symmetries of a geometric figure (the dihedral groups) to groups acting on various sets constructed from the group itself. This rich subject includes the notion of a transitive subgroup of a permutation group that plays an important role in Galois theory. The Sylow Theorems are the focus of Chapter 2, where they are applied to determine if a group of a particular cardinality can be a simple group. These theorems are also a nice warmup to Galois theory, which makes the connection between permutations of sets and the eventual use of permutations of roots of a polynomial. I added two important theorems to this discussion, *Iwasawa's Lemma*, which gives a criterion for simplicity, and the

Burnside Transfer Theorem, which provides a sophisticated tool for studying whether a group of a given order is simple or not.

Chapter 3 is a warmup chapter on linear algebra. Most students who have had the first course in Modern Algebra have been introduced to linear algebra in an earlier course, and some may have had a second course in Linear Algebra as well. In this chapter the useful notion of a projection is introduced along with tensor products of vector spaces. I introduce Hermitian inner products on complex vector spaces in anticipation of their appearance in representation theory. I also consider matrix groups over fields to obtain another class of simple groups, $\mathrm{PSl}_n(\mathbb{F})$ for $n \geq 2$, which give finite simple groups when $\mathbb{F} = \mathbb{Z}/p\mathbb{Z}$ for p a prime. Chapter 4 focuses on representation theory, developed in the framework of finite groups acting on vector spaces. There is an alternative viewpoint that features modules over the group ring $\mathbb{C}G$. Sadly, this approach is left out in an effort to work with as few prerequisites as possible. Comments on this choice are found in the Epilogue. The beautiful cascade of constructions and results is pointed toward obtaining properties of a finite group that reveal deeper structure and the connections between subgroup data, quotient data, and the invariants of a given group. The example of A_5 is developed completely.

Chapter 5 treats fields and polynomials. This chapter may have the largest overlap with the first course. The basic notion of a field extension is introduced, including the key notion of a splitting field. The existence of algebraic closures is proved, and further properties of finite fields are proved including the Frobenius automorphism. With the existence of new finite fields, there come new finite simple groups, $\mathrm{PSl}_n(\mathbb{F}_q)$. The exercises in this chapter treat the classical problem of compass and straightedge constructions in geometry. Chapter 6 focuses on Galois theory, which is about the isomorphisms between fields. The emphasis up to this chapter on the actions of groups gives motivation for the path of development of Galois's (and Dedekind's) important ideas. I cover the classical topics of the Galois correspondence, cyclotomic extensions, solvability by radicals, including the cases of cubic and quartic polynomials. The case of quintic polynomials is discussed from the point of view of the inverse Galois problem – which transitive subgroups of the symmetric group on five letters are realized as a Galois group of an extension over the rationals? Galois theory also provides some control of algebraic integers, and I return to a celebrated result of Burnside, which combines representation theory and Galois theory. The Normal Basis Theorem provides another bridge between representation theory and Galois theory establishing the insight of Emmy Noether and Max Deuring that Galois extensions realize the regular representation of the Galois group.

Some sections of the book involve fairly sophisticated topics that can be skipped on first reading. These sections are marked with an asterisk.

In all my courses of the past few years I have introduced a regular active learning session, my *Workshop Wednesdays*, in which students work together on some problems. My goal is to offer additional opportunities to explore the topics of the week, as well as the opportunity to talk together about mathematics. If you have studied a foreign language, then these sorts of sessions build assurance in knowledge of vocabulary, syntax, and semantics. All of this is useful in learning new mathematics. I have built this idea into the book as sets of exercises usually labeled *Get to know* something. These sets of exercises come at points in the narrative where hands-on experience is important. I encourage the reader to keep a pad and writing instrument handy to try their hand in the moment.

How to use this book

First courses in Algebra vary widely. Depending upon the coverage of topics (Did the course focus only on groups? Were the Sylow Theorems part of the curriculum? How much ring theory and field theory were covered? Were vector spaces over an arbitrary field part of the linear algebra course?), the reader will find plenty that they might skip or review, especially in the warmup chapters. Alternatively, the warmup chapters may be assigned as supplemental reading, or as in-class short presentations. The heart of the matter lies in the even numbered chapters. A thirteen-week course might be arranged in the following (breakneck speed) breakdown by week:

1. The Jordan–Hölder Theorem, simple and solvable groups, and A_n is simple for $n \geq 5$.
2. Action of a group on a set, orbits, stabilizers, equivariant mappings, the Class Equation.
3. Transitive actions, p-groups, the Sylow Theorems and consequences.
4. The Fundamental Theorem of Linear Algebra via projections, tensor products.
5. Hermitian vector spaces, matrix groups, $PSl_n(\mathbb{F})$ is simple for most n.
6. Representations, subrepresentations, irreducible representations, Maschke's Theorem.
7. Schur's Lemma, finding irreducible representations, characters.
8. The orthogonality theorems for character tables. Lifts, kernels, linear representations.

9. Field extensions, field homomorphisms, splitting fields, Kronecker's Theorem.
10. Review irreducibility criteria, finite fields and the Frobenius automorphisms, existence of algebraic closures.
11. Field automorphisms, finite field extension example, Galois group, Galois correspondence.
12. Dedekind Independence Theorem, normal and Galois extensions, the Fundamental Theorem of Galois Theory.
13. Cyclotomic extensions, solvability by radicals and Galois' Theorem.

Making some of the topics into workshops can preserve the order while encouraging active learning. The proofs of many results may be skipped to present the architecture of the ideas more clearly, but certainly many should be presented to reveal the inner workings of the subjects. I am sure every reader and every instructor will make their own choices.

Acknowledgments

This book emerged from a course I gave a couple times at Vassar College. I thank my students for their willingness to go in new directions. I especially thank Emily Stamm, whose notes helped reconstruct the story, and Ben Costa, Baian Liu, Alex May, Jonas Pandit, Eli Polston, Dean Spyropoulos, and Yanjie Zhang in 2017, and Thanh Bui, Gabriel Duska, Hannah Eakin, Colburn Morrison, Corin Rose, and Noah Winold in 2019 for their attention and hard work. My thanks to Yusra Naqvi for her support in conceiving the representation theory portion of the course. Her notes were a wonderful guide.

My thanks to the anonymous readers who reviewed the proposal and recommended pursuing the project. There were not many pages on which to base their remarks. I hope I have achieved what you thought I might be able to do.

Thanks to Annette Morriss whose painting, $\mathbf{P}2I3{,}2I3{,}2I3{,}2I3$ (oil on canvas, 15×15, 2021) graces the cover – our third such effort. I thank the EMS Press for use of quotes from Peter Neumann's edition of Galois's papers [65], and Cambridge University Press for use of the quotes from Burnside [10].

Some folks have taken valuable time to read an earlier draft of the book. My thanks to Wyatt Milgrim, Steve Andrilli, Michael Bush, and Joe Kirtland for helpful comments. Discussions with Adam Lowrance and Ben Lotto have saved me from some self-inflicted detours. Thanks also to Gerard Furey for egging me on.

The folks at Cambridge University Press have been patient, supportive, and great to work with. Many thanks to Kaitlin Leach and Johnathan Fuentes. Also thanks to Kęstutis Vizbaras from VTeX for patient help with the look of the final version.

For support of the most important kinds in getting this done, thanks to Carlie Graves, my best companion through so many adventures.

1

Warmup: More Group Theory

...put operations into groups, class them according to their difficulty
and not according to their form; that is, according to me, the mission of
future geometers ...

ÉVARISTE GALOIS from *Preface for two memoirs*

1.1 Isomorphism Theorems

The basic notions of Modern Algebra are modeled on the theory of groups.
Fundamental concepts include subgroups, normal subgroups, quotient groups,
homomorphisms, and isomorphisms. An important result that combines these
ideas is the *First Isomorphism Theorem*. The ingredients are a homomorphism
$\phi\colon G_1 \to G_2$ between two groups, and the subgroups

$\ker \phi = \{g \in G_1 \mid \phi(g) = e\} \subset G_1$, the *kernel* of ϕ, and

$\phi(G_1) = \{h \in G_2 \mid \text{there is } g \in G_1 \text{ with } h = \phi(g)\} \subset G_2$, *image* of ϕ.

The First Isomorphism Theorem *If $\phi\colon G_1 \to G_2$ is a homomorphism of
groups, then the kernel of ϕ is a normal subgroup of G_1, and the image of ϕ,
$\phi(G_1)$, is isomorphic to the quotient group $G_1/\ker \phi$.*

Recall that a subgroup $N \subset G$ is *normal* if, for all $g \in G$, $gNg^{-1} = N$, or
equivalently, $gN = Ng$. This theorem has wide-reaching consequences. There
are analogues of the theorem for ring homomorphisms, linear transformations,
and many other structures and their mappings. When we speak of *fundamental
concepts*, we focus on these key notions.

If there is a *first* such theorem, then what are the subsequent statements? To
state the next isomorphism theorem, let us consider a particular situation inside

a group: Let H and N be subgroups of G with N a normal subgroup. Define

$$HN = \{hn \in G \mid h \in H, n \in N\},$$

the set of products of an element in H with an element in N, in that order. Observe that HN is a subgroup of G: if $h_1 n_1$ and $h_2 n_2$ are elements of HN, then, because N is normal in G, we have $h_2^{-1} N h_2 = N$, and so

$$(h_1 n_1)(h_2 n_2) = (h_1 h_2)\big(h_2^{-1} n_1 h_2\big) n_2 = (h_1 h_2)(n_3 n_2).$$

Thus HN is closed under multiplication. Also,

$$(hn)^{-1} = n^{-1} h^{-1} = h^{-1}\big(hn^{-1} h^{-1}\big) = h^{-1} n',$$

and HN contains inverses. These arguments show that $HN = NH$. With these constructions, we can prove

The Second Isomorphism Theorem *Suppose H and N are subgroups of G with N a normal subgroup. Then HN is a subgroup of G containing N, and*

$$HN/N \cong H/(H \cap N).$$

Proof First notice that N is a normal subgroup of HN because N is normal in G. Consider the mapping $\pi \colon H \to HN/N$ given by $\pi(h) = hN$. Then $\pi(h_1 h_2) = h_1 h_2 N = (h_1 N)(h_2 N) = \pi(h_1)\pi(h_2)$, and π is a homomorphism. By the First Isomorphism Theorem, $\pi(H) \cong H/\ker \pi$. Given $hn \in HN$, the coset $hnN = hN$, and so the homomorphism π is surjective. The kernel of π consists of elements of H for which $hN = N$, that is, $h \in H \cap N$. It follows that $\pi(H) = HN/N \cong H/(H \cap N)$. $\square$

Corollary 1.1 *For G a finite group, $\#HN = (\#H)(\#N)/\#(H \cap N)$.*

The subgroups of a group G are partially ordered by inclusion. For small groups, this ordering can be pictured in the *Hasse diagram* of G, which is the graph that depicts this partially ordered set. A vertex is a subgroup, and an edge denotes an inclusion. For example, the Hasse diagrams for Σ_3, the group of permutations of $\{1, 2, 3\}$, and A_4, the group of even permutations of $\{1, 2, 3, 4\}$, take the form as in Fig. 1.1. Here $\langle g \rangle$ denotes the cyclic subgroup generated by g. Subgroups of the same cardinality are arranged horizontally.

Each element of a group G determines a homomorphism by conjugation: If $g \in G$, then the mapping *conjugation by g*, $c_g \colon G \to G$, is defined by $c_g(h) = ghg^{-1}$.

Proposition 1.2 *Conjugation by g is an isomorphism called an inner automorphism of G.*

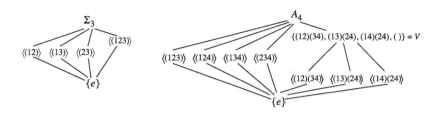

Figure 1.1

Proof We have $c_g(hk) = g(hk)g^{-1} = (ghg^{-1})(gkg^{-1}) = c_g(h)c_g(k)$, and c_g is a homomorphism. If g^{-1} is the inverse of g, then $c_{g^{-1}}$ satisfies $c_{g^{-1}}(c_g(h)) = g^{-1}(ghg^{-1})g = h$, and c_g is an isomorphism. □

Two elements h, k of G are said to be *conjugates* if there is an element $g \in G$ such that $c_g(h) = k$, that is, $k = ghg^{-1}$. This relation is an equivalence relation on G: $exe^{-1} = x$, so x is a conjugate of x; if x is a conjugate of y, that is, $y = c_g(x)$, then $x = c_{g^{-1}}(y)$, and y is a conjugate of x; finally, if x is conjugate to y and y is conjugate to z, then $y = c_g(x)$ and $z = c_h(y)$ for some $g, h \in G$. It follows that $z = c_h(c_g(x)) = h(gxg^{-1})h^{-1} = (hg)x(hg)^{-1} = c_{hg}(x)$ and x is conjugate to z.

Definition 1.3 The equivalence classes under conjugation of elements of G are called *conjugacy classes*. We denote a conjugacy class by $[g] = \{xgx^{-1} \mid x \in G\}$, and the set of equivalence classes is denoted by $\mathrm{Cl}(G)$.

The set $\mathrm{Cl}(G)$ tells us something about the binary operation on G. For example, if G is abelian, then $ghg^{-1} = gg^{-1}h = h$, and so the conjugacy class of h is a singleton set $\{h\}$. When G is nonabelian, the set $Z(G)$ of elements whose conjugacy classes are singletons, that is, $[g] = \{g\}$, determines a subgroup of G called the *center* of G: Suppose $x, y \in Z(G)$. Then $c_g(x) = x$ and $c_g(y) = y$ for all $g \in G$; $c_g(xy) = c_g(x)c_g(y) = xy$ for all $g \in G$, and so $xy \in Z(G)$; if $x \in Z(G)$, then $c_g(x^{-1}) = gx^{-1}g^{-1} = (g^{-1}xg)^{-1} = x^{-1}$ for all $g \in G$.

Let us explore the important example of conjugacy classes in Σ_n, the *symmetric group on n letters*, which is the group of all permutations of the set $[n] = \{1, 2, \ldots, n\}$. Recall that a *permutation* is a one-to-one correspondence of $[n]$ with itself. There are various notations for permutations. For example, if $\sigma \in \Sigma_7$, then we can write

$$\sigma = \begin{pmatrix} 1 & 2 & 3 & 4 & 5 & 6 & 7 \\ 2 & 3 & 4 & 1 & 5 & 7 & 6 \end{pmatrix} = (1, 2, 3, 4)(6, 7).$$

In the first case, we present the function values explicitly; reading downward $\sigma(1) = 2, \sigma(2) = 3$, etc. In the second case, we use *cycle notation*: $(1, 2, 3, 4)$ means $1 \mapsto 2 \mapsto 3 \mapsto 4 \mapsto 1$. Notice that 5 does not appear in the cycle notation because it is fixed by σ, $\sigma(5) = 5$. Every permutation has a unique presentation as a composite of disjoint cycles. Simply apply σ repeatedly to an element until it cycles back to itself. This gives one cycle. Apply σ to any remaining elements, leaving off any that are fixed, and repeat until every element fits into a cycle or is fixed. We denote the identity permutation $\text{Id}(j) = j$ for all j by ( ), the empty cycle. A given cycle has as many presentations as elements in its cycle; for example, $(1, 2, 3) = (2, 3, 1) = (3, 1, 2)$. The pattern of the bijection is fixed by the cyclic order of the elements.

The binary operation on Σ_n is a composition, and so when we compute the product of permutations, we compute the resulting cycle decomposition of the product **reading from right to left**. For example,

$$(3, 4, 5) \circ (1, 3, 5, 2, 4) = (1, 4)(2, 5).$$

We call a 2-cycle (a, b) a *transposition* because only two elements move and they are interchanged. Every permutation can be expressed as a product of transpositions. For example, $(a, b, c) = (a, b)(b, c)$. We say that a permutation is an *odd permutation* (*even permutation*) if it has a presentation as a product of an odd (even) number of transpositions. It is left as an exercise to prove that this is independent of the choice of presentation. Products of even permutations are even, and so they form a subgroup A_n of Σ_n, called the *alternating group* on n letters.

In 1844 [11], AUGUSTIN CAUCHY (1789–1857) first developed some of the key properties of symmetric groups. In particular, he proved the following:

Cauchy's Formula *If $\sigma \in \Sigma_n$ and $(a_1, a_2, \ldots, a_k)$ is a k-cycle in Σ_n, then*

$$\sigma \circ (a_1, a_2, \ldots, a_k) \circ \sigma^{-1} = \big(\sigma(a_1), \sigma(a_2), \ldots, \sigma(a_k)\big).$$

Proof   If $(a_1, \ldots, a_k)$ fixes $i \in [n]$, then $\sigma \circ (a_1, \ldots, a_k) \circ \sigma^{-1}$ fixes $\sigma(i)$:

$$\sigma \circ (a_1, \ldots, a_k) \circ \sigma^{-1}\big(\sigma(i)\big) = \sigma \circ (a_1, \ldots, a_k)(i) = \sigma(i).$$

For $1 \leq j \leq k$, we have

$$\sigma \circ (a_1, \ldots, a_k) \circ \sigma^{-1}\big(\sigma(a_j)\big) = \sigma \circ (a_1, \ldots, a_k)(a_j) = \sigma(a_{j+1}),$$

where we understand $a_{k+1} = a_1$. Because we have accounted for everything in $[n]$, the permutation $\sigma \circ (a_1, \ldots, a_k) \circ \sigma^{-1} = (\sigma(a_1), \ldots, \sigma(a_k))$. $\square$

Since conjugation is a homomorphism,

$$c_\sigma\big((a_1, \ldots, a_k)(b_1, \ldots, b_l)\big) = c_\sigma\big((a_1, \ldots, a_k)\big) \circ c_\sigma\big((b_1, \ldots, b_l)\big)$$
$$= \big(\sigma(a_1), \ldots, \sigma(a_k)\big)\big(\sigma(b_1), \ldots, \sigma(b_l)\big).$$

Cauchy's Formula extends to arbitrary products of cycles.

When a permutation $\sigma \in \Sigma_n$ is written as a product of disjoint cycles involving m_1 elements fixed by σ, m_2 2-cycles, m_3 3-cycles, and so on, we write the *type* of σ as $(1^{m_1}, 2^{m_2}, 3^{m_3}, \ldots, n^{m_n})$. Here $m_1 + 2m_2 + \cdots + nm_n = n$. For example, $\sigma = (1, 2, 3, 4)(6, 7) \in \Sigma_7$ has type $(1^1, 2^1, 3^0, 4^1, 5^0, 6^0, 7^0)$.

Corollary 1.4 *Two permutations in Σ_n are conjugate if and only if they share the same type.*

Proof By Cauchy's Formula conjugation preserves the length of a cycle. If σ and τ are conjugate, then they have the same type. Suppose conversely that σ and τ have the same type. We use the functional presentation of a permutation. Suppose the type shared by σ and τ is $(1^{m_1}, 2^{m_2}, \ldots, n^{m_n})$. Write the domain $[n]$ as an ordered set $1 < 2 < \cdots < n$ and place parentheses according to the type giving a permutation $\alpha \in \Sigma_n$. For example, in Σ_7, a type $(1^1, 2^1, 3^0, 4^1, 5^0, 6^0, 7^0)$ gives us $\alpha = (1)(2, 3)(4, 5, 6, 7)$. Write σ and τ as products of disjoint cycles that follow the type in the manner of α. For example, $\sigma = (1, 2, 3, 4)(6, 7)$ can be written as $(5)(6, 7)(1, 2, 3, 4)$. This gives us an ordering of $[7]$ by ignoring the parentheses. We obtain a permutation θ by stacking the natural order over the order given by σ:

$$\theta = \begin{pmatrix} 1 & 2 & 3 & 4 & 5 & 6 & 7 \\ 5 & 6 & 7 & 1 & 2 & 3 & 4 \end{pmatrix}.$$

Cauchy's Formula tells us that $\theta \circ \alpha \circ \theta^{-1} = \theta \circ (1)(2, 3)(4, 5, 6, 7) \circ \theta^{-1} = \sigma$. Reversing the conjugation by θ, we get $\alpha = \theta \sigma \theta^{-1}$. For τ of the same type, we get another permutation ζ with $\alpha = \zeta \tau \zeta^{-1}$, and so $\theta \sigma \theta^{-1} = \zeta \tau \zeta^{-1}$; it follows that σ and τ are conjugate. $\square$

The restriction that the type satisfies $m_1 + 2m_2 + \cdots + nm_n = n$ with $m_i \geq 0$ corresponds to a *partition* of the integer n, that is, an expression of n as a sum of positive integers in nondecreasing order. For example, $4 = 1 + 1 + 1 + 1 = 1 + 1 + 2 = 1 + 3 = 2 + 2$. Each partition of 4 corresponds to a type: $4 \leftrightarrow (1^0, 2^0, 3^0, 4^1)$, $1 + 1 + 1 + 1 \leftrightarrow (1^4, 2^0, 3^0, 4^0)$, $1 + 1 + 2 \leftrightarrow (1^2, 2^1, 3^0, 4^0)$, $1 + 3 \leftrightarrow (1^1, 2^0, 3^1, 4^0)$, and $2 + 2 \leftrightarrow (1^0, 2^2, 3^0, 4^0)$. The analysis above shows that the number of conjugacy classes filling Σ_n is $p(n)$, the *number of partitions of n*. The function $p(n)$ was introduced by Leibniz and is the subject of some remarkable mathematics (see, for example, [1]).

Get to know conjugation

1. For a group G, an element $g \in G$, and a subgroup H of G, show that gHg^{-1} is a subgroup of G that is isomorphic to H.
2. What are the conjugacy classes of the groups Σ_3 and Σ_4? What is the cardinality of each conjugacy class?
3. Show that a group G is abelian if and only if all of its conjugacy classes are singletons.
4. What is the partition into conjugacy classes for the group $A_4 \subset \Sigma_4$ of even permutations of four objects? What is the partition into conjugacy classes for the dihedral group of eight elements, D_8, the set of symmetries of a square?

   The *dihedral groups* D_{2n} are the groups of symmetries of a regular n-gon in the plane. There is a symmetry by rotating the n-gon through $2\pi/n$ radians. Denote this rotation by r and notice that $r^n = \text{Id}$. If we situate the n-gon in the plane with its center of gravity at the origin and one vertex at the point $(1, 0)$, then there is a symmetry of the plane obtained by (x, y) going to $(x, -y)$ (complex conjugation?). Denote this reflection by f and notice that $f^2 = \text{Id}$. These two elements *generate* the dihedral group, that is, every symmetry of the regular n-gon in the plane is a finite product of the form $f^{a_1} r^{b_1} f^{a_2} r^{b_2} \cdots$. We simplify using $f^2 = \text{Id} = r^n$ whenever they occur. There is also another relation that follows from the geometry: $fr = r^{n-1}f$. (Can you picture this?) We write these data as a *presentation* of the group:

$$D_{2n} = \langle f, r \mid r^n = \text{Id}, f^2 = \text{Id}, fr = r^{n-1}f \rangle.$$

   The dihedral group D_{2n} has order $2n$.

   If G is a group and $S \subset G$ is a subset of G, then $\langle S \rangle$ is the smallest subgroup of G that contains S, that is, $\langle S \rangle = \bigcap_{S \subset H, \text{subgroup}} H$.

5. Suppose that N is a normal subgroup of G. Show that N is the union of the conjugacy classes of its elements. If $[g]$ is a conjugacy class of G and $H = \langle [g] \rangle$ is the subgroup of G generated by the elements of $[g]$, then show that H is normal in G.

For a subgroup of a group, the conjugacy classes of elements in the subgroup need not coincide with those of the group. For example, if $H = \langle g \rangle$ is a cyclic subgroup generated by $g \in G$, then the conjugacy classes in H are singletons.

If G is nonabelian, then there are conjugacy classes of cardinality greater than one.

In a finite group, what is the cardinality of a conjugacy class? The *centralizer* of an element g in any group G is the subset $C_G(g) = \{x \in G \mid xgx^{-1} = g\}$. In other words, the centralizer consists of those elements x of G that commute with g.

Proposition 1.5 *For $g \in G$ a group, the centralizer $C_G(g)$ is a subgroup of G. Furthermore, if G is finite, then #[g], the cardinality of the conjugacy class of g, is given by the index of $C_G(g)$ in G, $[G : C_G(g)] = \#G/\#C_G(g)$.*

Proof If $x, y \in C_G(g)$, then $(xy)g(xy)^{-1} = x(ygy^{-1})x^{-1} = xgx^{-1} = g$. So $C_G(g)$ is closed under the binary operation. Because $g = xgx^{-1}$ implies $x^{-1}gx = g$, $C_G(g)$ is closed under inverses, and $C_G(g)$ is a subgroup of G. Recall that $G/C_G(g)$ denotes the set of left cosets of $C_G(g)$ in G.

Consider the function of sets $f : G/C_G(g) \to [g]$ given by $f : xC_G(g) \mapsto xgx^{-1}$. If $xC_G(g) = yC_G(g)$, then $y^{-1}x \in C_G(g)$ from which it follows that $(y^{-1}x)g(y^{-1}x)^{-1} = g$. This implies that $xgx^{-1} = ygy^{-1}$ and the mapping f is well-defined. It is clearly surjective. To see that f is injective, suppose $f(x) = f(y)$. Then $xgx^{-1} = ygy^{-1}$ and $(y^{-1}x)g(y^{-1}x)^{-1} = g$, that is, $y^{-1}x \in C_G(g)$, and so $xC_G(g) = yC_G(g)$. The bijection implies that $\#[g] = \#(G/C_G(g)) = [G : C_G(g)]$. $\qquad\square$

For a subgroup H of G and $h \in H$, we have $C_H(h) = \{k \in H \mid khk^{-1} = h\} = C_G(h) \cap H$.

Let us examine in detail the case of A_n, the subgroup of Σ_n consisting of even permutations. Suppose $\sigma \in A_n$. In Σ_n the conjugates of σ are permutations of the same type as σ. If τ is an odd permutation, then $\tau\sigma\tau^{-1}$ is an even permutation that might not be in the conjugacy class of σ in A_n. It can be that $\tau\sigma\tau^{-1} \neq \alpha\sigma\alpha^{-1}$ for all $\alpha \in A_n$. How can we recognize when this happens? Following Proposition 1.5,

$$\#[\sigma]_{A_n} = \left[A_n : C_{A_n}(\sigma)\right] = \left[A_n : A_n \cap C_{\Sigma_n}(\sigma)\right].$$

Because A_n is normal in Σ_n, the Second Isomorphism Theorem implies $C_{\Sigma_n}(\sigma)A_n/A_n \cong C_{\Sigma_n}(\sigma)/A_n \cap C_{\Sigma_n}(\sigma)$. Cross multiply to obtain $\#A_n/\#(A_n \cap C_{\Sigma_n}(\sigma)) = \#C_{\Sigma_n}(\sigma)A_n/\#A_n$, and

$$\left[A_n : A_n \cap C_{\Sigma_n}(\sigma)\right] = \left[C_{\Sigma_n}(\sigma)A_n : C_{\Sigma_n}(\sigma)\right].$$

Suppose there is an odd permutation that commutes with σ. Then $C_{\Sigma_n}(\sigma)$ contains an odd permutation, and so $C_{\Sigma_n}(\sigma)A_n = \Sigma_n$. From this identity,

$$\#[\sigma]_{A_n} = \left[A_n : A_n \cap C_{\Sigma_n}(\sigma)\right] = \left[C_{\Sigma_n}(\sigma)A_n : C_{\Sigma_n}(\sigma)\right] = \left[\Sigma_n : C_{\Sigma_n}(\sigma)\right]$$
$$= \#[\sigma]_{\Sigma_n}.$$

When there are no odd permutations that commute with σ, we have $C_{\Sigma_n}(\sigma) = C_{A_n}(\sigma)$, and $\#[\sigma]_{A_n} = [A_n : C_{\Sigma_n}(\sigma)]$, but then

$$\#[\sigma]_{\Sigma_n} = \left[\Sigma_n : C_{\Sigma_n}(\sigma)\right] = [\Sigma_n : A_n]\left[A_n : C_{\Sigma_n}(\sigma)\right] = 2\#[\sigma]_{A_n}.$$

So if σ commutes only with even permutations, then the conjugacy class of σ in Σ_n splits into two equal pieces in A_n.

For example, in Σ_3, $[(1,2,3)]_{\Sigma_3} = \{(1,2,3),(1,3,2)\}$, whereas $[(1,2,3)]_{A_3} = \{(1,2,3)\}$. It would nice to be able to tell directly from σ when $C_{\Sigma_n}(\sigma) = C_{A_n}(\sigma)$.

Proposition 1.6 *Suppose* $\sigma \in A_n$ *has type* $(1^{m_1}, 2^{m_2}, \ldots, n^{m_n})$. *Then* $C_{\Sigma_n}(\sigma) = C_{A_n}(\sigma)$ *and* $\#[\sigma]_{\Sigma_n} = 2\#[\sigma]_{A_n}$ *if and only if, for all* i, $m_{2i} = 0$ *and* $m_{2i+1} = 0$ *or* 1.

For example, in Σ_3, the type of $(1,2,3)$ is $(1^0, 2^0, 3^1)$.

Proof If $m_{2i} > 0$, then $\sigma = \alpha\beta$ with α a $2i$-cycle and β a product of cycles disjoint from α. This implies $\alpha\beta = \beta\alpha$, and multiplying by α on the left gives

$$\alpha\sigma = \alpha\alpha\beta = \alpha\beta\alpha = \sigma\alpha.$$

A $2i$-cycle is an odd permutation and α is in $C_{\Sigma_n}(\sigma)$ but not in A_n.

If $m_{2i} = 0$ but $m_{2i+1} > 1$ for some i, then, after renumbering, we have $\sigma = (1, 2, \ldots, 2i+1)(1', 2', \ldots, (2i+1)')\nu$, where the numbers 1 through $2i+1$ and $1'$ through $(2i+1)'$ are taken from $[n]$ and are disjoint from one another. The permutation ν is also disjoint from these numbers. Let $\zeta = (1,1')(2,2')\cdots(2i+1,(2i+1)')$, an odd permutation. Then $\zeta\sigma\zeta^{-1} = \sigma$ by Cauchy's Formula, and so $\zeta \in C_{\Sigma_n}(\sigma)$. This proves one direction of the proposition.

To prove the other direction, suppose the type of σ has $m_{2i} = 0$ and $m_{2i+1} = 0$ or 1 for all i. Then σ is a product of disjoint k-cycles with k odd. Since the type determines the conjugacy class in Σ_n, we can count the number of elements in $[\sigma]_{\Sigma_n}$ combinatorially. Take any of the $n!$ orderings of $[n]$ and form the permutation θ as in the proof of Proposition 1.5 by removing the first m_1 entries and introducing parentheses according to the type of σ. Wherever $m_{2i+1} = 1$, the corresponding $(2i+1)$-cycle can start from any of its entries and so is counted $2i+1$ times. We divide $n!$ by $2i+1$ for each $m_{2i+1} = 1$ to adjust for the overcounting. Thus $\#[\sigma] = n!/t$ with t an odd number. This implies $\#C_{\Sigma_n}(\sigma) = t$, a subgroup of Σ_n of odd order, and every permutation in $C_{\Sigma_n}(\sigma)$ has odd order. The order of a permutation is the least common multiple of the lengths of each constituent disjoint k-cycle. Since the order is odd, no $2m$-cycle appears in the disjoint cycle presentation of any element of $C_{\Sigma_n}(\sigma)$. Then $C_{\Sigma_n}(\sigma) \subset A_n$ and $C_{\Sigma_n}(\sigma) = C_{A_n}(\sigma)$. $\square$

For example, in A_5 the permutation $\sigma = (1, 2, 3, 4, 5)$ has type $(1^0, 2^0, 3^0, 4^0, 5^1)$, and so $C_{\Sigma_n}(\sigma) = C_{A_n}(\sigma)$ and $\#[\sigma]_{\Sigma_n} = 2\#[\sigma]_{A_n}$. On the other hand, the permutation $\tau = (1, 2, 3)$ has type $(1^2, 2^0, 3^1, 4^0, 5^0)$, and so the conjugacy class of τ satisfies $[(1, 2, 3)]_{\Sigma_5} = [(1, 2, 3)]_{A_5}$ and contains all 3-cycles.

1.2 The Jordan–Hölder Theorem

When $\phi\colon G_1 \to G_2$ is a homomorphism, we obtain subgroups of G_1 by taking the *inverse image* of subgroups of G_2: let $K \subset G_2$ denote a subgroup of G_2, and let

$$\phi^{-1}(K) = \{g \in G_1 \mid \phi(g) \in K\}.$$

If g and g' are in $\phi^{-1}(K)$, then $\phi(g) = k$ and $\phi(g') = k'$, both in K. Since K is a subgroup, $\phi(gg') = \phi(g)\phi(g') = kk'$ is in K, and so $gg' \in \phi^{-1}(K)$. For a subgroup $H \subset G_1$, the image $\phi(H) = \{\phi(h) \in G_2 \mid h \in H\}$ is a subgroup of G_2. Hence we can go to and fro between the collections of subgroups in this manner. In fact, we get a strong connection between these collections.

The Correspondence Theorem *Let $\phi\colon G_1 \to G_2$ be a surjective group homomorphism. Then there is a bijection*

$$\Phi\colon \mathcal{S}_\phi = \{H, \text{ a subgroup of } G_1, \text{ with } \ker\phi \subset H\}$$
$$\to \mathcal{T}_{G_2} = \{K, \text{ a subgroup of } G_2\}$$

given by $\Phi(H) = \phi(H)$. Furthermore, Φ and its inverse take normal subgroups to normal subgroups.

Proof We prove Φ is a bijection by presenting its inverse. For a subgroup K of G_2, let $\Phi^{-1}(K) = \phi^{-1}(K)$. In general, notice that $\phi^{-1}(\phi(H)) = (\ker\phi)H$ as subgroups of G_1: let $l \in \phi^{-1}(\phi(H))$. Then $\phi(l) \in \phi(H)$, and we can write $\phi(l) = \phi(h)$ for some $h \in H$. Then $\phi(lh^{-1}) = e$ and $lh^{-1} = k \in \ker\phi$, but then $l = kh \in (\ker\phi)H$ and $\phi^{-1}(\phi(H)) \subset (\ker\phi)H$. For $nh \in (\ker\phi)H$, $\phi(nh) = \phi(n)\phi(h) = \phi(h) \in \phi(H)$, and so $(\ker\phi)H \subset \phi^{-1}(\phi(H))$. For any subgroup $H \in \mathcal{S}_\phi$, we have $\ker\phi \subset H$, and so $H = (\ker\phi)H = \phi^{-1}(\phi(H)) = \Phi^{-1}(\Phi(H))$. In the other direction, $\Phi(\Phi^{-1}(K)) =$

$\phi(\phi^{-1}(K))$. Because ϕ is a surjection, $\phi(\phi^{-1}(K)) = K$, and so $\Phi \circ \Phi^{-1}$ is the identity mapping on $\mathcal{T}_{G_2}$.

If N is a normal subgroup, $N \lhd G_1$, and $k \in G_2$, then we can write $k\Phi(N)k^{-1} = \phi(g)\phi(N)\phi(g^{-1})$ because ϕ is surjective. It follows then that $\phi(g)\phi(N)\phi(g^{-1}) = \phi(gNg^{-1}) = \phi(N)$, and so $\Phi(N)$ is normal in G_2. For a normal subgroup M of G_2, and $h \in G_1$, $\phi(h\phi^{-1}(M)h^{-1}) = \phi(h)M\phi(h^{-1}) = M$, and $h\phi^{-1}(M)h^{-1} = \phi^{-1}(M)$, that is, $\Phi^{-1}(M)$ is normal in G_1.   □

The Correspondence Theorem implies that a surjective homomorphism gives a correspondence between the Hasse diagram of the codomain and portions of the Hasse diagram of the domain. Correspondences will figure prominently in other parts of the book.

One way to carry out inductive sorts of arguments on finite groups is choosing a normal subgroup N of a finite group G that gives a surjection $\pi : G \to G/N$ to a smaller group G/N. Properties of G/N may lift to G over π as in the Correspondence Theorem. However, some groups are not susceptible to this strategy, lacking nontrivial normal subgroups.

Definition 1.7 A group G is called a *simple group* if whenever N is a normal subgroup of G, then $N = \{e\}$ or $N = G$.

For example, an abelian finite group G is simple if and only if G is isomorphic to $\mathbb{Z}/p\mathbb{Z}$ for a prime p. (Can you prove this?) Simple groups turn out to play a role similar to atoms in a molecule or to primes in the prime factorization of a positive integer. To make this analogy precise, we introduce the following notion.

Definition 1.8 Suppose G is a finite group. A proper subgroup N is a *maximal normal subgroup* if N is normal in G, and if $N \lhd H \lhd G$ for some subgroup H, then either $H = N$ or $H = G$.

Proposition 1.9 *For a group G and a normal subgroup N, the quotient group G/N is a simple group if and only if N is a maximal normal subgroup.*

Proof Let $\pi : G \to G/N$ be the quotient homomorphism, $\pi(g) = gN$. Suppose H is a normal subgroup of G/N. Then $\pi^{-1}(H)$ is a normal subgroup of G that contains $N = \ker \pi$. If N is a maximal normal subgroup, then $N \lhd \pi^{-1}(H) \lhd G$ implies either that $\pi^{-1}(H) = N$, and so $H = \{eN\} \subset G/N$, or that $\pi^{-1}(H) = G$ and $H = G/N$. Hence G/N is a simple group.

In the case that G/N is a simple group, any normal subgroup M of G with $N \lhd M \lhd G$ is mapped to $\pi(M)$ a normal subgroup of G/N. This implies that $\pi(M) = \{eN\}$ or $\pi(M) = G/N$, which implies that $M = N$ or $M = G$ and N is a maximal normal subgroup of G.   □

The subgroup $V = \{(\ ), (1, 2)(3, 4), (1, 3)(2, 4), (1, 4)(2, 3)\}$ of the alternating group A_4, the *Viergruppe* of Klein, has quotient A_4/V isomorphic to a simple group $\mathbb{Z}/3\mathbb{Z}$. Therefore V is a maximal normal subgroup of A_4.

Definition 1.10 A *normal series* of a group G is a finite sequence of subgroups

$$\{e\} = G_0 \lhd G_1 \lhd \cdots \lhd G_{n-1} \lhd G_n = G,$$

for which G_{k-1} is normal in G_k for all $0 < k \leq n$. A normal series determines a sequence of quotient groups $G/G_{n-1}, G_{n-1}/G_{n-2}, \ldots, G_2/G_1, G_1$, whose nontrivial members are called *subquotients* or *composition factors*. A normal series is a *composition series* if the subquotients are all nontrivial simple groups.

For example, the group A_4 has a composition series

$$\big\{(\ )\big\} \lhd \big\langle (1, 2)(3, 4) \big\rangle \lhd V \lhd A_4.$$

The composition factors are nontrivial simple groups

$$A_4/V \cong \mathbb{Z}/3\mathbb{Z}, \quad V/\big\langle (1, 2)(3, 4) \big\rangle \cong \mathbb{Z}/2\mathbb{Z}, \text{ and } \big\langle (1, 2)(3, 4) \big\rangle / \big\{(\ )\big\} \cong \mathbb{Z}/2\mathbb{Z}.$$

Given a finite group $G = G_n$, we can construct a composition series by choosing a maximal normal proper subgroup G_{n-1}. This guarantees that G_n/G_{n-1} is a simple group. As G_{n-1} is smaller than G_n, we can proceed by induction to construct a composition series for G_{n-1} and extend it by attaching G, giving a composition series for G. However, it may be possible to choose different maximal normal subgroups. For example, consider $G = \mathbb{Z}/30\mathbb{Z} = \langle \mathbf{1} \rangle$. Here are four composition series for G and their subquotients. A bold digit **n** represents the equivalence class (mod 30) of n:

$$\{\mathbf{0}\} \lhd \langle \mathbf{15} \rangle \lhd \langle \mathbf{3} \rangle \lhd \langle \mathbf{1} \rangle = G, \quad \mathbb{Z}/3\mathbb{Z}, \mathbb{Z}/5\mathbb{Z}, \mathbb{Z}/2\mathbb{Z};$$

$$\{\mathbf{0}\} \lhd \langle \mathbf{6} \rangle \lhd \langle \mathbf{3} \rangle \lhd \langle \mathbf{1} \rangle = G, \quad \mathbb{Z}/3\mathbb{Z}, \mathbb{Z}/2\mathbb{Z}, \mathbb{Z}/5\mathbb{Z};$$

$$\{\mathbf{0}\} \lhd \langle \mathbf{6} \rangle \lhd \langle \mathbf{2} \rangle \lhd \langle \mathbf{1} \rangle = G, \quad \mathbb{Z}/2\mathbb{Z}, \mathbb{Z}/3\mathbb{Z}, \mathbb{Z}/5\mathbb{Z};$$

$$\{\mathbf{0}\} \lhd \langle \mathbf{10} \rangle \lhd \langle \mathbf{2} \rangle \lhd \langle \mathbf{1} \rangle = G \quad \mathbb{Z}/2\mathbb{Z}, \mathbb{Z}/5\mathbb{Z}, \mathbb{Z}/3\mathbb{Z}.$$

Get to know composition series

1. Give an example of a group G together with subgroups K and N with $K \lhd N$ and $N \lhd G$ but K not normal in G.
2. What is the set of subquotients for the composition series for $\mathbb{Z}/12\mathbb{Z} = \{\mathbf{0}, \mathbf{1}, \ldots, \mathbf{11}\}$ given by

$$\{\mathbf{0}\} \lhd \langle \mathbf{6} \rangle \lhd \langle \mathbf{3} \rangle \lhd \langle \mathbf{1} \rangle = \mathbb{Z}/12\mathbb{Z}?$$

Can you give another composition series for $\mathbb{Z}/12\mathbb{Z}$? Contrast the sets of composition factors you obtain and compare them with the set of composition factors for A_4.

3. Which composition series are possible for $G = \mathbb{Z}/pq\mathbb{Z}$ where p and q are distinct primes? Suppose H is an abelian group of order p^n for $n \geq 1$. What are the composition factors of H? Which composition series are possible for $n = 3$?

4. The Fundamental Theorem of Finite Abelian Groups [31] states that a finite abelian group A is isomorphic to a direct product of products of cyclic groups whose orders are powers of prime integers coming from the prime factorization of the order of A. What does the Fundamental Theorem imply about the composition series of a finite abelian group A? Illustrate your answer with an example.

5. Suppose N is a normal subgroup of H and of K, subgroups of a group G, and that $H \lhd K$. Show that H/N is normal in K/N. Consider the mapping $\psi \colon K \to (K/N)/(H/N)$ given by $\psi(k) = kN(H/N)$. Show that ψ is a surjective homomorphism with kernel H. Show that this proves the *Third Isomorphism Theorem*, namely $K/H \cong (K/N)/(H/N)$.

In fact, the order of our choices of maximal normal subgroups does not matter.

The Jordan–Hölder Theorem *Given two composition series for a finite group G,*

$$G_{\bullet} \colon \{e\} = G_0 \lhd G_1 \lhd \cdots \lhd G_n = G \text{ and } H_{\bullet} \colon \{e\} = H_0 \lhd H_1 \lhd \cdots \lhd H_k = G,$$

we have $n = k$, and there is a permutation σ of $\{1, \ldots, n\}$ such that, for all i, $H_i/H_{i-1} \cong G_{\sigma(i)}/G_{\sigma(i)-1}$.

Proof We follow the proof of [63] and proceed by induction on the order of G. If $\#G = 1$, then the theorem is trivially true. Suppose $\#G = n$ and the theorem is true for all groups of order less than n. With each composition series, associate the ordered list of nontrivial simple groups given by the subquotients

$$(G/G_{n-1}, G_{n-1}/G_{n-2}, \ldots, G_2/G_1, G_1) \text{ and}$$
$$(G/H_{k-1}, H_{k-1}/H_{k-2}, \ldots, H_2/H_1, H_1).$$

If $G_{n-1} = H_{k-1}$, then since $\#G_{n-1} < \#G$, the theorem holds by induction for the composition series ending at G_{n-1} and at H_{k-1} and hence for G.

Let us assume that $G_{n-1} \neq H_{k-1}$. The theorem applies to the smaller groups G_{n-1} and H_{k-1}. Consider the composition series $\{e\} = G_0 \lhd G_1 \lhd \cdots \lhd G_{n-1}$ and $\{e\} = H_0 \lhd H_1 \lhd H_2 \lhd \cdots \lhd H_{k-1}$. Because $H_{k-1} \lhd G$, $G_{n-1} \cap H_{k-1}$ is normal in G_{n-1}, and, by the Second Isomorphism Theorem, the group $K = G_{n-1} \cap H_{k-1}$ satisfies

$$G_{n-1}/K = G_{n-1}/(G_{n-1} \cap H_{k-1}) \cong G_{n-1}H_{k-1}/H_{k-1} = G/H_{k-1}.$$

The last equality holds because $H_{k-1} \lhd G_{n-1}H_{k-1} \lhd G$, and G/H_{k-1} is simple: since $G_{n-1} \neq H_{k-1}$, $G_{n-1}H_{k-1} = G$. It follows that K is a maximal normal subgroup of G_{n-1} and, symmetrically, of H_{k-1} with $H_{k-1}/K \cong G/G_{n-1}$. Let $K_\bullet$ denote any composition series for K, $K_\bullet$: $\{e\} = K_0 \lhd K_1 \lhd \cdots \lhd K_{r-1} \lhd K_r = K$. Then $K_\bullet \lhd G_{n-1}$ and $K_\bullet \lhd H_{k-1}$ are also composition series for G_{n-1} and H_{k-1}, respectively. Thus the sequences of subquotients $(G_{n-1}/G_{n-2}, G_{n-2}/G_{n-3}, \ldots)$ and $(G_{n-1}/K, K/K_{r-1}, \ldots)$ are the same up to permutation. Similarly, $(H_{k-1}/H_{k-2}, H_{k-2}/H_{k-3}, \ldots)$ and $(H_{k-1}/K, K/K_{r-1}, \ldots)$ are the same up to permutation. By induction, $n - 1 = r + 1 = k - 1$ and $n = k$. Introducing G to the subquotients, we find

$$(G/G_{n-1}, \overbrace{G_{n-1}/G_{n-2}, G_{n-2}/G_{n-3}, \ldots})$$

$$\Big\downarrow \cong \quad \text{by induction, the same up to order}$$

$$(H_{k-1}/K, \overbrace{G_{n-1}/K, K/K_{r-1}, \ldots})$$

$$\Longleftrightarrow \quad \text{switch first two entries}$$

$$(G_{n-1}/K, H_{k-1}/K, K/K_{r-1}, \ldots)$$

$$\Big\downarrow \cong$$

$$(G/H_{k-1}, H_{k-1}/K, K/K_{r-1}, \ldots)$$

$$\text{by induction, the same up to order}$$

$$(G/H_{k-1}, \overbrace{H_{k-1}/H_{k-2}, H_{k-2}/H_{k-3}, \ldots}).$$

The composite of these permutations of indexed subquotients proves the theorem. □

The Jordan–Hölder Theorem suggests a strategy for analyzing finite groups. Because the set of subquotients is a collection of simple groups, by inductively building the subgroups in the composition series from the simple groups in the collection of composition factors, we can arrive at any group. Begin with a simple group G_1 and a subquotient G_2/G_1, from which we find a group G_2 and a surjection $G_2 \to G_2/G_1$ whose kernel is isomorphic to G_1. Then repeat with G_2 and G_3/G_2 to construct G_3, and continue.

In general, we can put these data together into a diagram called a *short exact sequence*:

$$\{e\} \to K \to G \to Q \to \{e\},$$

a sequence of homomorphisms for which the kernel of each outgoing homomorphism is the image of the incoming homomorphism. With K and Q known, we call G an *extension of Q by K*. In this parlance, G_2 is an extension of G_2/G_1 by G_1.

There is plenty of room for variation. For example, if $K \cong \mathbb{Z}/2\mathbb{Z} \cong Q$, then there are two extensions of Q by K, the *trivial* extension $G = K \times Q \cong \mathbb{Z}/2\mathbb{Z} \times \mathbb{Z}/2\mathbb{Z}$ and $G = \mathbb{Z}/4\mathbb{Z}$ with the surjection $\mathbb{Z}/4\mathbb{Z} \to \mathbb{Z}/2\mathbb{Z}$ given by $1 \mapsto 1$, whose kernel is the subgroup generated by 2. Another nontrivial example is the nonabelian group Σ_3 together with the surjection sgn: $\Sigma_3 \to \{\pm 1\} \cong \mathbb{Z}/2\mathbb{Z}$ given by

$$\mathrm{sgn}(\sigma) = \text{the sign of the permutation } \sigma.$$

This homomorphism has the kernel $A_3 = \langle (1, 2, 3) \rangle \cong \mathbb{Z}/3\mathbb{Z}$. Thus an extension of an abelian group by an abelian group need not be abelian.

Given a pair of groups K and Q, the general problem of which extensions G of Q by K are possible is quite difficult. (See Chapter 17 of [24] for an introduction to homological algebra.) However, the first step toward a method of classification is finding the irreducible parts of the construction, the finite simple groups – they are the atoms, and finite groups the molecules.

A particular family of groups, whose importance will be made clear in later chapters, is the class of *solvable groups*.

Definition 1.11 A finite group G is *solvable* if G has a composition series all of whose subquotients are abelian.

Since the subquotients are simple and abelian, they are each isomorphic to some $\mathbb{Z}/p_i\mathbb{Z}$ for a prime p_i.

Proposition 1.12 *Subgroups and quotients of solvable groups are solvable. Given a short exact sequence* $\{e\} \to K \to G \to Q \to \{e\}$, *the extension G is solvable if and only if K and Q are solvable.*

Proof Suppose G is solvable, and let $\{e\} = G_0 \triangleleft G_1 \triangleleft \cdots \triangleleft G_{n-1} \triangleleft G_n = G$ be a composition series for G. If K is any subgroup of G, then the series

$$\{e\} = (G_0 \cap K) \triangleleft (G_1 \cap K) \triangleleft \cdots \triangleleft (G_{n-1} \cap K) \triangleleft (G_n \cap K) = K$$

is a normal series for K. The canonical mapping $\pi: G_i \cap K \rightarrow G_i/G_{i-1}$, given by $\pi(g) = gG_{i-1}$, has the kernel given by $G_{i-1} \cap K$. Therefore the quotient $(G_i \cap K)/(G_{i-1} \cap K)$ is isomorphic to a subgroup of G_i/G_{i-1} and hence is abelian. Furthermore, the image of π is a normal subgroup of the abelian group G_i/G_{i-1}, which is simple; either π is surjective and $(G_i \cap K)/(G_{i-1} \cap K)$ is isomorphic to G_i/G_{i-1}, or π is the trivial mapping and $G_{i-1} \cap K = G_i \cap K$. By eliminating repeats in the normal series for K we obtain a composition series with all subquotients abelian, and K is solvable.

For Q, we have a surjective mapping $\phi: G \rightarrow Q$. The Correspondence Theorem gives a bijection between subgroups of G that contain $K \simeq \ker\psi$ and subgroups of Q. This allows us to send the composition series of G to a normal series for Q given by $Q_i = \phi(G_i)$. However, $\phi(G_i)$ is isomorphic to $G_i/\ker\phi \cong G_i/K$. By the Third Isomorphism Theorem $Q_i/Q_{i-1} \cong (G_i/K)/(G_{i-1}/K) \cong G_i/G_{i-1}$, and so Q is also solvable.

If we assume that K and Q are solvable with composition series $\{e\} = Q_0 \lhd Q_1 \lhd \cdots \lhd Q_{r-1} \lhd Q_r = Q$ and $\{e\} = K_0 \lhd K_1 \lhd \cdots \lhd K_{s-1} \lhd K_s = K$, then consider the preimage in G of the subgroups Q_i, $G_i = \phi^{-1}(Q_i) = \{g \in G \mid \phi(g) \in Q_i\}$. Then $G_0 = \ker\phi = K$. Concatenate the series to get the normal series

$$\{e\} = K_0 \lhd K_1 \lhd \cdots \lhd K_{s-1} \lhd K_s = K \lhd G_1 \lhd \cdots \lhd G_{r-1} \lhd G_r = G,$$

for which $G_i/G_{i-1} \cong Q_i/Q_{i-1}$ and K_j/K_{j-1} are simple and abelian. Also, $G_1/K \cong \phi(G_1) = Q_1$ is also simple and abelian. Hence G is solvable. $\square$

1.3 Some Simple Groups

Not all groups are solvable. There exist nonabelian simple groups. Let us meet an important family of such groups, the alternating groups A_n for $n \geq 5$.

Lemma 1.13 *For any $n > 2$, A_n is generated by the set of 3-cycles in A_n.*

Proof. The alternating group consists of all permutations that can be expressed as a product of an even number of transpositions. In general we have $(a, b)(c, d) = (a, b, c)(b, c, d)$ and $(a, b)(b, c) = (a, b, c)$, so each product of a pair of transpositions can be replaced with a 3-cycle or a product of two 3-cycles. The set of 3-cycles generates A_n. $\square$

Theorem 1.14 *For $n \geq 5$, the alternating group A_n is simple.*

Proof Notice that $A_3 = \langle (1, 2, 3) \rangle \cong C_3$ is an abelian simple group and that A_4 contains V as a normal subgroup. In A_5 the types of the even permutations of five objects are $(1^5, 2^0, 3^0, 4^0, 5^0)$, $(1^1, 2^2, 3^0, 4^0, 5^0)$, $(1^2, 2^0, 3^1, 4^0, 5^0)$, and $(1^0, 2^0, 3^0, 4^0, 5^1)$. This gives us five conjugacy classes with the following cardinalities:

$[g_i]$	$[()]$	$[(1,2)(3,4)]$	$[(1,2,3)]$	$[(1,2,3,4,5)]$	$[(2,1,3,4,5)]$
$\#[g_i]$	1	15	20	12	12

By Proposition 1.6 we know that $[(1,2)(3,4)]_{A_5} = [(1,2)(3,4)]_{\Sigma_5}$ and $[(1,2,3)]_{A_5} = [(1,2,3)]_{\Sigma_5}$, making the counting easy. The type $(1^0, 2^0, 3^0, 4^0, 5^1)$ in Σ_5 splits into two equal pieces, each of cardinality 12.

Suppose N is a nontrivial normal subgroup of A_5. A normal subgroup is the union of conjugacy classes of the elements in the subgroup, so N can be written as a disjoint union of conjugacy classes. Of course, $() \in N$ implies $[()] \subset N$. It is also the case that $\#N$ divides 60. However, every sum of the cardinalities of conjugacy classes of A_5, 20, 15, 12, and 12, with 1 for $[()]$, gives a number that does not divide 60, except $\#N = 60$. Hence $N = A_5$, and A_5 is simple.

To establish the shape of an induction argument, we next prove that A_6 is simple. Let $St_i = \{\sigma \in A_6 \mid \sigma(i) = i\}$ denote the subgroup of A_6 that fixes $i \in \{1, 2, \ldots, 6\}$. For each i, St_i is isomorphic to A_5.

Suppose that N is a normal subgroup of A_6 and $N \cap St_i \neq \{()\}$ for some i. Then $N \cap St_i$ is normal in St_i. Since A_5 is simple, $N \cap St_i = St_i$. Then N contains a 3-cycle (a, b, c) for a, b, c not equal to i. Because $(b, x)(c, y)(a, b, c) \times (b, x)(c, y) = (a, x, y)$ and $(w, z, a)(a, x, y)(a, z.w) = (x, y, z)$, every 3-cycle is in N. By Lemma 1.13, $N = A_6$.

If $N \cap St_i = \{()\}$ for all i, then N must contain only permutations that have no fixed values. Since 6-cycles are odd, for some labeling, either $\eta = (1, 2)(3, 4, 5, 6)$ or $\gamma = (1, 2, 3)(4, 5, 6)$ must be in N. However, η^2 fixes 1 and 2, and so $\eta^2 \in St_1 \cap N$. If $\gamma \in N$ and you compute $\tau = ((2, 3, 4)\gamma \times (4, 3, 2))\gamma^{-1}$, then $\tau \in N$, and τ fixes 6. Hence $N \cap St_6 \neq \{()\}$. These computations contradict $N \cap St_i = \{()\}$ for all i and imply that $N = A_6$ and A_6 is simple.

By induction, suppose A_k is simple for $6 \leq k < n$. Suppose $N \neq \{()\}$ is a normal subgroup of A_n. Let $\tau \in N$ with $\tau(i) = j \neq i$, and let k and l be integers that are not equal to i or j. Let $\beta = (j, k, l)$. Notice that $\tau\beta(i) = \tau(i) = j$ and $\beta\tau(i) = \beta(j) = k$. Since $j \neq k$, we have $\tau\beta \neq \beta\tau$, and the commutator $\tau\beta\tau^{-1}\beta^{-1} \neq ()$. It is also the case that $\tau(\beta\tau^{-1}\beta^{-1})$ is in N because N is normal in A_n. However, $\tau\beta\tau^{-1} = (\tau(j), \tau(k), \tau(l))$, and so $\tau\beta\tau^{-1}\beta^{-1} = (\tau(j), \tau(k), \tau(l))(l, k, j)$.

Let $B = \{\sigma \in A_n \mid \sigma(m) = m$ for all $m \neq j, k, l, \tau(j), \tau(k), \tau(l)\}$. If $\tau(j), \tau(k)$, and $\tau(l)$ are distinct from j, k, and l, then B is isomorphic to A_6.

If they are not distinct, then simply choose other values from $[n]$ to make B isomorphic to A_6. We then have that $N \cap B$ is normal in B and $\tau \beta \tau^{-1} \beta^{-1} \in N \cap B$. Since $B \cong A_6$ and $\{(\ )\} \neq N \cap B$, we have $N \cap B = B$. Then N contains a 3-cycle $(\tau(j), \tau(k), \tau(l))$, and so every 3-cycle is in N. Because the 3-cycles generate A_n, $N = A_n$, and A_n is simple. $\qquad\qquad\square$

Corollary 1.15 *The symmetric group on n letters Σ_n is not solvable for $n \geq 5$.*

Proof The composition series for Σ_n is $\{(\ )\} \lhd A_n \lhd \Sigma_n$ with composition factors $\mathbb{Z}/2\mathbb{Z}$ and A_n. $\qquad\qquad\square$

In the chapters that follow, we will explore the questions when a finite group G is simple and when is G solvable.

Exercises

1.1 Suppose H and K are subgroups of a group G. Form the set

$$HK = \{hk \mid h \in H, k \in K\}.$$

Show that HK is a subgroup of G if and only if $HK = KH = \{kh \mid k \in K, h \in H\}$. Furthermore, show that if H and K are normal subgroups of G, then HK is also a normal subgroup of G.

1.2 Work out a group table for the dihedral groups D_8 and D_{10}. What are the conjugacy classes in these groups? What are the conjugacy classes in general for D_{2n}?

1.3 What is a composition series for the group D_{2n}? How does it depend upon n?

1.4 Suppose a group G has a normal series (not necessarily a composition series), $\{e\} = G_0 \lhd G_1 \lhd \cdots \lhd G_{n-1} \lhd G_n = G$ for which the subquotients are all abelian. Prove that G is a solvable group.

Definition 1.16 For elements g and h in a group G, the *commutator* $[g, h]$ is defined by $[g, h] = ghg^{-1}h^{-1}$. The *commutator subgroup* of G is defined by

$$G' = [G, G] = \langle [g, h] \mid g, h \in G \rangle$$

$$= \text{the subgroup generated by all commutators in } G.$$

The *derived* series of a group is given by letting $G^{(0)} = G$ and $G^{(n)} = (G^{(n-1)})'$ for $n > 0$. This gives $G = G^{(0)} \supset G^{(1)} \supset G^{(2)} \supset \cdots \supset G^{(n)} \supset \cdots$.

1.5   Show that G' is a normal subgroup of G and that G/G' is abelian.

1.6   Suppose that N is a normal subgroup of G. Suppose further that G/N is abelian. Show that N must contain G', the commutator subgroup of G.

1.7   By Exercise 1.5, the derived series is a normal series:

$$G = G^{(0)} \rhd G^{(1)} \rhd G^{(2)} \rhd \cdots \rhd G^{(n)} \rhd \cdots .$$

Show that a finite group G is solvable if and only if $G^{(n)} = \{e\}$ for some $n > 0$.

1.8   Let us extend the collection of isomorphism theorems to add the *Zassenhaus Lemma*: Given four subgroups $N \lhd N^*$ and $M \lhd M^*$ of a group G, prove that

$$N\left(N^* \cap M\right) \lhd N\left(N^* \cap M^*\right), \quad M\left(M^* \cap N\right) \lhd M\left(M^* \cap N^*\right),$$

$$\text{and} \quad \frac{N(N^* \cap M^*)}{N(N^* \cap M)} \cong \frac{M(M^* \cap N^*)}{M(M^* \cap N)}.$$

1.9   A *refinement* of a normal series $\{e\} = G_0 \lhd G_1 \lhd \cdots \lhd G_{n-1} \lhd G_n = G$ is another normal series $\{e\} = H_0 \lhd H_1 \lhd \cdots \lhd H_{k-1} \lhd H_k = G$ for which each G_i is equal to $H_{K(i)}$ for some $0 \le K(i) \le k$. Prove the *Schreier Refinement Theorem*: Any two normal series for a group G have refinements that share the same subquotients. (Hint: You can insert normal subgroups between $G_{i-1} \lhd G_i$ by letting $G_{ij} = G_{i-1}(G_i \cap H_j)$. The Second Isomorphism Theorem and the Zassenhaus Lemma are handy tools.) Show that this leads to another proof of the Jordan–Hölder Theorem.

1.10 Show that the order of an extension of finite groups K and Q is $(\#K)(\#Q)$. Suppose $\#K = p$ and $\#Q = q$, where p and q are distinct primes. If G is an extension of K and Q, what are the possible composition series of G? More generally, show that two solvable groups of the same order have the same composition factors.

2

Groups Acting on Sets

All these propositions have been deduced from the theory of permutations.

ÉVARISTE GALOIS, *Analysis of a memoir on the algebraic solution of equations*

2.1 Basic Definitions

The finite groups most familiar to students of a first course in Modern Algebra arise out of arithmetic (the cyclic groups, $\langle g \rangle \cong C_n \cong \mathbb{Z}/n\mathbb{Z}$), geometry (the dihedral groups, D_{2n}), and combinatorics (the symmetric groups, Σ_n). New examples can be manufactured from abstract constructions, such as products and quotients, subgroups like the commutator subgroup, the center, and so on.

 Some mathematical objects come equipped with a group of symmetries that preserve the nature of the object. The symmetries of a square give one of the simplest examples. Let D_8 denote the collection of invertible linear mappings of the plane to itself that take a square with center at the origin onto itself. Label the corners of the square counterclockwise 1, 2, 3, 4. Rotation counterclockwise by 90° gives the permutation of our labels (in cycle notation) $r = (1, 2, 3, 4)$. Notice that $r^4 = \text{Id}$; reflection across the vertical axis through the origin gives the permutation of the corners $f = (1, 4)(2, 3)$ and $f^2 = \text{Id}$. A computation shows $fr = r^3 f$ (rotating and flipping is the same as flipping and rotating three times). We can list the elements of the group $D_8 = \{\text{Id}, r, r^2, r^3, f, rf, r^2 f, r^3 f\}$.

A richer example is the group $\mathbf{Doc}^+$ of rotational symmetries of a regular dodecahedron. (This group is denoted by $[5, 3]^+$ in Coxeter's notation [17].) Label the set of the 20 vertices of the dodecahedron by a through t.

19

Each rotation gives a bijection of this set to itself, that is, a permutation of $A = \{a, \ldots, t\}$. The group of rotational symmetries of the dodecahedron determines a subgroup of $\Sigma(A)$, the symmetric group on A. Because rotations preserve distances, only certain permutations of A are possible.

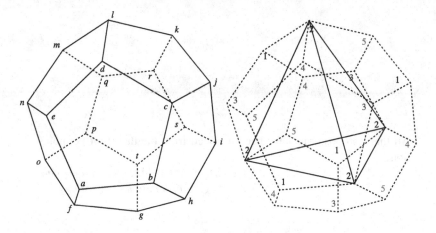

Figure 2.1

To get a better handle on the structure of **Doc**$^+$, we make a geometric observation: The vertices of a regular dodecahedron can be partitioned into five subsets of four vertices, each forming a regular tetrahedron. Rotations of the dodecahedron preserve all distances, and so the tetrahedra are permuted among each other by a rotation. With the labels a through t the tetrahedra are $\mathbf{1} = \{a, j, m, t\}$, $\mathbf{2} = \{b, l, o, s\}$, $\mathbf{3} = \{c, g, n, r\}$, $\mathbf{4} = \{d, f, i, q\}$, and $\mathbf{5} = \{e, h, k, p\}$.

Rotations in $\mathbb{R}^3$ are around an axis that passes through the center of the dodecahedron. Particular elements of **Doc**$^+$ are pictured in Fig. 2.2. An axis can join the center of a face to the center of the opposite face, giving a rotation σ through $2k\pi/5$ radians with k an integer; or the axis is through the center of an edge to the center of the opposite edge, giving a rotation τ through π; or the axis can be between a vertex and its opposite vertex, giving a rotation ρ through $2l\pi/3$ for an integer l. For example, rotating the face $abcde$ around its center through $2\pi/5$ gives the permutation σ of a through t:

$$\sigma = \begin{pmatrix} a & b & c & d & e & f & g & h & i & j & k & l & m & n & o & p & q & r & s & t \\ b & c & d & e & a & h & i & j & k & l & m & n & o & f & g & t & p & q & r & s \end{pmatrix}.$$

In cycle notation we have $\sigma = (a, b, c, d, e)(f, h, j, l, n)(g, i, k, m, o)(p, t, s, r, q)$. Notice that this computation shows that all other face–face axis rota-

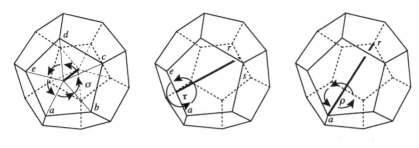

Figure 2.2

tions are powers of a product of four 5-cycles. Restricting to the tetrahedra, this permutation induces $(1, 2, 3, 4, 5)$, an element of Σ_5, the symmetric group on $[5] = \{1, 2, 3, 4, 5\}$.

The rotation τ around the axis joining the midpoint of the edge ae and the midpoint of rs through π radians has the cycle expression

$$\tau = (a, e)(b, n)(c, o)(d, f)(g, l)(h, m)(i, q)(j, p)(k, t)(r, s).$$

On the set of tetrahedra, τ induces $(1, 5)(2, 3)$, another element of Σ_5. Finally, the rotation ρ around the axis joining a and r through angle $2\pi/3$ has the form

$$\rho = (b, e, f)(c, n, g)(d, o, h)(i, l, p)(j, m, t)(k, q, s).$$

On the tetrahedra, we get the permutation $(2, 5, 4)$.

Notice that $(1, 2, 3, 4, 5)$, $(1, 2)(3, 4)$, and $(2, 5, 4)$ all lie in the subgroup A_5 of Σ_5 of even permutations of $[5]$. After developing a more general framework, we will be able to identify $\mathbf{Doc}^+$ with a subgroup of A_5.

In the examples of the square and the dodecahedron, we have a group of symmetries of a set of vertices determined by the geometry. More generally, let X denote a set and define the *symmetry group on X* as

$$\Sigma(X) = \{\text{bijections } f \colon X \to X\}.$$

A binary operation on $\Sigma(X)$ is given by composing bijections. With this operation, $\Sigma(X)$ is a group because the composite of bijections is a bijection; the identity mapping $\mathrm{Id} \colon X \to X$ acts as an identity for composition; and the inverse of a bijection is a bijection. Being functions, composition is associative.

When $X = [n] = \{1, 2, \ldots, n\}$, we write $\Sigma([n]) = \Sigma_n$, the *symmetric group on n letters*. The order of Σ_n is $n!$.

Definition 2.1 A group G *acts on a set X* if each $g \in G$ determines a bijection $\rho(g) \colon X \to X$ in such a way that for e, the identity element in G, $\rho(e)(x) =$

$\text{Id}(x) = x$, and $\rho(gh) = \rho(g) \circ \rho(h)$ for all $g, h \in G$. If G acts on a set X, then we call X a *G-set*.

In the examples, D_8 acts on the set of the vertices of a square, and **Doc**$^+$ similarly acts on the set of vertices of the dodecahedron by rotations; **Doc**$^+$ also acts on the set of tetrahedra inside the dodecahedron. More generally, D_{2n} acts on the set of vertices of a regular n-gon.

Proposition 2.2 *A group G acts on a set X precisely when there is a homomorphism $\rho \colon G \to \Sigma(X)$, called the action of G on X.*

Proof Because $\text{Id} = \rho(e) = \rho(gg^{-1}) = \rho(g) \circ \rho(g^{-1})$, the function $\rho(g)$ has an inverse, and $\rho(g) \in \Sigma(X)$. It follows that $\rho(g)^{-1} = \rho(g^{-1})$. Because $\rho(gh) = \rho(g) \circ \rho(h)$, $\rho \colon G \to \Sigma(X)$ is a homomorphism.

Conversely, any homomorphism $\rho \colon G \to \Sigma(X)$ determines an action G on X: Each $\rho(g)$ is a bijection $X \to X$, and $\rho(gh) = \rho(g) \circ \rho(h)$ because the operation on the symmetric group $\Sigma(X)$ is composition. A homomorphism takes $e \in G$ to the identity element of $\Sigma(X)$, which is $\text{Id} \colon X \to X$. $\square$

With this formulation, we can obtain new group actions from a given action $\rho \colon G \to \Sigma(X)$. If $\phi \colon H \to G$ is any group homomorphism, then $\rho \circ \phi \colon H \to \Sigma(X)$ determines an action of H on X. In particular, for any subgroup H of G, the inclusion homomorphism of H into G leads to the *restriction* of ρ to H, $\rho|_H \colon H \to \Sigma(X)$. For example, the subgroup generated by the rotation $\langle r \rangle \subset D_{2n}$ acts on the set of vertices of a regular n-gon.

The examples D_8 and **Doc**$^+$ begin with a set X, and geometry determines a group action on X. If we are given a group G, then are there natural choices of sets X on which G acts? A rich source of group actions is provided by the group itself.

(a) If H is a subgroup of a group G, let $X = G/H = \{gH \mid g \in G\}$, the set of left cosets of H. For $x \in G$, define $\rho(x)(gH) = xgH$. Notice that $\rho(xy)(gH) = (xy)gH = x(ygH) = \rho(x)(ygH) = \rho(x)(\rho(y)(gH)) = (\rho(x) \circ \rho(y))(gH)$. Hence $\rho(xy) = \rho(x) \circ \rho(y)$. Also, $\rho(e)(gH) = egH = gH = \text{Id}(gH)$, so $\rho \colon G \to \Sigma(G/H)$ is a homomorphism, and G acts on G/H. When $H = \{e\}$, this action is on $X = G = G/\{e\}$, and it is referred to as the *left action* of G on itself for which $\rho(g)(h) = gh$.

(b) When we let $X = \mathcal{T}_G = \{H, \text{a subgroup of } G\}$, the collection of all subgroups of G, we can define $\rho_c(g)(H) = gHg^{-1}$, the *conjugate* of H by g. It is a simple exercise to show that gHg^{-1} is also a subgroup of G. We check that ρ_c defines an action of G on $\mathcal{T}_G$:

$$\rho_c(xy)(H) = (xy)H(xy)^{-1} = (xy)H\left(y^{-1}x^{-1}\right) = x\left(yHy^{-1}\right)x^{-1}$$
$$= \rho_c(x)\left(yHy^{-1}\right) = \rho_c(x)\left(\rho_c(y)(H)\right),$$

so $\rho_c(xy) = \rho_c(x) \circ \rho_c(y)$. Furthermore, $\rho_c(e)(H) = eHe^{-1} = H$. Thus G acts on the set of its subgroups by conjugation.

Similarly, G acts on itself by conjugation $\rho_c \colon G \to \Sigma(G)$ given by $\rho_c(g)(x) = gxg^{-1}$.

(c) Let $X = \binom{G}{k} = \{A \subset G \mid \#A = k\}$, the set of subsets of G containing k elements. If $x \in G$ and $A \in \binom{G}{k}$, then define $\rho(x)(A) = xA = \{xa \mid a \in A\}$. When G is a finite group and $k \leq \#G$, then $\#xA = \#A$ because the group operation has left cancelation, that is, $xa = xb$ implies $a = b$. To see that G acts on X, notice that $\rho(xy)(A) = (xy)A = \rho(x)(yA) = \rho(x)(\rho(y)(A))$, and so $\rho(xy) = \rho(x) \circ \rho(y)$. Clearly, $\rho(e)(A) = A$, and G acts on $\binom{G}{k}$ by left multiplication.

Get to know group actions

1. A source of group actions on infinite sets comes from matrices acting on $\mathbb{R}^n$. An example is the set $UT = \{\left[\begin{smallmatrix} a & b \\ 0 & c \end{smallmatrix}\right] \mid a, b, c \in \mathbb{R}, ac \neq 0\}$ acting on $\mathbb{R}$ by $\rho(\left[\begin{smallmatrix} a & b \\ 0 & c \end{smallmatrix}\right])(r) = (ar + b)/c$. Show that UT is a group and that $\rho \colon UT \to \Sigma(\mathbb{R})$ is an action of UT on $\mathbb{R}$.

2. A bracelet has n beads on it. Of course, you can rotate it or put it on after reflecting it across a diameter. Thus the group D_{2n} acts on any bracelet. However, if we can choose identical beads from a small set of colors, then some of the rotations and flips give the same bracelet. Work out what happens if there are five beads of two colors to make the bracelet. Which operations of D_{10} change the bracelet, and which do not?

3. What is the cardinality of the set $\binom{G}{k}$ in terms of $\#G$?

4. Suppose $\rho \colon G \to \Sigma(X)$ is an action of G on X and $f \colon X \to Y$ is a bijection. Show that f induces an isomorphism of groups $f_* \colon \Sigma(X) \to \Sigma(Y)$ by $f_*(\sigma) = f \circ \sigma \circ f^{-1}$. Composition with f_* determines a function ρ_f from G to $\Sigma(Y)$ by $g \mapsto \rho_f(g) = f_*(\rho(g))$. Show that $\rho_f \colon G \to \Sigma(Y)$ is an action of G on Y.

Definition 2.3 Two actions of $\rho_1 \colon G \to \Sigma(X_1)$ and $\rho_2 \colon G \to \Sigma(X_2)$ are *equivalent* if there is a bijection $f \colon X_1 \to X_2$ for which $f_* \circ \rho_1 = \rho_2$.

5. Show that the action of D_8 on the square and the action of D_8 on the set $\{1, 2, 3, 4\}$ given by $\rho_2(r) = (1, 2, 3, 4)$ and $\rho_2(f) = (1, 4)(2, 3)$ are equivalent.

2.2 Fundamental Concepts for G-Sets

For the rest of the chapter, we assume that G is a finite group and that G acts on a finite set X via a homomorphism $\rho \colon G \to \Sigma(X)$. With such an action we can associate some useful subgroups of G and certain subsets of X.

Definition 2.4 For $x \in X$, the *stabilizer* of x is given by

$$G_x = \big\{g \in G \mid \rho(g)(x) = x\big\} \subset G,$$

that is, the elements g of the group G that fix x under the bijection $\rho(g)$.

For example, the *trivial action* $\rho \colon G \to \Sigma(X)$ for which $\rho(g) = \mathrm{Id}_X$ for all g has $G_x = G$ for all $x \in X$. For D_8 acting on the labels $\{1, 2, 3, 4\}$ of the corners of a square, $(D_8)_1 = (D_8)_3 = \langle fr^3 \rangle$ and $(D_8)_2 = (D_8)_4 = \langle fr \rangle$, subgroups of order two.

Lemma 2.5 *The stabilizer G_x is a subgroup of G.*

Proof Suppose $g, h \in G_x$. Then $\rho(gh)(x) = \rho(g)(\rho(h)(x)) = \rho(g)(x) = x$, so $gh \in G_x$. Also, for $g \in G_x$, $\rho(g^{-1})(x) = \rho(g^{-1})(\rho(g)(x)) = \rho(g^{-1}g)(x) = \rho(e)(x) = \mathrm{Id}_X(x) = x$, and $g^{-1} \in G_x$. $\qquad\square$

For example, when G acts by conjugation on $\mathcal{T}_G$, the set of subgroups of G, the stabilizer of a particular subgroup is given by

$$G_H = \big\{g \in G \mid gHg^{-1} = H\big\} = N_G(H),$$

the *normalizer* of H in G. This subgroup contains H, and H is a normal subgroup of the normalizer. In fact, $N_G(H)$ is the largest subgroup of G in which H is a normal subgroup.

Definition 2.6 For $x \in X$, let $\mathcal{O}_x = \{\rho(g)(x) \mid g \in G\} \subset X$ denote the *orbit* of x.

The orbits of the trivial action are the singleton subsets of X. If $X = G/H$ is the set of left cosets of a subgroup H of G, with the action $\rho(g)(xH) = gxH$, the stabilizer subgroup of a coset xH is given by $G_{xH} = \{g \in G \mid gxH =$

xH}. However, $gxH = xH$ implies $x^{-1}gxH = H$, that is, $x^{-1}gx \in H$. Thus G_{xH} is the set of $g \in G$ for which conjugation by x^{-1} lands in H, $G_{xH} = \{g \in G \mid x^{-1}gx \in H\} = xHx^{-1}$.

The orbit of xH can be determined by noticing that $gH = (gx^{-1})xH = \rho(gx^{-1})(xH)$, and so $gH \in \mathcal{O}_{xH}$ for any $gH \in G/H$. Hence, for any $xH \in G/H$, $\mathcal{O}_{xH} = G/H = X$.

For the conjugation action of G on $\mathcal{T}_G$, the set of subgroups of G, the orbits are given by $\mathcal{O}_H = \{gHg^{-1} \mid g \in G\}$, the set of conjugates of H.

Proposition 2.7 *For any action of G on X, $\rho: G \to \Sigma(X)$, and $x, y \in X$, either $\mathcal{O}_x = \mathcal{O}_y$ or $\mathcal{O}_x \cap \mathcal{O}_y = \emptyset$.*

Proof Suppose $\mathcal{O}_x \cap \mathcal{O}_y \neq \emptyset$ and $z \in \mathcal{O}_x \cap \mathcal{O}_y$. Then $z = \rho(g)(x) = \rho(h)(y)$ for some $g, h \in G$. Then $y = \rho(h^{-1})(\rho(h)(y)) = \rho(h^{-1})(\rho(g)(x)) = \rho(h^{-1}g)(x)$, and $y \in \mathcal{O}_x$. Also, $\rho(k)(y) = \rho(k) \circ \rho(h^{-1}g)(x)$ for any $k \in G$, so $\mathcal{O}_y \subset \mathcal{O}_x$. Reversing the roles of x and y, we have $\mathcal{O}_x \subset \mathcal{O}_y$. Hence $\mathcal{O}_x = \mathcal{O}_y$. □

It follows from the proposition that X is a disjoint union of distinct orbits, $X = \mathcal{O}_{x_1} \cup \mathcal{O}_{x_2} \cup \cdots \cup \mathcal{O}_{x_m}$, where $\mathcal{O}_{x_i} \cap \mathcal{O}_{x_j} = \emptyset$ for $i \neq j$.

For a G-set X, what is the cardinality of an orbit $\mathcal{O}_x$ or of a stabilizer subgroup G_x? How are these cardinalities related to the order of G and the cardinality of X? There is a fundamental relation among these quantities.

The Orbit-Stabilizer Theorem *If G is a finite group acting on a finite set X, then for any $x \in X$, $\#\mathcal{O}_x = [G : G_x] = \#(G/G_x)$.*

Proof Consider the mapping $\phi: G/G_x \to \mathcal{O}_x$ given by $\phi(hG_x) = \rho(h)(x)$. Notice that ϕ is well-defined: if $gG_x = hG_x$, then $h^{-1}g \in G_x$, and so $\rho(h^{-1}g)(x) = x$, that is, $\rho(g)(x) = \rho(h)(x)$. For $\rho(g)(x) \in \mathcal{O}_x$, we have $\rho(g)(x) = \phi(gG_x)$, and so ϕ is surjective. If $\phi(gG_x) = \phi(hG_x)$, then $\rho(g)(x) = \rho(h)(x)$, which implies $h^{-1}g \in G_x$, and so $gG_x = hG_x$, and ϕ is injective. Hence ϕ is a bijection between G/G_x and $\mathcal{O}_x$. □

Get to know stabilizers and orbits

1. Let us consider the example of the dodecahedron. The set $A = \{a, b, \ldots, t\}$ of vertices is acted on by the group of rotational symmetries **Doc$^+$**. Given a vertex, say a, show that the orbit of a is all of A, that is, any two vertices are related by a rotation.

2. The rotations σ, τ, and ρ in **Doc**$^+$ are defined using the face $\{a, b, c, d, e\}$, the edge $\{a, e\}$, and the vertex a, respectively. Use the representations of these rotations to argue that the stabilizer $G_a = $ **Doc**$_a^+$ has three elements, $\{\text{Id}, \rho, \rho^2\}$.

3. We associated elements of **Doc**$^+$ with elements of A_5 by thinking of a rotation as a permutation of the tetrahedra inside the dodecahedron. Show that this association is injective, that is, if r_1 and r_2 are rotations of the dodecahedron that induce the same permutation of tetrahedra, then $r_1 = r_2$.

4. A consequence of the Orbit-Stabilizer Theorem $\#\mathcal{O}_x = [G : G_x] = \#G/\#G_x$ is that $\#G = \#G_x \cdot \#\mathcal{O}_x$. Use the previous exercises to show that $\#\textbf{Doc}^+ = 60 = \#A_5$. We showed that **Doc**$^+$ can be treated as a subgroup of A_5, so it follows that **Doc**$^+$ is isomorphic to A_5.

5. Carry out a similar analysis for D_8 acting on the vertices of a square. Remember to consider the reflections. Let **Cu**$^+$ denote the group of rotations of a cube. Use the tools developed so far to determine the structure of **Cu**$^+$.

Corollary 2.8 *If G is a finite group acting on a finite set X, then for any $x \in X$, $\#\mathcal{O}_x$ divides the order of G.*

Proof   The value of the index $[G : G_x]$ is a divisor of the order of G.   □

Divisibility, used carefully, can lead to some powerful consequences. The two extremes of the size of an orbit have a special designations.

Definition 2.9 If G acts on the set X, then $x \in X$ is a *fixed element* if $\mathcal{O}_x = \{x\}$, that is, $\rho(g)(x) = x$ for all $g \in G$. An action is called *transitive* if $\mathcal{O}_y = X$ for all $y \in X$, that is, for all $y, z \in X$, there is $h \in G$ such that $\rho(h)(y) = z$.

It follows from the definition that for a fixed element x, $G_x = G$, that is, G acts as the identity on $\{x\}$.

When the action is transitive, $\#X = [G : G_y] = \#G/\#G_y$, and so $\#G = \#X \cdot \#G_y$, and $\#X$ must divide the order of G. An example of a transitive action is given by G acting by left multiplication on G/H for H a subgroup of G.

Let us explore some of the fundamental algebraic concepts for actions of a group G on sets. If X is a G-set, then a subset $Y \subset X$ is *G-invariant* if for all $y \in Y$ and $g \in G$, $\rho(g)(y) \in Y$. The action ρ on X is also an action of G on

Y in this case, and Y is a G-set. For example, let

$$X^G = \{x \in X \mid \rho(g)(x) = x \text{ for all } g \in G\}$$

denote the *set of fixed elements* of the G-set X. In the case $X^G \neq \emptyset$, X^G is G-invariant, and the action of G on X^G induced by ρ is the trivial action.

The orbits of any action are G-invariant subsets: $\mathcal{O}_x = \{\rho(g)(x) \mid g \in G\}$, so $\rho(h)(\rho(g)(x)) = \rho(hg)(x) \in \mathcal{O}_x$ for all $h \in G$, and $\mathcal{O}_x$ is G-invariant. It is also the case that the action of G on $\mathcal{O}_x$ is transitive: if $\rho(g)(x)$ and $\rho(h)(x)$ are in $\mathcal{O}_x$, then $\rho(h)(x) = \rho(hg^{-1})(\rho(g)(x))$. Transitive actions are *irreducible* in the sense that they do not contain nontrivial G-invariant subsets: If $Y \subset X$ is a nontrivial G-invariant subset and $y \in Y$, then $\rho(g)(y)$ is in Y for all $g \in G$. Because the action is transitive, this includes every element of X, and so $Y = X$.

The orbit decomposition $X = \mathcal{O}_{x_1} \cup \mathcal{O}_{x_2} \cup \cdots \mathcal{O}_{x_n}$ where the orbits are distinct is a partition of X into a disjoint union of G-invariant subsets on which G acts transitively, that is, a decomposition into irreducible pieces.

Given two G-sets (X, ρ_X) and (Y, ρ_Y), we can combine them to produce a new G-set. The formal definition of the *disjoint union* of sets X and Y is given as the union

$$X \sqcup Y = (X \times \{0\}) \cup (Y \times \{1\})$$

for $0 \neq 1$. Since $(X \times \{0\}) \cap (Y \times \{1\}) = \emptyset$, this union is disjoint. Let G act on $X \sqcup Y$ by $\rho: G \to \Sigma(X \sqcup Y)$, $\rho(g)((x, 0)) = (\rho_X(g)(x), 0)$ and $\rho(g)((y, 1)) = (\rho_Y(g)(y), 1)$. This construction gives a binary operation $(X, Y) \mapsto X \sqcup Y$ on the collection of all G-sets, a kind of addition of G-sets. The action of G on $X \times Y$ given by $\rho(g)(x, y) = (\rho_X(x), \rho_Y(y))$ determines a multiplication of G-sets.

The orbit decomposition of (X, ρ) can be expressed as such a sum. If $Y \subset X$ is any G-invariant subset, then the complement $X - Y$ is also G-invariant. To see this, suppose $z \in X - Y$ and $\rho(g)(z) = y \in Y$. Then $z = \rho(g^{-1})(y)$, and so z is in Y, a contradiction. Thus $X = Y \cup (X - Y)$, a disjoint union of G-sets, whenever Y is a G-invariant subset of X. Iterating this procedure leads to the orbit decomposition of X.

The stabilizer subgroup of $x \in X$, $G_x = \{g \in G \mid \rho(g)(x) = x\}$, is the subgroup of elements of G that fix x. Reversing the roles of G and X, we can define the *fixed set* associated with $g \in G$ by

$$\text{Fix}(g) = \{x \in X \mid \rho(g)(x) = x\}.$$

The fixed sets and the orbits of an action are intimately related:

Burnside's Orbit Counting Theorem *If a finite group G acts on a finite set X, then*

$$\frac{1}{\#G} \sum_{g \in G} \#\mathrm{Fix}(g) = \# \text{ orbits of } G \text{ acting on } X.$$

Proof The theorem follows by a counting argument in which we determine the cardinality of a set in two different ways. Consider the set

$$W = \{(g, x) \in G \times X \mid \rho(g)(x) = x\}.$$

The number of pairs (g_0, x) in W for a particular $g_0 \in G$ is $\#\mathrm{Fix}(g_0)$. The number of pairs (g, x_0) for a particular $x_0 \in X$ is the order of the stabilizer of x_0, $\#G_{x_0}$.

To tame the counting, notice that when x and y are in the same orbit, G_x and G_y have the same cardinality.

Lemma 2.10 *If $\rho \colon G \to \Sigma(X)$ is a group action, then for all $g \in G$ and $x \in X$, we have $G_{\rho(g)(x)} = gG_x g^{-1}$.*

Proof If $h \in G_{\rho(g)(x)}$, then $\rho(h)(\rho(g)(x)) = \rho(g)(x)$. This implies that $\rho(g^{-1}) \circ \rho(hg)(x) = x$, that is, $g^{-1}hg \in G_x$ and $h \in gG_x g^{-1}$. In the other direction, if $k \in G_x$, then

$$\rho(gkg^{-1})(\rho(g)(x)) = \rho(gk)(x) = \rho(g)(\rho(k)(x)) = \rho(g)(x),$$

that is, $gkg^{-1} \in G_{\rho(g)(x)}$.   □

By Lemma 2.10 we compute $\#W$ by summing over $x \in X$, giving

$$\#W = \sum_{x \in X} \#G_x = \sum_{\text{orbits}} \#\mathcal{O}_x \cdot \#G_x = \sum_{\text{orbits}} [G : G_x] \cdot \#G_x = \#G \cdot \#\text{orbits}.$$

Computing $\#W$ by summing over $g \in G$ gives $\#W = \sum_{g \in G} \#\mathrm{Fix}(g)$, and Burnside's Theorem follows.   □

For the action of G on itself by conjugation, the orbits are conjugacy classes of elements of G. The orbit counting theorem implies that

$$\#\text{conjugacy classes in } G = \frac{1}{\#G} \sum_{g \in G} \#C_G(g),$$

where $C_G(g) = \{h \in G \mid hg = gh\}$ is the centralizer of g in G.

The relevant mappings between G-sets are the *G-equivariant* mappings:

Definition 2.11 A function $f \colon (X, \rho_X) \to (Y, \rho_Y)$ is *G-equivariant* when, for all $g \in G$ and $x \in X$, $f(\rho_X(g)(x)) = \rho_Y(g)(f(x))$, that is, $f \circ \rho_X = \rho_Y \circ f$.

A G-equivariant mapping $f : (X, \rho_X) \to (Y, \rho_Y)$ is a *G-equivalence* if f is a bijection.

An interesting example arising from any G-set X is the mapping $St : X \to \mathcal{T}_G$ given by $St(x) = G_x$, assigning to each $x \in X$ its stabilizer. We let G act on $\mathcal{T}_G$ by conjugation. We claim that St is G-equivariant: By Lemma 2.10 we have

$$St\big(\rho(g)(x)\big) = G_{\rho(g)(x)} = gG_x g^{-1} = \rho_c(g)(G_x) = \rho_c(g)\big(St(x)\big),$$

so $St \circ \rho = \rho_c \circ St$, and $St : X \to \mathcal{T}_G$ is a G-equivariant mapping.

Another example of an equivalence comes from a transitive action $\rho : G \to \Sigma(X)$. For $x \in X$, $\mathcal{O}_x = X$. The proof of the Orbit-Stabilizer Theorem gives a mapping $\phi : X \to G/G_x$ by $\phi(\rho(g)(x)) = gG_x$. The action of G on G/G_x determined by left multiplication, $\rho(h)(gG_x) = (hg)G_x$, makes ϕ a G-equivariant map since $\phi(\rho(h)(\rho(g)(x))) = (hg)G_x = \rho(h)(gG_x) = \rho(h)(\phi(\rho(g)(x))$. Thus $\phi \circ \rho(h) = \rho(h) \circ \phi$ for all $h \in G$. Because ϕ is a bijection, it is an equivalence of G-sets.

An action $\rho : G \to \Sigma(X)$ is a homomorphism, and so it has a kernel

$$\ker \rho = \big\{ g \in G \mid \rho(g) = \mathrm{Id}_X : X \to X \big\},$$

the subgroup of elements that act as the identity. This is the *kernel of the action* ρ.

When $\ker \rho = \{e\}$, that is, ρ is injective, we say that the action of G on X is *faithful*. Let G act on itself by left multiplication, $\rho(g)(x) = gx$. This action is faithful: $\rho(g)(x) = x$ means $gx = x$, and by cancellation, $g = e$. Therefore $\rho : G \to \Sigma(G)$ is injective, and we can identify G with the subgroup $\rho(G)$ of the permutation group $\Sigma(G)$. This observation gives a proof of a theorem of ARTHUR CAYLEY (1821–1895) in 1854 [12].

Cayley's Theorem *Every group G may be identified with a subgroup of the symmetric group $\Sigma(G)$. If $\#G = n$, then G may be identified with a subgroup of Σ_n.*

Since $\ker \rho$ is a normal subgroup of G, a group action with a nontrivial kernel cannot be simple. A simple example is when G acts on a "small" set.

Proposition 2.12 *Suppose a finite group G acts on a set X for which $\#G$ does not divide $(\#X)!$. Then $\ker \rho \ne \{e\}$ and the action is not faithful.*

Proof If $\ker \rho = \{e\}$, then $G \cong \rho(G)$, a subset of $\Sigma(X)$. By Lagrange's Theorem $\#G$ must divide $\#\Sigma(X) = (\#X)!$, a contradiction. $\square$

More subtle observations are possible for particular actions.

Lemma 2.13 *The conjugation action* $\rho_c \colon G \to \Sigma(G)$ *has the kernel given by* $\ker \rho_c = Z(G)$, *the center of* G.

Proof Recall that the center of G is defined as the subset of G of elements that commute with every element of G. Then

$$Z(G) = \{x \in G \mid \text{for all } g \in G, gx = xg\}$$
$$= \{x \in G \mid \text{for all } g \in G, xgx^{-1} = g\}$$
$$= \{x \in G \mid \text{for all } g \in G, \rho_c(x)(g) = g\} = \{x \in G \mid \rho_c(x) = \text{Id}\}.$$

This is the kernel of the conjugation action.   □

For $h \in G$, the orbit of the conjugation action is given by $\mathcal{O}_h = \{ghg^{-1} \mid g \in G\} = [h]$, the conjugacy class of h. The stabilizer subgroup of h is the centralizer of h in G,

$$G_h = \{g \in G \mid ghg^{-1} = h\} = \{g \in G \mid gh = hg\} = C_G(h),$$

and $\#\mathcal{O}_h = \#[h] = [G : C_G(h)]$. Because $X = G$ for the conjugation action, we can write $\#G = \#\mathcal{O}_{g_1} + \#\mathcal{O}_{g_2} + \cdots + \#\mathcal{O}_{g_m}$ for appropriately chosen $g_1, \ldots, g_m$. It follows immediately that, for the conjugation action,

(1) $\#\mathcal{O}_h = 1$ if and only if $h \in Z(G)$.

To see this, $\#\mathcal{O}_h = 1$ if and only if $\mathcal{O}_h = \{h\}$ if and only if $ghg^{-1} = h$ for all $g \in G$ if and only if $hg = gh$ for all $g \in G$, that is, $h \in Z(G)$.

(2) $\#G = \#Z(G) + \#\mathcal{O}_{y_1} + \cdots + \#\mathcal{O}_{y_s}$, where $\#\mathcal{O}_{y_i} > 1$ for $i = 1, 2, \ldots, s$.

Because $\#[y] = \#G/\#C_G(y)$, we have established a result first proved by FERDINAND GEORG FROBENIUS (1849–1917) in 1887 [28].

The Class Equation *For a finite group* G,

$$\#G = \#Z(G) + \sum_j \#G/\#C_G(y_j),$$

where the sum is taken over conjugacy classes of G *that are not singletons.*

For certain special classes of groups, the Class Equation can reveal some of the deeper structure of the group:

Definition 2.14 If p is a prime number, then a *p-group* is a finite group P of order p^n from some integer $n \geq 1$.

An example of a 2-group is Q, the *quaternion group*, given by

$$Q = \{\pm 1, \pm i, \pm j, \pm k\} \text{ satisfying } i^2 = j^2 = k^2 = -1, ij = k,$$
$$jk = i, \text{ and } ki = j.$$

From these relations you can prove $ji = -k$ and the other pairwise products, $ik = -j$ and $kj = -i$. The group Q has order $8 = 2^3$.

In the case of p-groups, divisibility properties lead to fixed elements of a group action.

Lemma 2.15 *For a prime p, let P be a p-group that acts on a finite set X for which p does not divide #X. Then there is at least one fixed element of the action of P in X.*

Proof Generally, $\#X = \#\mathcal{O}_{x_1} + \#\mathcal{O}_{x_2} + \cdots + \#\mathcal{O}_{x_m}$ for a choice of $x_1, \ldots,$ $x_m \in X$. Because P is a p-group, each $\#\mathcal{O}_{x_i} = [P : P_{x_i}]$ is a power of p. If $\#\mathcal{O}_{x_i} > 1$ for all i, then p would divide the cardinality of X. By the assumption that p does not divide #X, there is some x_k with $\mathcal{O}_{x_k} = \{x_k\}$, and x_k is a fixed element of the action. $\qquad\square$

Proposition 2.16 *For a prime p and a p-group P, the center of P is not the trivial subgroup, $Z(P) \neq \{e\}$.*

Proof If $Z(P) = \{e\}$, then $\#Z(P) = 1$, and the Class Equation implies that the rest of the conjugacy classes $[y_i]$ have cardinality p^k for some $k > 1$. We have $\#P = p^N = 1 + pm$ for some integer m, which is arithmetically impossible. So $\#Z(P) > 1$, and the center of P is nontrivial. $\qquad\square$

Corollary 2.17 *For a prime p, a group of order p^2 is abelian.*

Proof Let P be a group of order p^2. Then $Z(P) \neq \{e\}$ implies that $\#Z(P) = p$ or p^2. If $\#Z(P) = p^2 = \#P$, then $P = Z(P)$ is abelian. Suppose $p = \#Z(P)$ and $x \in P$ with $x \notin Z(P)$. The centralizer of x in P, $C_P(x) = \{g \in P \mid gx = xg\}$, contains $Z(P)$. (Can you prove this?) Since $x \notin Z(P)$, $\#C_P(x) > \#Z(P)$. However, $C_P(x)$ is a subgroup of P, so $\#C_P(x) = p^2$, that is, $P = C_P(x)$. This means that x commutes with every element of P, and so $x \in Z(P)$, a contradiction to our assumption. Hence $\#Z(P) = p^2$, and P is abelian. $\qquad\square$

For the conjugation action, $Z(P) = \ker \rho_c$, and the center $Z(P)$ is a normal subgroup of P. By the proposition, a p-group that is not isomorphic to $\mathbb{Z}/p\mathbb{Z}$ is not a simple group: Either P is abelian and P contains a normal cyclic subgroup of order p, or P is nonabelian and $Z(P) \neq P$ is a nontrivial normal subgroup of P.

Get to know Frobenius groups

1. Let $\mathbb{F}_p$ denote the finite field of order p and consider the collection of 2×2 matrices

$$F_{p(p-1)} = \left\{ \begin{bmatrix} a & b \\ 0 & 1 \end{bmatrix} \mid a, b \in \mathbb{F}_p, a \neq 0 \right\}.$$

Let $X = \mathbb{F}_p \times \{1\} \subset \mathbb{F}_p^{\times 2}$ denote the set of 2-vectors with 1 in the second entry. Then show that $F_{p(p-1)}$ is a group that acts transitively on X by the usual matrix multiplication on 2-vectors.

2. Let $H = (F_{p(p-1)})_0$, the stabilizer of $\begin{bmatrix} 0 \\ 1 \end{bmatrix} \in X$. Show that H has order $p-1$ and for other elements of X, $\begin{bmatrix} j \\ 1 \end{bmatrix}$, has a stabilizer subgroup a conjugate of H.

3. Suppose $H = (F_{p(p-1)})_0$, and let A_1 and A_2 be in $F_{p(p-1)}$. Consider $A_1 H A_1^{-1}$ and $A_2 H A_2^{-1}$. Show that either $A_1 H A_1^{-1} = A_2 H A_2^{-1}$ or $(A_1 H A_1^{-1}) \cap (A_2 H A_2^{-1}) = \{\text{Id}\}$.

4. Prove that the normalizer of $H = (F_{p(p-1)})_0$ in $F_{p(p-1)}$ is H.

Definition 2.18 An action of a group G on a set X of cardinality at least two, $\rho \colon G \to \Sigma(X)$, is called a *Frobenius action* if the action is transitive but not *regular*, that is, there are elements $x \in X$ whose stabilizer is not the trivial subgroup, and whenever $x \neq y$, $G_x \cap G_y = \{e\}$.

Definition 2.19 A group G is called a *Frobenius group* if there is a nontrivial subgroup $H \neq G$ for which $H = N_G(H)$ and if $g_1 H g_1^{-1} \neq g_2 H g_2^{-1}$, then $g_1 H g_1^{-1} \cap g_2 H g_2^{-1} = \{e\}$. The subgroup H is called the *Frobenius complement* in G.

5. Show that $F_{p(p-1)}$ with $X = \mathbb{F}_p \times \{1\}$ is a Frobenius action. Show that the dihedral group D_{2n} is a Frobenius group if and only if $n \geq 3$ is odd.

6. Show that if $\rho \colon G \to \Sigma(X)$ is a Frobenius action, then $H = G_x \neq \{e\}$ is a Frobenius group.

A stricter version of a transitive action is a *k-transitive action* of G on a set X.

Definition 2.20 The action of a group G on a set X is *k-transitive* if for any pair of k-tuples $(x_1, x_2, \ldots, x_k)$ and $(y_1, y_2, \ldots, y_k)$ in $X^{\times k}$ with $x_i \neq x_j$ and

$y_i \neq y_j$ for all $i \neq j$, there is an element $g \in G$ for which $\rho(g)(x_i) = y_i$ for all $1 \leq i \leq k$. When $k = 2$, we say that G acts on X *doubly transitively*.

For example, the symmetric group Σ_n acts on $[n] = \{1, 2, \ldots, n\}$ n-transitively: given $(x_1, \ldots, x_n)$ and $(y_1, \ldots, y_n)$ with all entries distinct, the ordering determines the permutations

$$\sigma = \begin{pmatrix} 1 & 2 & \cdots & n \\ x_1 & x_2 & \cdots & x_n \end{pmatrix}, \quad \tau = \begin{pmatrix} 1 & 2 & \cdots & n \\ y_1 & y_2 & \cdots & y_n \end{pmatrix}.$$

The product $\tau \circ \sigma^{-1}$ takes $x_i \mapsto y_i$ for all i.

The alternating group A_n acts $(n-2)$-transitively on $[n]$. To see this, consider the set of $(n-2)$-tuples $(x_1, \ldots, x_{n-2}) \in [n]^{\times n-2}$ for which $x_i \neq x_j$ for all $i \neq j$. Let Σ_n act on such $(n-2)$-tuples by $\rho(\sigma)(x_1, x_2, \ldots, x_{n-2}) = (\sigma(x_1), \sigma(x_2), \ldots, \sigma(x_{n-2}))$. Let $\mathbf{x} = (1, 2, \ldots, n-2)$. The stabilizer $(\Sigma_n)_{\mathbf{x}}$ of this $(n-2)$-tuple consists of $\{\mathrm{Id}, (n-1, n)\}$, and so the orbit of the action has the cardinality $n!/2 = \#A_n$. Restricting the action of Σ_n to the subgroup A_n, the stabilizer of $\mathbf{x}$ in A_n is trivial (check on any 3-cycle), and so the orbit is all of the orbit of Σ_n. Thus A_n acts transitively on this set of $(n-2)$-tuples, that is, A_n acts $(n-2)$-transitively on $[n]$.

An example of a doubly transitive action is the group of invertible linear transformations of a vector space to itself acting on the set of lines through the origin in the vector space. This example will be developed in Chapter 3. Having a doubly transitive action restricts the properties of the group.

Proposition 2.21 *Suppose the group G acts doubly transitively on a set X. For $x \in X$, the stabilizer subgroup G_x is a maximal subgroup of G, that is, if H is another subgroup of G and $G_x \subset H \subset G$, then either $H = G_x$ or $H = G$.*

Proof Suppose H is a subgroup of G with $G_x \subset H \subset G$. Suppose there is $h \in H$ and $h \notin G_x$. Then $\rho(h)(x) = y \neq x$. Furthermore, if $H \neq G$, then there is $g \in G$ such that $g \notin H$. Then $\rho(g)(x) = z \neq x$. Because G acts doubly transitively on X, there is $k \in G$ such that $\rho(k)(x) = x$ and $\rho(k)(y) = z$. Notice that $k \in G_x$. Then $\rho(k)(\rho(h)(x)) = \rho(g)(x)$, from which it follows that $\rho(g^{-1}kh)(x) = x$ and $g^{-1}kh \in G_x$. Then $g^{-1} \in G_x h^{-1} k^{-1} \subset G_x H G_x = H$, because $G_x \subset H$. By assumption $g \notin H$, and we get a contradiction. Thus $H = G_x$ or $H = G$. $\qquad \square$

The Orbit Stabilizer Theorem implies a divisibility condition for the existence of a doubly transitive action.

Proposition 2.22 *If G acts doubly transitively on a set X with $\#X = n$, and $x, y \in X$ with $x \neq y$, then $\#G = n(n-1)\#(G_x \cap G_y)$.*

Proof First notice that if $x \in X$, then the stabilizer subgroup G_x acts on the set $X - \{x\}$. Furthermore, when the action is doubly transitive, this action is transitive: for $y, z \in X - \{x\}$, there is $g \in G$ such that $\rho(g)(x) = x$ and $\rho(g)(y) = z$, so $g \in G_x$ and takes y to z. The stabilizer of $y \in X - \{x\}$ for the G_x-action is the subgroup of elements that fix y and x, that is, $(G_x)_y = G_x \cap G_y$. The Orbit Stabilizer Theorem gives

$$\#G_x = \#(X - \{x\}) \cdot \#(G_x)_y = (n-1) \cdot \#(G_x \cap G_y).$$

Since $\#G = \#G_x \cdot \#\mathcal{O}_x$, and a doubly transitive action is also transitive, $\#\mathcal{O}_x = \#X = n$, and so $\#G = n\#G_x = n(n-1)\#(G_x \cap G_y)$. $\qquad\square$

An immediate consequence is that p-groups cannot act doubly transitively on sets of cardinality greater than two.

The condition that a group G acts on a set X doubly transitively, together with other conditions, leads to a proof that the group is simple, a theorem of KENKICHI IWASAWA (1917–1998) from 1941 [47]. Recall that the commutator subgroup is denoted by G' or $[G, G]$.

Iwasawa's Lemma *Suppose a group G acts doubly transitively on a set X via $\rho\colon G \to \Sigma(X)$. Suppose further that $G = G'$, that is, G is equal to its commutator subgroup, and there exists an element $x \in X$ for which G_x contains a normal abelian subgroup $H \triangleleft G_x$ for which the conjugates of H in G generate G. Then $G/\ker \rho$ is a simple group.*

Proof Saying that the set of conjugates of H generates G means that the subgroup of G made up of all possible finite products of elements of $T = \bigcup_{g \in G} gHg^{-1}$ is all of G.

By the Correspondence Theorem we can show that $\ker \rho$ is a maximal normal subgroup of G to prove that $G/\ker \rho$ is simple. Suppose N is a normal subgroup of G with $\ker \rho \subset N \subset G$. For all $x \in X$, we know that G_x is a maximal subgroup of G by Proposition 2.21. Consider the subgroup $G_x N$ of G. Then $G_x \subset G_x N \subset G$, and so either $G_x N = G_x$ or $G_x N = G$.

Restricting the action of G to the subgroup N, we claim that N either acts transitively or trivially on X. Suppose N acts nontrivially. Then there are some $x \in X$ and $n \in N$ such that $\rho(n)(x) \neq x$. Suppose $y \neq y'$ are elements of X. Because G acts doubly transitively on X, there is an element $g \in G$ such that $\rho(g)(x) = y$ and $\rho(g)(\rho(n)(x)) = y'$. It follows that

$$y' = \rho(g)\big(\rho(n)(x)\big) = \rho(g)\big(\rho(n)\rho(g^{-1})(y)\big) = \rho\big(gng^{-1}\big)(y).$$

Since N is normal in G, $gng^{-1} \in N$, and because $y \neq y'$ determines an arbitrary pair in X, N acts transitively on X.

If $G_x N = G_x$, then $N \subset G_x$, and N fixes x. Then N does not act transitively on X. We showed that if N acts nontrivially on X, then N acts transitively on X. Hence, if $G_x N = G_x$, then N acts trivially on X, and $N \subset \ker \rho$. Therefore $N = \ker \rho$.

If $G_x N = G$, then we call on the abelian normal subgroup H of G_x. Since $H \lhd G_x$ implies $NH \lhd NG_x = G$, we have $gHg^{-1} \subset gNHg^{-1} = NH$ for all $g \in G$, and so NH contains all conjugates of H. By assumption the conjugates of H generate G, and so $NH = G$. The Second Isomorphism Theorem implies

$$G/N = NH/N \cong H/(H \cap N),$$

which is an abelian group. The fact that G/N is abelian implies that $G' \subset N$. However, $G = G'$, so $G \subset N$ and $G = N$. Therefore $\ker \rho$ is a maximal normal subgroup, and $G/\ker \rho$ is a simple group. $\qquad \square$

We can apply this result immediately to prove again that A_5 is simple. In fact, A_5 acts 3-transitively on [5], but a k-transitive action is m-transitive for all $m \leq k$. The stabilizer $(A_5)_1$ is isomorphic to A_4, which contains the abelian normal subgroup $V = \{( ), (1, 2)(3, 4), (1, 3)(2, 4), (1, 4)(2, 3)\}$. Conjugating V with $\sigma = (1, 2, 3, 4, 5)$ repeatedly, we obtain all two-fold products of distinct transpositions from which we can obtain any two-fold product of transpositions. These products generate A_5, and so we only need to show that the commutator subgroup $(A_5)' = A_5$. The relation

$$(b, c)(d, e)(a, b, c)(b, c)(d, e)(a, c, b) = (a, c, b)(a, c, b) = (a, b, c)$$

shows that every 3-cycle is a commutator, and the result follows from the fact that A_5 is generated by its 3-cycles.

2.3 The Sylow Theorems

Lagrange's Theorem tells us that the order of a subgroup of a finite group divides the order of the group. Does the converse hold, that is, if k divides the order of a group G, then is there a subgroup of G of order k? A careful accounting of the subgroups of the alternating group A_4 of order 12 shows that there is no subgroup of order 6. (You might want to check this.) The converse of Lagrange's Theorem fails. A partial converse to Lagrange's Theorem was proved by Cauchy [11].

Cauchy's Theorem *For a finite group G, if p is a prime that divides #G, then there is an element $g \in G$ of order p.*

Proof   We base our proof on group actions following [61]. For a prime p, consider the set $X = \{(g_1, g_2, \ldots, g_p) \in G^{\times p} \mid g_1 g_2 \cdots g_p = e\}$. To determine the cardinality of this set, choose, in order, $g_1, g_2, \ldots, g_{p-1}$ from G. Then let $g_p = (g_1 g_2 \cdots g_{p-1})^{-1}$. Such p-tuples are in X, and so $\#X = (\#G)^{p-1}$.

Denote the cyclic group of order p by $C_p = \langle x \rangle$ for a generator x. Then C_p acts on X by $\rho(x)(g_1, \ldots, g_p) = (g_2, g_3, \ldots, g_p, g_1)$. Since $g_1 = (g_2 \cdots g_p)^{-1}$, we have $g_1(g_2 \cdots g_p) = e = (g_2 \cdots g_p)g_1$, and the permuted p-tuple is an element in X.

The partition of X into disjoint orbits $X = \mathcal{O}_{x_1} \cup \cdots \cup \mathcal{O}_{x_m}$ includes the singleton orbit $\mathcal{O}_{(e,\ldots,e)} = \{(e, e, \ldots, e)\}$. Every orbit has the cardinality given by the index $[C_p : (C_p)_{x_i}]$, which is either 1 or p. Because p divides $(\#G)^{p-1} = \#X$, there must be other orbits of cardinality one. Such an orbit represents a fixed element in $G^{\times p}$ under the action of C_p. However, the permutation $\rho(x)$ fixes only p-tuples of the form $(h, h, \ldots, h)$, and so we have found an element $h \neq e$ with $h^p = e$. $\square$

The Norwegian mathematician LUDWIG SYLOW (1832–1918) [83] extended Cauchy's Theorem considerably. For a prime p dividing the order of a group G, there is a subgroup of order p in G. Sylow considered other subgroups of G that are also p-groups. In Cauchy's and Sylow's arguments, the group G was taken to be a subgroup of a symmetric group on some $[n]$. The first proof of Sylow's Theorems without this assumption is due to Frobenius [28].

Definition 2.23   Let p be a prime number, and let G be a finite group. Suppose $\#G = p^e m$ where p does not divide m. A *Sylow p-subgroup* of G is a subgroup P that is a p-group of order p^e.

A Sylow p-subgroup of G has the largest possible pth power order since p does not divide the index $[G : P] = m$.

The Sylow Theorems *Suppose p is a prime and G is a finite group for which p divides #G. Then*

a.   *G contains a Sylow p-subgroup.*

b.   *The Sylow p-subgroups of G are conjugates of one another, that is, if P and Q are Sylow p-subgroups of G, then there is an element $h \in G$ such that $Q = hPh^{-1}$. Also, every p-subgroup of G is contained in a Sylow p-subgroup.*

c. *If #G = $p^e m$ with m not divisible by p, and s_p denotes the number of Sylow p-subgroups of G, then s_p divides m, and $s_p \equiv 1 \,(\mathrm{mod}\; p)$, that is, $s_p = kp + 1$ for some integer k.*

This structure theorem is very strong. For example, any group G of order $45 = 3^2 \cdot 5$ has at least one Sylow 3-subgroup. Since 5 is prime, there can be one or four Sylow 3-subgroups, but 4 does not divide 5, so there is only one Sylow 3-subgroup, call it P_3. Because all Sylow 3-subgroups are conjugate, P_3 is normal in G – each of its conjugates equals itself. It follows that $G/P_3 \cong \mathbb{Z}/5\mathbb{Z}$.

To prove the Sylow Theorems, we recall a little number theory. Let $\binom{n}{k}$ for $0 \le k \le n$ denote the *binomial coefficient* associated with $(1 + x)^n = \sum_{k=0}^n \binom{n}{k} x^k$. Recall that $\binom{p}{k} \equiv 0 \,(\mathrm{mod}\; p)$ for p prime and $1 \le k < p$.

Lemma 2.24 *If the prime p does not divide $m \ge 1$, then, for $e \ge 1$, $\binom{p^e m}{p^e} \equiv \binom{m}{1} = m \,(\mathrm{mod}\; p)$. In particular, p does not divide $\binom{p^e m}{p^e}$.*

Proof This lemma is an instance of a theorem of É. LUCAS (1842–1891) [59]. Fermat's Little Theorem [23] can be written $a^p \equiv a \,(\mathrm{mod}\; p)$ for any integer a and a prime p. This means that

$$(x + 1)^{pk} \equiv \left(x^p + 1\right)^k \,(\mathrm{mod}\; p).$$

Expanding the binomials gives $\sum_{i=0}^{pk} \binom{pk}{i} x^i \equiv \sum_{j=0}^{k} \binom{k}{j} x^{pj} \,(\mathrm{mod}\; p)$. It follows that $\binom{pk}{i} \equiv 0 \,(\mathrm{mod}\; p)$ unless $i = pj$ for some $0 \le j \le k$ and that $\binom{pk}{pj} \equiv \binom{k}{j} \,(\mathrm{mod}\; p)$. The lemma follows by induction on the exponent e. □

Proof of the Sylow Theorems This is a story of four group actions. The first has G acting on the set $\binom{G}{p^e} = \{A \subset G \mid \#A = p^e\}$ by left multiplication. Recall that this means $\rho(g)(A) = gA = \{ga \mid a \in A\}$. The set $\binom{G}{p^e}$ contains $\binom{p^e m}{p^e}$ elements.

The lemma tells us that p does not divide $\#\binom{G}{p^e}$, since $\binom{p^e m}{p^e} \equiv m \,(\mathrm{mod}\; p)$. Partitioning the set $\binom{G}{p^e}$ into the orbits of this action, we get the count:

$$\binom{p^e m}{p^e} = \#\binom{G}{p^e} = \underbrace{\sum \#\mathcal{O}_{A_i}}_{\text{orbits}} = \underbrace{\sum [G : G_{A_i}]}_{\text{orbits}}.$$

Because p does not divide $\binom{p^e m}{p^e}$, at least one $[G : G_{A_i}]$ is not divisible by p. Let us denote such a subset A_i by B, and so p does not divide $[G : G_B] = p^e m/\#G_B$, which implies that p^e divides $\#G_B$.

The second group action of the proof is the action of G_B on B by left multiplication. By definition

$$G_B = \{g \in G \mid \rho(g)(B) = B\} = \{g \in G \mid gb \in B \text{ for all } b \in B\},$$

and B is stable under the action of G_B. Because $B \subset G$, this action is faithful, and the orbit of any $b \in B$ under the action of G_B is the right coset $G_B b$. It follows that B is a union of cosets of G_B, and so $\#G_B$ divides $\#B = p^e$. We know that p does not divide $[G : G_B]$, and so $\#G_B = p^e$. We have found a Sylow p-subgroup, G_B. This proves part a of the theorem.

Our third group action is the left multiplication action of G on the set of cosets G/P where P is a Sylow p-subgroup. Suppose K is any other p-group that is a subgroup of G. The action of G on G/P is given by $\rho(g)(xP) = (gx)P$. It restricts to an action of K on G/P, $\rho|_K : K \to \Sigma(G/P)$, with the same definition. The order of K is p^d, and the cardinality of G/P is m. Since p does not divide m, by Lemma 2.15 there is a fixed element of the action of K on G/P, that is, a coset yP such that $(ky)P = yP$ for all k in K. However, $kyP = yP$ implies $y^{-1}kyP = P$ for all k in K, that is, $y^{-1}ky$ is in P for all $k \in K$, and so $y^{-1}Ky \subset P$. Thus any p-subgroup of G is conjugate to a subgroup of a Sylow p-subgroup. Of course, this applies to any other Sylow p-subgroup of G, and hence all the Sylow p-subgroups of G are conjugate to one another. This proves part b of the theorem.

The last group action is the conjugation action of G on $\mathcal{S}_p = \{$Sylow p-subgroups of $G\}$, namely, $\rho : G \to \Sigma(\mathcal{S}_p)$, $\rho(g)(P) = g^{-1}Pg$. Restrict this action to a choice P_0 of Sylow p-subgroup of G, $\rho|_{P_0} : P_0 \to \Sigma(\mathcal{S}_p)$. Certainly, $x^{-1}P_0 x = P_0$ for all $x \in P_0$, and so the orbit of P_0 under the action of P_0 on $\mathcal{S}_p$ is a singleton, $\mathcal{O}_{P_0} = \{P_0\}$.

If Q is another Sylow p-subgroup for which the action of P_0 on $\mathcal{S}_p$ gives a singleton orbit $\mathcal{O}_Q = \{Q\}$, then $x^{-1}Qx = Q$ for all $x \in P_0$. In G the stabilizer of Q under the conjugation action of G on $\mathcal{S}_p$ is $G_Q = \{g \in G \mid g^{-1}Qg = Q\} = N_G(Q)$, the normalizer of Q in G. Because the orbit under P_0 of Q is $\{Q\}$, we find $P_0 \subset G_Q = N_G(Q)$. Because $Q \subset N_G(Q)$, p^e divides $\#N_G(Q)$ by Lagrange's Theorem. It follows that P_0 and Q are Sylow p-subgroups of $N_G(Q)$, and so they are conjugate to one another in $N_G(Q)$ by part b of the theorems. However, Q is fixed under conjugation in $N_G(Q)$. If $y^{-1}P_0 y = Q$ for some $y \in N_G(Q)$, then $P_0 = yQy^{-1} = Q$, and so $Q = P_0$. Thus the orbits of the action of P_0 on $\mathcal{S}_p$ are (1) $\mathcal{O}_{P_0}$, a singleton, and (2) $\mathcal{O}_{Q_i}$ of cardinality greater than 1. Because $\#\mathcal{O}_{Q_i} = [P_0; (P_0)_{Q_i}] > 1$, it follows that $\#\mathcal{O}_{Q_i}$ is a power of p. Hence

$$s_p = \#\mathcal{S}_p = \#\mathcal{O}_{P_0} + \#\mathcal{O}_{Q_1} + \cdots + \#\mathcal{O}_{Q_t} = 1 + kp \equiv 1 \pmod p.$$

Because any Sylow p-subgroup P is conjugate to any other Sylow p-subgroup, the conjugation action of G on $\mathcal{S}_p$ is transitive, that is, $\mathcal{O}_P = \mathcal{S}_p$ for any $P \in \mathcal{S}_p$. The stabilizer of P is $G_P = \{g \in G \mid g^{-1}Pg = P\} = N_G(P)$, which contains P. Hence $\#N_G(P) = p^e d$ for some positive integer d. It follows that

$$s_p = \#\mathcal{S}_p = \#\mathcal{O}_P = [G : G_P] = \big[G : N_G(P)\big] = p^e m / p^e d = m/d,$$

and s_p divides m. This proves part c. $\qquad\square$

2.4 Small Groups

Corollary 2.25 *If P is the only Sylow p-subgroup in a finite group G, then P is a normal subgroup of G. More generally, $s_p = [G : N_G(P)]$ for any Sylow p-subgroup P of G.*

Proof The subgroup P is fixed under conjugation by all of G. $\qquad\square$

The Sylow Theorems provide powerful tools for opening up the structure of finite groups. To see these tools in action, let us consider the collection of all groups of order less than or equal to 200. Which of these groups have a chance of being simple groups knowing only their cardinality?

The groups of prime order are simple groups, and this accounts for 46 cases. Another 14 cases can be eliminated as p-groups with order p^e and $e > 1$: their centers are nontrivial normal subgroups.

The simplest application of the Sylow Theorems follows from Corollary 2.25.

Proposition 2.26 *If the order of a group G is $p^e m$ and neither p nor $1 + kp$ divides m for $k > 0$, then G is not a simple group.*

Proof Because $s_p = 1 + kp$ does not divide m if $k > 0$, we have $s_p = 1$. Therefore a Sylow p-subgroup P is a nontrivial normal subgroup of G. $\qquad\square$

This eliminates a great number of possible orders. For example, $28 = 2^2 \cdot 7$ ($s_7 = 1$), $40 = 2^3 \cdot 5$ ($s_5 = 1$), and $78 = 2 \cdot 3 \cdot 13$ ($s_{13} = 1$) cannot be the orders of simple groups. Applying this proposition leaves us with the list

$$30, 56, 60, 72, 90, 105, 108, 112, 120, 144, 150, 168, \text{ and } 180.$$

A more subtle condition is given by the following:

Proposition 2.27 *If G is a simple group of order greater than 2 and G acts on a set X with $\#X = n$, then G is isomorphic to a subgroup of A_n.*

Proof Denote the action by $\rho: G \to \Sigma(X)$. If we label the elements of X by 1 through n, then we have that $\Sigma(X) \cong \Sigma_n$. We identify $\Sigma(X)$ with Σ_n. Since G is assumed to be simple, $\ker \rho = \{e\}$, and so $\rho(G)$ is a subgroup of Σ_n isomorphic to G. Consider the homomorphism $\pi: \Sigma_n \to \Sigma_n/A_n \cong \{\pm 1\}$. Compose π with ρ to get a homomorphism $\pi \circ \rho: G \to \{\pm 1\}$. If the order of G is greater than two, then this homomorphism has a nontrivial kernel if it is surjective. Since G is simple, we must have $\ker(\pi \circ \rho) = G$, which implies that $\rho(G) \subset A_n$. $\square$

Corollary 2.28 *If G acts on a set X of cardinality $\#X = n > 1$ and $\#G$ does not divide $n!/2$, then G is not simple.*

This condition can be used in different ways. For example, if P is a Sylow p-subgroup of a group G of order $p^e m$, then G acts on G/P, a set of cardinality m. If $\#G$ does not divide $m!/2 = \#A_m$, then G is not simple. For example, if G has order $80 = 2^4 \cdot 5$, then a Sylow 2-subgroup has order 16 and 80 does not divide $5!/2 = 60$. Using the action of G on $X = \mathcal{S}_p$, the set of Sylow p-subgroups, the corollary may apply. For example, suppose G is a group of order $72 = 2^3 \cdot 3^2$. If G were simple, then $s_3 \neq 1$, and $s_3 = 4$ is the only other possible value. However, 72 does not divide $4!/2 = 12$, so the corollary tells us that G cannot be simple.

If $\#G = 120 = 2^3 \cdot 3 \cdot 5$, then in order for G to be isomorphic to a subgroup of A_n, n must be greater than 5. This implies $s_5 = 6$, $s_2 = 15$, and $s_3 = 10$ or 40. Using the action of G on the set of Sylow 5-subgroups, we can view G as a subgroup of A_6. Then A_6 acts on the set A_6/G of left cosets of G. Such an action may be expressed as $\rho: A_6 \to \Sigma_3$, which would have a nontrivial kernel. Since A_6 is simple, no such injective homomorphism of G into A_6 exists, and G is not a simple group.

This leaves $30, 56, 60, 90, 105, 144, 168$, and 180. An argument by counting eliminates some cases. For example, $30 = 2 \cdot 3 \cdot 5$. If G were simple, then $s_5 = 6$ and $s_3 = 10$ by the Sylow Theorems. The nonidentity elements in Sylow 5-subgroups are disjoint from the elements in the Sylow 3-subgroups. This accounts for 24 elements of order 5 and 20 elements of order 3. Because all of these elements must be distinct, there are too many elements in a group of order 30. Hence G cannot be simple. A similar argument works for 56 and 105.

To apply the Sylow Theorems to groups of orders 90, 144, and 180, we consider a more subtle counting argument. Each group has order divisible by 3^2. If the Sylow 3-subgroups shared only the identity element in each case, then a counting argument with other divisors leads to a contradiction. So there are Sylow 3-subgroups P and Q of order 9 and intersection $P \cap Q$ of order 3. Consider the normalizer of $P \cap Q$, $N_G(P \cap Q)$. Since P and Q are abelian, both

P and Q are subgroups of $N_G(P \cap Q)$. Thus the set of products PQ is a subset of $N_G(P \cap Q)$. By the argument for the Second Isomorphism Theorem $\#PQ = \dfrac{\#P \# Q}{\#P \cap Q} = \dfrac{81}{3} = 27$, and so $\#N_G(P \cap Q) \geq 27$. Checking divisors, we find that the index $[G : N_G(P \cap Q)]$ takes values among 1, 2, or 4. However, if 3^2 divides $n!/2$, then $n \geq 6$. Thus G of order 90, 144, or 180 cannot be simple.

This leaves only 60 and 168. The reader will want to apply the arguments above to see that they do not work to eliminate groups of order 60 or 168 from being simple groups. Of course, $\#A_5 = 60$, so no such argument is possible. As we will see in Chapter 3, no such argument is possible for $\#G = 168$.

Get to know the Sylow Theorems

1. Suppose $\#G = pq$ with $p < q$, p and q primes. Suppose further that $q \not\equiv 1 \pmod{p}$. Show that G is isomorphic to $\mathbb{Z}/p\mathbb{Z} \times \mathbb{Z}/q\mathbb{Z}$. To fashion such a proof, show that there are a unique Sylow p-subgroup P and a unique Sylow q-subgoup Q of G such that $P \cap Q = \{e\}$. Since P and Q are cyclic, suppose $P = \langle a \rangle$ and $Q = \langle b \rangle$. Prove that $aba^{-1}b^{-1} = e$ and so $ab = ba$. Finally, show that the mapping $\phi \colon P \times Q \to G$ given by $\phi(a^r, b^s) = a^r b^s$ is a group isomorphism.
2. Suppose N is a normal subgroup of a finite group G and $[G : N]$ is prime to p. Show that N contains every Sylow p-subgroup of G.
3. Suppose P is a Sylow p-subgroup of a finite group G. Let $N_G(P)$ be the normalizer of P in G. Show that P is a Sylow p-subgroup of $N_G(P)$ and that it is the only Sylow p-subgroup of $N_G(P)$.
4. Suppose G has order $p^2 q$ where p and q are distinct primes. Show that G contains a normal Sylow subgroup for at least one of p or q.
5. Suppose n is an odd integer. Show that every Sylow p-subgroup of D_{2n} is cyclic.
6. Extend the list of values n for which there is no simple group of order n, say for $n \leq 300$.

2.5 Burnside's Transfer Theorem*

The inclusion of a subgroup H into a group G is always a homomorphism. Under what conditions can we guarantee that there is also a homomorphism $G \to H$? For finite groups and certain conditions on the subgroup H, this will turn out to be possible. Since $\#H < \#G$ in this case, such a homomorphism has a kernel, and hence G is not a simple group. To make this construction, we use quite a few aspects of elementary group theory, none of which are difficult, although they may appear in unexpected ways.

The set of (left) cosets of H may be presented as $G/H = \{H, t_2 H, \ldots, t_m H\}$ for a choice of elements $\{e = t_1, t_2, \ldots, t_m\}$ called a *transversal* for H in G where $[G : H] = m$. The action of G on G/H, $\rho(g)(kH) = (gk)H$, determines a homomorphism $\rho: G \to \Sigma(G/H) \cong \Sigma_m$ given by $\rho(g)(i) = \sigma(i)$ when $g t_i H = t_{\sigma(i)} H$. This mapping is well defined because G is the disjoint union $H \sqcup t_1 H \sqcup \cdots \sqcup t_m H$ and every $g \in G$ falls into exactly one coset. When $g t_i \subset t_{\sigma(l)} H$, then $g t_i = t_{\sigma(i)} h_i^g$ for some $h_i^g \in H$.

Under the condition that H is an abelian subgroup of G, we construct a homomorphism $T: G \to H$ called the *transfer*.

Definition 2.29 For an abelian subgroup H of a group G of index m, the *transfer* is a homomorphism $T: G \to H$ defined by choosing a transversal of H in G, $\{e = t_1, t_2, \ldots, t_m\}$, and letting $g t_i = t_{\sigma(i)} h_i^g$. Then $T(g) = h_1^g h_2^g \cdots h_m^g$.

Proposition 2.30 *The transfer is well-defined and a homomorphism.*

Proof Suppose $\{u_1, u_2, \ldots, u_m\}$ is another transversal of H in G. The left cosets are determined by H, so we can renumber the transversal if needed to get $t_i H = u_i H$. This means that $u_i = t_i v_i$ for some $v_i \in H$. Because $g t_i = t_{\sigma(i)} h_i^g$, the factors of $T(g)$ are $h_i^g = t_{\sigma(i)}^{-1} g t_i$. The corresponding elements for the other transversal are $k_i^g = u_{\sigma(i)}^{-1} g u_i$. (The same permutation appears because $u_i H = t_i H$.) Then $k_i^g = (t_{\sigma(i)} v_{\sigma(i)})^{-1} g(t_i v_i)$, and, because H is abelian,

$$k_1^g k_2^g \cdots k_m^g = \prod_i v_{\sigma(i)}^{-1} \left(t_{\sigma(i)}^{-1} g t_i\right) v_i = \prod_i t_{\sigma(i)}^{-1} g t_i \prod_i v_{\sigma(i)}^{-1} \prod_i v_i$$

$$= T(g) \prod_i v_{\sigma(i)}^{-1} \prod_i v_i.$$

Because σ is a permutation of $[m]$, each v_i has its inverse among the $v_{\sigma(j)}^{-1}$, and so the latter two products cancel leaving $T(g)$, and T is well-defined.

Suppose $\rho: G \to \Sigma(G/H) \cong \Sigma_m$ is the action, and let $a, b \in G$ with $\rho(a) = \sigma$ and $\rho(b) = \tau$. Then $\rho(ab) = \sigma \circ \tau$. We can compute

$$T(ab) = \prod_i t_{\sigma \circ \tau(i)}^{-1} (ab) t_i = \prod_i t_{\sigma \circ \tau(i)}^{-1} a t_{\tau(i)} t_{\tau(i)}^{-1} b t_i$$

$$= \prod_i t_{\sigma(\tau(i))}^{-1} a t_{\tau(i)} \prod_i t_{\tau(i)}^{-1} b t_i = T(a) T(b).$$

Thus T is a homomorphism. □

Recall that the *centralizer of a subgroup* H of G is defined by $C_G(H) = \{g \in G \mid gh = hg$ for all $h \in H\}$. It follows immediately that $C_G(H)$ is a subgroup of the normalizer of H in G, which is $N_G(H) = \{g \in G \mid gH = Hg\}$. When H is abelian, H is a subgroup of $C_G(H)$. The following important result was proved by WILLIAM BURNSIDE (1852–1927) [10].

Burnside's Transfer Theorem *Suppose G is a finite group and P is a Sylow p-subgroup of G for which $C_G(P) = N_G(P)$. Then there is a normal subgroup N of G such that $N \cap P = \{e\}$ and $G = NP$. We call N a normal complement of P.*

Proof Suppose $\#G = p^e m$ where p does not divide m. Let P denote a Sylow p-subgroup of G for which $C_G(P) = N_G(P)$. Notice that P is a subgroup of its normalizer $N_G(P)$, and so if $N_G(P) = C_G(P)$, then P is abelian, and we can define the transfer $T: G \to P$. The theorem will follow if the transfer is surjective. The argument is based on some basic properties of permutations and numbers.

Let left multiplication by $g \in G$ determine the permutation $\sigma \in \Sigma_m$ for which $g t_i = t_{\sigma(i)} h_i^g$. Then we can write σ as a product of disjoint cycles

$$\sigma = \big(j(1,1), \ldots, j\big(1, l(1)\big)\big)$$
$$\cdot \big(j(2,1), \ldots, j\big(2, l(2)\big)\big) \cdots \big(j(t,1), \ldots, j\big(t, l(t)\big)\big),$$

where $l(i)$ is the length of the ith cycle, which can be 1 if $j(i,1)$ is a fixed letter. The sum of the lengths is m, $\sum_i l(i) = m$.

Let us focus on a factor of $T(g)$: $h_i^g = t_{\sigma(i)}^{-1} g t_i$. Since $i \in [m]$, i appears in some disjoint cycle as $i = j(k,s)$, it follows that $\sigma(i) = j(k, s+1)$, where $s+1$ is treated mod $l(k)$. Then $g t_i = t_{\sigma(i)} h_i^g = t_{j(k,s+1)} h_{j(k,s)}^g$. Repeating gives $g^2 t_i = g t_{j(k,s+1)} h_{j(k,s)}^g = t_{j(k,s+2)} h_{j(k,s+1)}^g h_{j(k,s)}^g$. After $l(k)$ repetitions, we obtain

$$g t_{j(k,s)} = t_{j(k,s+1)} h_{j(k,s)}^g,$$
$$g^2 t_{j(k,s)} = t_{j(k,s+2)} h_{j(k,s+1)}^g h_{j(k,s)}^g,$$
$$\vdots$$
$$g^r t_{j(k,s)} = t_{j(k,s+r)} h_{j(k,s+r-1)}^g \cdots h_{j(k,s)}^g,$$
$$\vdots$$
$$g^{l(k)} t_{j(k,s)} = t_{j(k,s)} h_{j(k,s+l(k)-1)}^g \cdots h_{j(k,s)}^g$$
$$= t_{j(k,s)} h_{j(k,1)}^g h_{j(k,2)}^g \cdots h_{j(k,l(k))}^g.$$

It follows that $t_{j(k,s)}^{-1} g^{l(k)} t_{j(k,s)} = h_{j(k,1)}^g h_{j(k,2)}^g \cdots h_{j(k,l(k))}^g$. This leads to an expression for $T(g)$:

$$T(g) = \prod_i h_i^g = \prod_{k=1}^{t} \prod_{s=1}^{l(k)} h_{j(k,s)}^g = \prod_{k=1}^{t} t_{j(k,s_k)}^{-1} g^{l(k)} t_{j(k,s_k)}.$$

We next make a technical observation.

Lemma 2.31 *For P a Sylow p-subgroup of G and $u, v \in C_G(P)$, suppose $v = k^{-1}uk$ for some $k \in G$. Then there is $x \in N_G(P)$ such that $v = k^{-1}uk = x^{-1}ux$.*

Proof Let $C_G(v) = \{g \in G \mid gv = vg\}$ denote the centralizer of v in G. Having chosen $v \in C_G(P)$, P is a subgroup of $C_G(v)$ because every element of P commutes with v. It follows that, for $h \in P$,

$$(k^{-1}hk)v = (k^{-1}hk)(k^{-1}uk) = k^{-1}(hu)k = k^{-1}(uh)k$$
$$= (k^{-1}uk)(k^{-1}hk) = v(k^{-1}hk).$$

So $k^{-1}Pk \subset C_G(v)$. Therefore P and $k^{-1}Pk$ are Sylow p-subgroups of $C_G(v)$, and there is an element of $C_G(v)$, say w, with $P = w^{-1}k^{-1}Pkw$, which means that $kw \in N_G(P)$. Let $kw = x$. Then, because $w \in C_G(v)$, $x^{-1}ux = (kw)^{-1}u(kw) = w^{-1}(k^{-1}uk)w = w^{-1}vw = v$. $\square$

Suppose $g \in P$, which is a subgroup of $C_G(P)$. We can call on the lemma to replace $t_{j(k,s_k)}^{-1} g^{l(k)} t_{j(k,s_k)}$ with $x_{j(k,s_k)}^{-1} g^{l(k)} x_{j(k,s_k)}$ where $x_{j(k,s_k)}$ is in $N_G(P)$. By assumption, $N_G(P) = C_G(P)$, and so we have that $x_{j(k,s_k)}^{-1} g^{l(k)} x_{j(k,s_k)} = g^{l(k)}$. Therefore

$$T(g) = \prod_{k=1}^{t} t_{j(k,s_k)}^{-1} g^{l(k)} t_{j(k,s_k)} = \prod_{k=1}^{t} x_{j(k,s_k)}^{-1} g^{l(k)} x_{j(k,s_k)} = \prod_{k=1}^{t} g^{l(k)} = g^m.$$

With $\#G = p^e m$ and $\gcd(m, p^e) = 1$, we can find integers a and b with $ma + p^e b = 1$. Then, for $g \in P$, we have $g^{p^e} = e$, and so

$$T(g^a) = (g^m)^a = g^{ma}(g^{p^e})^b = g^{ma + p^e b} = g^1 = g.$$

It follows that $T: G \to P$ is surjective. Let $\ker T = N$, a normal subgroup of G. Because $G/N \cong P$, we know that the order of N is m. To see that $N \cap P = \{e\}$, suppose $h \in P \cap N$. Then $h^{p^e} = e$ and $h^m = T(h) = e$. This implies that $h = h^{ma + p^e b} = (h^m)^a (h^{p^e})^b = e$. The subgroup NP has the cardinality determined by the Second Isomorphism Theorem: $NP/N \cong P/(P \cap N) = P$, and so $\#NP = \#G$. Since G is a finite group, $G = NP$. $\square$

As an example of how the Burnside Transfer Theorem can be applied, let us consider whether a group of order 180 can be simple. Since $180 = 2^2 \cdot 3^2 \cdot 5$, the possible values of s_5, the number of Sylow 5-subgroups, are 1, 6, and 36. If $s_5 = 6$, then there is an injection $G \hookrightarrow A_6$. Since $\#A_6 = 360 = 2\#G$, the image of G would be normal in A_6, a simple group. Hence $s_5 \neq 6$. So $s_5 = 36 = [G : N_G(P)]$ for a Sylow 5-subgroup P. This implies that $\#N_G(P) = 5$, but P is a subgroup of $N_G(P)$, so $P = N_G(P)$. Since a group of order 5 is abelian, P is also a subgroup of $C_G(P)$, which is a subgroup of $N_G(P)$. So we deduce $C_G(P) = N_G(P)$. The Burnside Transfer Theorem implies that P has a normal complement $N \lhd G$ of order 36, and so G is not simple. The reader can obtain other examples of this reasoning by choosing products of primes to fit this pattern.

Exercises

2.1 The orbits of an action of G on a finite set X determine a partition of X into G-invariant subsets. Show that this leads to a factorization of the action $\rho\colon G \to \Sigma(X)$ through $\rho\colon G \to \Sigma(\mathcal{O}_{x_1}) \times \cdots \times \Sigma(\mathcal{O}_{x_s})$ for which $\rho(g)$ acts on each orbit separately. Relate this to the representation of a permutation as a product of disjoint cycles.

2.2 Suppose $g \in G$ is an element of even order. Show that $g^2 \notin [g]$, the conjugacy class of g. Suppose G is a group of odd order. Show that for all $g \neq e$ in G, $g^{-1} \notin [g]$.

2.3 Show that every subgroup of the quaternion group Q is normal in Q. Exhibit a subgroup of D_8 that is not normal in D_8. Hence these groups of order 8 are not isomorphic.

2.4 For $\sigma \in \Sigma_n$ of type $(1^{m_1}, 2^{m_2}, \ldots, n^{m_n})$, show that the order of σ is $\mathrm{lcm}\{j \mid 1 \leq j \leq n, m_j > 0\}$, the least common multiple of the set of indices j with $m_j > 0$.

2.5 Let us count the number of distinct bracelets with five beads each of which can be one of three colors. The dihedral group D_{10} acts on each bracelet giving an equivalent bracelet. Thus a distinct bracelet corresponds to an orbit of the action. Determine the number of distinct bracelets by applying Burnside's Orbit Counting Theorem.

2.6 The are five groups of order 12. Use the Sylow Theorems to classify them.

2.7 Suppose G is a simple group of order 60. If G has a subgroup H of index five, then the faithful and transitive action of G on G/H determines an inclusion of G into A_5, which must be an isomorphism. Show that the

Sylow Theorems imply that a simple group of order 60 must contain a subgroup of index five. (Hint: Consider the Sylow 2-subgroups. They number $s_2 = 1, 3, 5$, or 15. Observe that 1 and 3 are impossible. If $s_2 = 15$, then $\#N_G(P) = 2^2$ and P is abelian, so $P = N_G(P) = C_G(P)$. The Burnside Transfer Theorem implies that G would have a normal subgroup and not be simple. So $s_2 = 5 = [G : N_G(P)]$ for a Sylow 2-subgroup P, and G has a subgroup of index five.) Deduce further that any simple group of order 60 is isomorphic to A_5.

2.8 Give examples of groups G and nontrivial actions of G on a set X that fail to be transitive.

2.9 Find all subgroups of Σ_3 that act transitively on $[3] = \{1, 2, 3\}$. (There are two such groups up to isomorphism.) Then find all subgroups of Σ_4 that act transitively on $[4] = \{1, 2, 3, 4\}$. (There are five such groups up to isomorphism.)

2.10 Suppose G acts transitively on X via $\rho\colon G \to \Sigma(X)$. Show that the kernel of ρ satisfies

$$\ker \rho = \bigcap_{x \in X} G_x = \bigcap_{g \in G} g G_x g^{-1} \text{ for some } x \in X.$$

What does this tell us for the kernel of the left multiplication action of G on G/H when H is a subgroup of G? What does this say about the kernel of the action of G on the orbit of a subgroup H under conjugation?

2.11 Suppose $\#G = p^2 m$ and p does not divide m. If P is a Sylow p-subgroup, then P is abelian. However, P need not be cyclic. Suppose P and Q are Sylow p-subgroups and $P \cap Q \neq \{e\}$. Show that both P and Q are subgroups of the normalizer of $P \cap Q$. If G is simple, then $N_G(P \cap Q) \neq G$. Estimate $[G : N_G(P \cap Q)] \leq m$ and argue that G must be isomorphic to a subgroup of A_m. In certain cases (for example, try $\#G = 90$), this condition can lead to a proof that G is not simple.

2.12 Show that only 60 and 168 are *stubborn* orders for groups of order less than 200, that is, the methods of the Sylow Theorems and the transfer do not exclude groups of order 60 or 168 from being simple. (Of course, we know why for $\#G = 60$.)

2.13 Suppose G is a finite group and $H \subset Z(G)$ is a subgroup of the center of G. Show that the transfer $T\colon G \to H$ takes the form $T(g) = g^m$ where $m = [G : H]$. An example [74] is the group of units in $\mathbb{Z}/p\mathbb{Z}$, denoted $(\mathbb{Z}/p\mathbb{Z})^{\times}$. Let H denote the subgroup $\{\pm 1\} \subset (\mathbb{Z}/p\mathbb{Z})^{\times}$. The transfer in this case is $T\colon (\mathbb{Z}/p\mathbb{Z})^{\times} \to \{\pm 1\}$ defined as $T(x) =$

$x^{(p-1)/2} = \left(\dfrac{x}{p} \right)$, the *Legendre symbol* mod p. (See [38] for details about the Legendre symbol.)

2.14 Suppose $K \subset H \subset G$ is a chain of subgroups with abelian K and H. Then there are transfer homomorphisms $T_G^H : G \to H$, $T_H^K : H \to K$, and $T_G^K : G \to K$. Show that $T_G^K = T_K^H \circ T_G^H$.

2.15 Extend the definition of the transfer for a nonabelian subgroup H of a group G by composing the definition with the homomorphism $\pi : H \to H^{ab} = H/H'$, where H' is the commutator subgroup of H, and H^{ab} is the *abelianization* of H.

3

Warmup: Some Linear Algebra

Thus simplified the group has $(p + 1)p\dfrac{p-1}{2}$ permutations. But it is easy to see that it is not further decomposable properly unless $p = 2$ or $p = 3$.

ÉVARISTE GALOIS, *The testamentary letter of 29 May 1832*

The standard introductory course in linear algebra explores the properties of $\mathbb{R}^n$ as a vector space over the field $\mathbb{R}$. The effective tool of Gaussian elimination leads to the fundamental notions of linear independence, bases, change-of-basis, and determinants. The next step includes eigenvalues and eigenvectors and their wonderful applications. The "*algebra*" of linear algebra is based on addition and multiplication by a scalar. The main properties carry over to more general scalars and other forms of addition.

3.1 The Fundamentals

A *field* $\mathbb{F}$ is a commutative ring with unit for which every nonzero element has a multiplicative inverse. It is customary to assume that $1 \neq 0$ in $\mathbb{F}$. The well-known examples are the rational numbers $\mathbb{Q}$, the real numbers $\mathbb{R}$, the complex numbers $\mathbb{C}$, and the finite fields $\mathbb{F}_p = \mathbb{Z}/p\mathbb{Z}$. In a later chapter, we will explore fields more thoroughly.

Definition 3.1 A *vector space* over a field $\mathbb{F}$ is a set V together with an addition $+: V \times V \to V$ and a multiplication by a scalar, mbs: $\mathbb{F} \times V \to V$, for which $(V, +)$ is an abelian group. If we write $+(u, v) = u + v$, then

1. for all u, v in V, $u + v = v + u$;
2. for all u, v, w in V, $(u + v) + w = u + (v + w)$;
3. there is an element $\mathbb{O} \in V$ satisfying $\mathbb{O} + v = v + \mathbb{O} = v$ for all $v \in V$;

48

4. for each v in V, there is an element $-v \in V$ for which $v + (-v) = \mathbb{O}$.

If we write the multiplication by a scalar by $\mathrm{mbs}(a, v) = av$, then

5. for all $a, b \in \mathbb{F}$ and $v \in V$, $(a + b)v = av + bv$;
6. for all $a \in \mathbb{F}$ and $u, v \in V$, $a(u + v) = au + av$.
7. for all $a, b \in \mathbb{F}$ and $v \in V$, $a(bv) = (ab)v$;
8. for all $v \in V$, $1v = v$.

The reader is familiar with examples of vector spaces over $\mathbb{R}$ that provide the motivation for more general vector spaces.

For an arbitrary field $\mathbb{F}$, the ring $\mathbb{F}[x]$ of polynomials with coefficients in $\mathbb{F}$ forms a vector space with the usual addition (add like terms) and multiplication by a scalar given by

$$a\left(b_0 + b_1 x + b_2 x^2 + \cdots + b_n x^n\right) = (ab_0) + (ab_1)x + (ab_2)x^2 + \cdots + (ab_n)x^n.$$

For an arbitrary field $\mathbb{F}$, the analogue of $\mathbb{R}^n$ is the vector space of n-vectors with entries in $\mathbb{F}$,

$$\mathbb{F}^{\times n} = \left\{ \begin{bmatrix} a_1 \\ a_2 \\ \vdots \\ a_n \end{bmatrix} \;\middle|\; a_i \in \mathbb{F} \right\},$$

with addition and multiplication by a scalar given by

$$\begin{bmatrix} a_1 \\ a_2 \\ \vdots \\ a_n \end{bmatrix} + \begin{bmatrix} b_1 \\ b_2 \\ \vdots \\ b_n \end{bmatrix} = \begin{bmatrix} a_1 + b_1 \\ a_2 + b_2 \\ \vdots \\ a_n + b_n \end{bmatrix}, \quad a\begin{bmatrix} a_1 \\ a_2 \\ \vdots \\ a_n \end{bmatrix} = \begin{bmatrix} aa_1 \\ aa_2 \\ \vdots \\ aa_n \end{bmatrix}.$$

The proof of the vector space properties of $\mathbb{R}^n$ go over immediately to the case of $\mathbb{F}^{\times n}$.

Given any set X, we can form a vector space $\mathbb{F}X$ over $\mathbb{F}$ by taking all finite formal sums $a_1 x_1 + a_2 x_2 + \cdots + a_k x_k$ with $x_i \in X$ and $a_i \in \mathbb{F}$. The addition and multiplication by a scalar are given by

$$(a_1 x_1 + \cdots + a_n x_n) + (b_1 x_1 + \cdots + b_n x_n)$$
$$= (a_1 + b_1)x_1 + \cdots + (a_n + b_n)x_n;$$
$$a(a_1 x_1 + \cdots + a_n x_n) = (aa_1)x_1 + \cdots + (aa_n)x_n.$$

The identity element in $\mathbb{F}X$ is given by any sum with all coefficients zero. It follows that the additive inverse of the expression ax is given by $(-a)x$. This construction makes an important appearance in the next chapter.

A *subspace* of a vector space V is a nonempty subset $W \subset V$ for which, if $u, v \in W$ and $a \in \mathbb{F}$, then $au + v \in W$. The condition guarantees that $\mathbb{O} \in W$ because, for any $w \in W$, $(-1)w + w = \mathbb{O} \in W$. Taking $a = 1$ shows that W is closed under addition, and taking $v = \mathbb{O}$ implies that W is closed under multiplication by scalars. Thus W is a vector space with the addition and mbs inherited from V. Notice that $(W, +)$ is a subgroup of the abelian group $(V, +)$.

Given a subset $S \subset V$, we can construct a subspace called the *Span* of S in V as the subset of all finite linear combinations of elements of S:

$$\text{Span}(S) = \{a_1 s_1 + a_2 s_2 + \cdots + a_n s_n \in V \mid s_i \in S, a_i \in \mathbb{F}\}.$$

Because the addition is addition in V, the span may differ from the vector space of formal sums $\mathbb{F}S$: relations among vectors in V can lead to relations in $\text{Span}(S)$, and some nonzero formal sums may be zero.

$\text{Span}(S)$ is a subspace of V for any $S \subset V$ ($\text{Span}(\emptyset) = \{\mathbb{O}\}$). The proof is the same as the proof that $\mathbb{F}S$ is a vector space together with the condition that for any $u, v \in \text{Span}(S)$, $a \in \mathbb{F}$, we have $au + v \in \text{Span}(S)$. As the motto says, *"Linear combinations of linear combinations are linear combinations"* [56].

Definition 3.2 A subset $S \subset V$ of a vector space V is *spanning* if $\text{Span}(S) = V$. A subset S of V is *linearly independent* if the condition $a_1 s_1 + a_2 s_2 + \cdots + a_n s_n = \mathbb{O}$ with $s_i \in S$ distinct and $a_i \in \mathbb{F}$ implies $a_1 = 0, a_2 = 0, \ldots, a_n = 0$.

For example, the set of vectors

$$\mathbf{e}_1 = \begin{bmatrix} 1 \\ 0 \\ \vdots \\ 0 \end{bmatrix}, \mathbf{e}_2 = \begin{bmatrix} 0 \\ 1 \\ \vdots \\ 0 \end{bmatrix}, \ldots, \mathbf{e}_n = \begin{bmatrix} 0 \\ 0 \\ \vdots \\ 1 \end{bmatrix}$$

in $\mathbb{F}^{\times n}$ is both spanning and linearly independent. If we add any other vector to $\{\mathbf{e}_1, \ldots, \mathbf{e}_n\}$, then the set $S = \{\mathbf{e}_1, \ldots, \mathbf{e}_n, v\}$ is not linearly independent:

$$\text{let } \mathbf{v} = \begin{bmatrix} a_1 \\ a_2 \\ \vdots \\ a_n \end{bmatrix}, \quad (-a_1)\mathbf{e}_1 + (-a_2)\mathbf{e}_2 + \cdots + (-a_n)\mathbf{e}_n + 1\mathbf{v} = \mathbb{O},$$

where some of the coefficients are not zero (in particular, the coefficient 1 times **v**). A subset is *linearly dependent* when it is not linearly independent.

Theorem 3.3 *If a subset $S \subset V$ of a vector space V is both spanning and linearly independent, then for all $v \in V$, there is a unique expression*

$$v = a_1 s_1 + a_2 s_2 + \cdots + a_n s_n$$

with $s_i \in S$ and $a_i \in \mathbb{F}$.

Proof Because S is spanning, every vector v in V can be written $v = a_1 s_1 + a_2 s_2 + \cdots + a_n s_n$. Suppose v has another representation $v = b_1 s_1 + b_2 s_2 + \cdots + b_n s_n$. Then

$$\mathbb{O} = v - v = (a_1 - b_1)s_1 + (a_2 - b_2)s_2 + \cdots + (a_n - b_n)s_n.$$

Since S is linearly independent, all coefficients $a_j - b_j = 0$, and so $a_j = b_j$ for all j. □

Definition 3.4 A subset $S \subset V$ of a vector space V is a *basis* for V if it is linearly independent and spanning.

If we impose the partial order induced by inclusion, $A \subset B$, on the collection of all linearly independent subsets of a vector space V, then any ordered sequence of linearly independent subsets $A_1 \subset A_2 \subset A_3 \subset \cdots$ has an upper bound $A_\infty = \bigcup_n A_n$. Then $A_k \subset A_\infty$ for any k. Notice that A_∞ is linearly independent: any linear relation $c_1 v_1 + \cdots + c_m v_m = \mathbb{O}$ with $v_i \in A_\infty$ and $c_i \in \mathbb{F}$, not all zero, must occur in A_k for some k, which contradicts the linear independence of A_k. *Zorn's Lemma* [4] is a handy result from set theory. It states that any partially ordered set $(S, \leq)$ for which every totally ordered subset $s_1 \leq s_2 \leq \cdots$ has an upper bound has a *maximal element* $t \in S$, that is, if $t \leq s$ for any $s \in S$, then $s = t$. In our case, there is a maximal linearly independent subset $B \subset V$ for which if A is a linearly independent subset and $B \subset A$, then $B = A$. Consider the span of B. If $\text{Span}(B) \neq V$, then there is an element $v \in V$ that is not a finite linear combination of elements of B. Adding v to B gives a larger linearly independent subset, which contradicts the maximal condition. Hence B is both spanning and linearly independent and thus a basis for V. Every vector space has a basis.

Get to know bases

A vector space V is a *finite-dimensional vector space* if V has a finite basis. In the following, assume that all vector spaces are finite-dimensional.

1. Show that a subset of nonzero elements $v_1, v_2, \ldots, v_k$ is linearly dependent if there is a least index j, $1 < j \leq k$, with $v_j = a_1 v_1 + \cdots + a_{j-1} v_{j-1}$. Also show that any subset $S \subset V$ that contains the zero element is linearly dependent.

2. If $B = \{v_1, \ldots, v_n\}$ is a basis for V, show that $\mathrm{Span}(\{v_1, v_2, \ldots, v_j\}) \neq V$ for any $j < n$.

3. Suppose V is a finite-dimensional vector space and $B = \{v_1, v_2, \ldots, v_n\}$ and $B' = \{w_1, w_2, \ldots, w_m\}$ are bases for V. Since B is spanning, $w_1 = a_1 v_1 + \cdots + a_n v_n$. Suppose $a_j \neq 0$ and $a_k = 0$ for $j < k \leq n$. Then $v_j = \dfrac{-1}{a_j}(a_1 v_1 + \cdots + a_{j-1} v_{j-1} - w_1)$. If we denote the omission of an element from a set by $\{v_1, v_2, \ldots, \widehat{v_j}, \ldots, v_n\} = \{v_1, \ldots, v_{j-1}, v_{j+1}, \ldots, v_n\}$, then $v_j \in \mathrm{Span}(\{v_1, \ldots, \widehat{v_j}, \ldots, v_n, w_1\})$. Show that the set $\{v_1, \ldots, \widehat{v_j}, \ldots, v_n, w_1\}$ is also a basis for V. Continue this process and add w_2 and remove v_i to get a new basis $\{v_1, \ldots, \widehat{v_i}, \ldots, \widehat{v_j}, \ldots, v_n, w_1, w_2\}$. If we add, one at a time, all of the w_k to the remaining v_s, then show that this implies $n \geq m$. Then argue that if $n > m$, then we would obtain a linearly dependent set when all w_m are added. Thus $n \leq m$, and we have shown that $n = m$, that is, all bases of a finite-dimensional vector space have the same cardinality.

   We call $\#B$, the number of elements in a basis for V, the *dimension* of V, denoted $\dim V$. For example, $\dim \mathbb{F}^{\times n} = n$.

4. Show that any subset $S \subset V$ with $\#S > \dim V$ must be linearly dependent. Suppose S is a spanning set for V. Show that S contains a basis of V.

5. Suppose V is a finite-dimensional vector space. If $W \subset V$ is a subspace of V, then show that any basis for W can be extended to a basis for V, that is, if $\{u_1, \ldots, u_k\}$ is a basis for W, then there are vectors $\{v_1, \ldots, v_l\}$ in V with $\{u_1, \ldots, u_k, v_1, \ldots, v_l\}$ a basis for V. Show that W is also finite-dimensional.

We compare vector spaces over a field $\mathbb{F}$ via the structure-preserving functions between vector spaces.

Definition 3.5 A function $\phi \colon V \to W$ between vector spaces V and W over a field $\mathbb{F}$ is a *linear transformation* if ϕ satisfies

$$\text{for all } u, v \in V \text{ and } a \in \mathbb{F}, \ \phi(au + v) = a\phi(u) + \phi(v).$$

A linear transformation that is both injective and surjective is called an *isomorphism* between the vector spaces.

A familiar source of linear transformations are the mappings $T_M : \mathbb{R}^n \to \mathbb{R}^m$ given by multiplication by M, where M is an $m \times n$ matrix with real entries and $T_M(\mathbf{x}) = M\mathbf{x}$. If $M = (a_{ij})$, then

$$M\mathbf{x} = \begin{bmatrix} a_{11} & a_{12} & \cdots & a_{1n} \\ a_{21} & a_{22} & \cdots & a_{2n} \\ \vdots & \vdots & \cdots & \vdots \\ a_{m1} & a_{m2} & \cdots & a_{mn} \end{bmatrix} \begin{bmatrix} x_1 \\ x_2 \\ \vdots \\ x_n \end{bmatrix}$$

$$= x_1 \begin{bmatrix} a_{11} \\ a_{21} \\ \vdots \\ a_{m1} \end{bmatrix} + x_2 \begin{bmatrix} a_{12} \\ a_{22} \\ \vdots \\ a_{m2} \end{bmatrix} + \cdots + x_n \begin{bmatrix} a_{1n} \\ a_{2n} \\ \vdots \\ a_{mn} \end{bmatrix}.$$

The basic properties of matrix multiplication imply that T_M is a linear transformation.

For a finite-dimensional vector space V with a basis $B = \{v_1, \ldots, v_n\}$, a linear transformation $\phi : V \to W$ is determined by choosing vectors $w_1, \ldots, w_n$ in W and letting $\phi(v_1) = w_1, \ldots, \phi(v_n) = w_n$. Then ϕ extends to the rest of V by

$$\phi(a_1 v_1 + \cdots + a_n v_n) = a_1 w_1 + \cdots + a_n w_n.$$

Thus a function $B \to W$ determines a linear transformation $V \to W$.

Denote by $\text{Hom}_{\mathbb{F}}(V, W)$ the set of all linear transformations from V to W. This set supports an addition: for linear transformations $\phi, \psi : V \to W$, define $\phi + \psi : V \to W$ by $(\phi + \psi)(v) = \phi(v) + \psi(v)$. Then

$$(\phi + \psi)(av + w) = \phi(av + w) + \psi(av + w)$$
$$= a\phi(v) + \phi(w) + a\psi(v) + \psi(w)$$
$$= a(\phi + \psi)(v) + (\phi + \psi)(w),$$

and $\phi + \psi$ is also a linear transformation. When we let $(a\phi)(v) = a\phi(v)$, this defines a multiplication by scalars on $\text{Hom}_{\mathbb{F}}(V, W)$ making $\text{Hom}_{\mathbb{F}}(V, W)$ into a vector space over the field $\mathbb{F}$.

Proposition 3.6 *If V and W are finite-dimensional vector spaces over $\mathbb{F}$, then* $\dim \text{Hom}_{\mathbb{F}}(V, W) = (\dim V)(\dim W).$

Proof Let $B = \{v_1, \ldots, v_n\}$ be a basis for V, and let $B' = \{w_1, \ldots, w_m\}$ be a basis for W. For $1 \leq i \leq n$ and $1 \leq j \leq m$, let $L_{ij} : V \to W$ denote the linear transformation determined on the basis B by $L_{ij}(v_i) = w_j$ and $L_{ij}(v_k) = \mathbb{O}$ for $k \neq i$. A linear transformation is determined by its values on the elements of a basis and linearity, so given $T : V \to W$ in $\text{Hom}_{\mathbb{F}}(V, W)$, we have

$$T(v_i) = a_{i1}w_1 + a_{i2}w_2 + \cdots + a_{im}w_m$$
$$= a_{i1}L_{i1}(v_i) + a_{i2}L_{i2}(v_i) + \cdots + a_{im}L_{im}(v_i).$$

Let $T' = \sum_{i=1}^{n} \sum_{j=1}^{m} a_{ij}L_{ij}$. The coefficients are chosen to guarantee $T'(v_i) = T(v_i)$ for all i. By linearity, $T' = T$ because they agree on a basis. Thus $\{L_{ij}\}$ is a spanning set of $\text{Hom}_{\mathbb{F}}(V, W)$. To see that $\{L_{ij}\}$ is linearly independent, suppose $\sum_{i,j} c_{ij}L_{ij} = \mathbb{O}$. Then

$$\mathbb{O} = \mathbb{O}(v_i) = \sum_{i,j} c_{ij}L_{ij}(v_i) = c_{i1}w_1 + \cdots + c_{im}w_m.$$

Since B' is a basis, $c_{i1} = \cdots = c_{im} = 0$. This relation holds for all i, and $\{L_{ij}\}$ is a basis for $\text{Hom}_{\mathbb{F}}(V, W)$. $\square$

With a linear transformation $L : V \to W$, we associate two subspaces: $\ker L = \{v \in V \mid L(v) = \mathbb{O}\}$, the *kernel* (or *null space*) of L, and $L(V) = \{L(v) \in W \mid v \in V\}$, the *image* of L. For $v, v' \in \ker L$, we have $L(av+v') = aL(v) + L(v') = a\mathbb{O} + \mathbb{O} = \mathbb{O}$; for $w, w' \in L(V)$, we have $aw + w' = aL(v) + L(v') = L(av + v') \in L(V)$. Hence $\ker L$ is a subspace of V, and $L(V)$ is a subspace of W.

When $V = W$, the vector space $\text{Hom}_{\mathbb{F}}(V, V)$ has some extra structure: Given $T_1 : V \to V$ and $T_2 : V \to V$, we can form the composite $T_2 \circ T_1 : V \to V$ given by $T_2 \circ T_1(v) = T_2(T_1(v))$. This binary operation is associative and has an identity element $\text{Id} : V \to V$ given by $\text{Id}(v) = v$.

Proposition 3.7 *The inverse of an isomorphism is also an isomorphism.*

Proof Suppose $\phi : V \to W$ is an isomorphism. Because ϕ is bijective, ϕ has an inverse function $\phi^{-1} : W \to V$. It suffices to show that ϕ^{-1} is a linear transformation. Suppose $v, w \in W$ and $a \in \mathbb{F}$. Then $v = \phi(v')$ and $w = \phi(w')$ for $v', w' \in V$, and

$$\phi^{-1}(av + w) = \phi^{-1}\big(a\phi(v') + \phi(w')\big) = \phi^{-1}\big(\phi(av' + w')\big)$$
$$= av' + w' = a\phi^{-1}(v) + \phi^{-1}(w).$$

Thus ϕ^{-1} is a bijective linear transformation. $\square$

The proposition implies that the subset of $\text{Hom}_{\mathbb{F}}(V, V)$ given by

$$\text{Gl}(V) = \{T : V \to V \mid T \text{ is a bijective linear transformation}\}$$

with composition as binary operation is a group, the *general linear group* associated with the vector space V. When $V = \mathbb{F}^{\times n}$, we can identify $\mathrm{Gl}(\mathbb{F}^{\times n})$ with $\mathrm{Gl}_n(\mathbb{F})$, the group of $n \times n$ invertible matrices with entries in $\mathbb{F}$. With the choice of a basis for V, we can represent any linear transformation $T : V \to V$ by an $n \times n$ matrix.

Theorem 3.8 *For any vector space V of dimension n over the field $\mathbb{F}$, V is isomorphic to $\mathbb{F}^{\times n}$. Furthermore, the groups $\mathrm{Gl}(V)$ and $\mathrm{Gl}_n(\mathbb{F})$ are isomorphic.*

Proof Let $B = \{v_1, v_2, \ldots, v_n\}$ be a basis for V. Consider the mapping $c_B \colon V \to \mathbb{F}^{\times n}$, the coordinate mapping with respect to D, given by

$$c_B(v) = c_B(a_1 v_1 + a_2 v_2 + \cdots + a_n v_n) = \begin{bmatrix} a_1 \\ a_2 \\ \vdots \\ a_n \end{bmatrix}.$$

It follows from $av + w = a(a_1 v_1 + \cdots + a_n v_n) + (b_1 v_1 + \cdots + b_n v_n) = (aa_1 + b_1)v_1 + \cdots + (aa_n + b_n)v_n$ that c_B is a linear transformation. The linear independence of B implies that c_B is injective: if $c_B(v) = c_B(v')$, then $c_B(v - v') = \mathbb{O}$, which implies that the coefficients of each linear combination are the same. That c_B is surjective follows by using the entries of a vector as the coefficients of a linear combination.

For a linear transformation $\phi \colon V \to V$, we associate a matrix with ϕ by considering the following diagram:

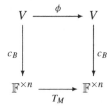

where $c_B \colon V \to \mathbb{F}^{\times n}$ is the coordinate isomorphism with respect to the basis B. When ϕ is an isomorphism, that is, $\phi \in \mathrm{Gl}(V)$, the composite $T_M = c_B \circ \phi \circ c_B^{-1} \colon \mathbb{F}^{\times n} \to \mathbb{F}^{\times n}$ is an isomorphism and so in $\mathrm{Gl}(\mathbb{F}^{\times n})$. The matrix, given by columns,

$$M = \begin{bmatrix} c_B \circ \phi \circ c_B^{-1}(\mathbf{e}_1) & c_B \circ \phi \circ c_B^{-1}(\mathbf{e}_2) & \cdots & c_B \circ \phi \circ c_B^{-1}(\mathbf{e}_n) \end{bmatrix}$$

satisfies $c_B \circ \phi \circ c_B^{-1}(\mathbf{v}) = M\mathbf{v}$. This gives us the mapping $\Theta_B \colon \mathrm{Gl}(V) \to \mathrm{Gl}_n(\mathbb{F})$ defined by $\Theta_B(\phi) = M$. For a pair of linear isomorphisms ϕ, ψ, we have

$$c_B \circ (\phi \circ \psi) \circ c_B^{-1} = \left(c_B \circ \phi \circ c_B^{-1} \right) \circ \left(c_B \circ \psi \circ c_B^{-1} \right).$$

Since composition of linear transformations corresponds to matrix multiplication, we have $\Theta_B(\phi \circ \psi) = \Theta_B(\phi)\Theta_B(\psi)$, and the mapping is a group homomorphism. The inverse is given by $\Theta_B^{-1}(M) = c_B^{-1} \circ M \circ c_B$, and so $\mathrm{Gl}(V)$ and $\mathrm{Gl}_n(\mathbb{F})$ are isomorphic groups. $\qquad\qquad\square$

Corollary 3.9 *If V and W are finite-dimensional vector spaces of the same dimension, then V is isomorphic to W.*

Proof If V and W are finite-dimensional vector spaces of the same dimension over $\mathbb{F}$, each is isomorphic to $\mathbb{F}^{\times n}$ and hence to each other. $\qquad\square$

When we consider isomorphic finite-dimensional vector spaces as equivalent, the dimension is a complete invariant of an equivalence class.

We can combine two vector spaces V and W over $\mathbb{F}$ into a new vector space, the *direct sum*

$$V \oplus W = \big\{(v, w) \mid v \in V, w \in W\big\}.$$

The addition on $V \oplus W$ is given by $(v_1, w_1) + (v_2, w_2) = (v_1+v_2, w_1+w_2)$, as in the case of the product of two abelian groups. The multiplication by a scalar is given by $a(v, w) = (av, aw)$. The zero in $V \oplus W$ is $(\mathbb{O}_V, \mathbb{O}_W)$. Given a basis $B = \{v_1, \ldots, v_n\}$ for V and a basis $C = \{x_1, \ldots, x_m\}$ for W, we can form a basis for $V \oplus W$ given by $\mathcal{B} = \{(v_1, \mathbb{O}_W), \ldots, (v_n, \mathbb{O}_W), (\mathbb{O}_V, x_1), \ldots, (\mathbb{O}_V, x_m)\}$. It follows that $\dim(V \oplus W) = \dim(V) + \dim(W)$. For example, the direct sum $\mathbb{F}^{\times n} \oplus \mathbb{F}^{\times m}$ has the same dimension as $\mathbb{F}^{\times(n+m)}$. Hence they are isomorphic. Notice that the subsets $V \oplus \{\mathbb{O}\} = \{(v, \mathbb{O}) \mid v \in V\}$, which is isomorphic to V, and $\{\mathbb{O}\} \oplus W \cong W$ are subspaces of $V \oplus W$. (Can you prove this?) The direct sum is the domain of two linear transformations $\mathrm{pr}_1 \colon V \oplus W \to V$ and $\mathrm{pr}_2 \colon V \oplus W \to W$ given by $\mathrm{pr}_1(v, w) = v$ and $\mathrm{pr}_2(v, w) = w$.

Given linear transformations $f_1 \colon W_1 \to V_1$ and $f_2 \colon W_2 \to V_2$, there is a linear mapping $f_1 \oplus f_2 \colon W_1 \oplus W_2 \to V_1 \oplus V_2$ defined by $(f_1 \oplus f_2)(w_1, w_2) = (f_1(w_1), f_2(w_2))$. The reader will want to check that $f_1 \oplus f_2$ is also a linear transformation, and if f_1 and f_2 are both isomorphisms, then so is $f_1 \oplus f_2$.

Get to know linear transformations

1. Let $T \colon \mathbb{R}^2 \to \mathbb{R}^2$ be given by $T\begin{bmatrix} x \\ y \end{bmatrix} = \begin{bmatrix} x \\ 0 \end{bmatrix}$. Let $\mathcal{B} = \{\mathbf{e}_1, \mathbf{e}_2\}$ denote the standard basis for $\mathbb{R}^2$, and let $\mathcal{B}' = \{\begin{bmatrix} 2 \\ 1 \end{bmatrix}, \begin{bmatrix} -1 \\ 1 \end{bmatrix}\}$. Represent T as a matrix with respect to the basis $\mathcal{B}$, as a matrix with respect to the basis $\mathcal{B}'$, and as a mapping $T \colon (\mathbb{R}^2, \mathcal{B}) \to (\mathbb{R}^2, \mathcal{B}')$.

2. Two $n \times n$ matrices M and N are *similar* if there is an invertible matrix P with $N = PMP^{-1}$. Show that this relation is an equivalence relation. Show further that we can view $P = c_B$ as a change-of-basis matrix for some basis B for $\mathbb{F}^{\times n}$. How can you relate c_B to P?

3. If $V = \mathbb{F}^{\times 2}$ and $T_1, T_2: V \to V$ are linear transformations represented by matrices M_1 and M_2, then $T_1 \oplus T_2: V \oplus V \to V \oplus V$ is a linear transformation. In terms of M_1 and M_2, what is the matrix representing $T_1 \oplus T_2$?

4. Let $M_{n \times n}$ denote the set of $n \times n$ matrices with entries in a field $\mathbb{F}$. Show that $M_{n \times n}$ is a vector space over $\mathbb{F}$. If we fix a matrix $B \in M_{n \times n}$, then show that the mapping $T: M_{n \times n} \to M_{n \times n}$ given by $T(A) = AB - BA$ is a linear mapping. Interpret its kernel (null space).

5. Let $T: V \to V$ be a linear transformation, and let $U: V \to W$ be a linear isomorphism. Consider the mapping $U^*: \mathrm{Hom}_{\mathbb{F}}(V, V) \to \mathrm{Hom}_{\mathbb{F}}(W, W)$ given by $U^*(T) = U \circ T \circ U^{-1}$. Show that U^* induces an isomorphism of vector spaces.

Suppose W_1 and W_2 are subspaces of a vector space V. We say that V is an *internal direct sum* if $V \cong W_1 \oplus W_2$. How can we recognize when V is an internal direct sum of subspaces?

Proposition 3.10 *A vector space V is an internal direct sum of subspaces W_1 and W_2 if (1) for all $v \in V$, there are $w_1 \in W_1$ and $w_2 \in W_2$ such that $v = w_1 + w_2$ and (2) $W_1 \cap W_2 = \{\mathbb{0}_V\}$.*

Proof To exhibit an isomorphism, consider the mapping $\phi: W_1 \oplus W_2 \to V$ given by $\phi(w_1, w_2) = w_1 + w_2$. By assumption (1), ϕ is surjective. To see that ϕ is linear, simply check

$$\phi\big(a(w_1, w_2) + (w_1', w_2')\big)$$
$$= \phi\big(aw_1 + w_1', aw_2 + w_2'\big) = aw_1 + w_1' + aw_2 + w_2'$$
$$= a(w_1 + w_2) + (w_1' + w_2') = a\phi(w_1, w_2) + \phi(w_1', w_2').$$

To prove that ϕ is injective, suppose $\phi(w_1, w_2) = \phi(w_1', w_2')$. Then $w_1 + w_2 = w_1' + w_2'$, which implies that $w_1 - w_1' = w_2' - w_2$. Since $w_1 - w_1' \in W_1$ and $w_2' - w_2 \in W_2$, both sides are in $W_1 \cap W_2 = \{\mathbb{0}_V\}$. So $w_1 = w_1'$ and $w_2 = w_2'$, and thus ϕ is injective. $\qquad\square$

The following lemma gives another condition from which we can recognize when V is an internal direct sum.

Lemma 3.11 *Suppose there is a linear transformation* $P \colon V \to V$ *satisfying* $P \circ P = P$, *that is,* $P(P(v)) = P(v)$ *for all* $v \in V$. *Then* V *is an internal direct sum of* $P(V)$, *the image of* P, *and* $\ker P$, $V \cong P(V) \oplus \ker P$.

Proof Consider the expression $v - P(v)$. Then $P(v - P(v)) = P(v) - P(P(v)) = P(v) - P(v) = \mathbb{O}$ and $v - P(v) \in \ker P$. Every v in V may be written as $v = P(v) + v - P(v)$ with $P(v) \in P(V)$ and $v - P(v) \in \ker P$. Suppose $u \in P(V) \cap \ker P$. Then

$$u = P(v) = P\big(P(v)\big) = P(u) = \mathbb{O},$$

and so $P(V) \cap \ker P = \{\mathbb{O}\}$. By Proposition 3.10, $V \cong P(V) \oplus \ker P$. $\square$

An important consequence of the lemma is the following result.

Fundamental Theorem of Linear Algebra *Let* V *be a finite-dimensional vector space over* $\mathbb{F}$. *Given a linear transformation* $\phi \colon V \to W$, *the domain* V *is isomorphic to* $\ker \phi \oplus \phi(V)$, *the direct sum of the kernel of* ϕ *and the image of* ϕ.

Proof The kernel of ϕ is a subspace of V and hence finite-dimensional. Let $\{u_1, \ldots, u_k\}$ be a basis for $\ker \phi$. Extend the basis for $\ker \phi$ to a basis for V, $B = \{u_1, \ldots, u_k, v_{k+1}, \ldots, v_n\}$. Define $P \colon V \to V$ on the basis by $P(u_i) = u_i$ and $P(v_j) = \mathbb{O}$. The linear extension to V is given for $v = a_1 u_1 + \cdots + a_k u_k + a_{k+1} v_{k+1} + \cdots a_n v_n$ by

$$P(v) = P(a_1 u_1 + \cdots a_k u_k + a_{k+1} v_{k+1} + \cdots + a_n v_n) = a_1 u_1 + \cdots + a_k u_k.$$

Notice that $P(P(v)) = P(v)$ for all $v \in V$. By Lemma 3.11, V is isomorphic to $P(V) \oplus \ker P$. The image of P is given by the span of $\{u_1, \ldots, u_k\}$, which is $\ker \phi$. The kernel of P is seen to be $\mathrm{Span}\{v_{k+1}, \ldots, v_n\}$. The theorem follows by showing that $\mathrm{Span}\{v_{k+1}, \ldots, v_n\} \cong \phi(V)$.

We claim that $\{\phi(v_{k+1}), \ldots, \phi(v_n)\}$ is a basis for $\phi(V)$. Suppose $w \in \phi(V)$. Then, for some $v \in V$,

$$
\begin{aligned}
w = \phi(v) &= \phi(a_1 u_1 + \cdots + a_k u_k + a_{k+1} v_{k+1} + \cdots + a_n v_n) \\
&= \phi(a_1 u_1 + \cdots + a_k u_k) + a_{k+1}\phi(v_{k+1}) + \cdots + a_n\phi(v_n) \\
&= a_{k+1}\phi(v_{k+1}) + \cdots + a_n\phi(v_n).
\end{aligned}
$$

Thus $\{\phi(v_{k+1}), \ldots, \phi(v_n)\}$ spans $\phi(V)$. Suppose $\mathbb{O} = c_{k+1}\phi(v_{k+1}) + \cdots + c_n\phi(v_n) = \phi(c_{k+1}v_{k+1} + \cdots + c_n v_n)$. Then $c_{k+1}v_{k+1} + \cdots + c_n v_n$ is in the kernel of ϕ and can be written $c_{k+1}v_{k+1} + \cdots + c_n v_n = b_1 u_1 + \cdots + b_k u_k$, giving the relation

$$b_1 u_1 + \cdots + b_k u_k - c_{k+1}v_{k+1} - \cdots - c_n v_n = \mathbb{O}.$$

Because B is a basis for V, all the coefficients are zero, and so $c_{k+1} = \cdots = c_n = 0$, and thus $\{\phi(v_{k+1}), \ldots, \phi(v_n)\}$ is a basis for $\phi(V)$.

The mapping $\widehat{\phi}\colon \ker P \to \phi(V)$ given on the basis by $\widehat{\phi}(v_j) = \phi(v_j)$ is an isomorphism because it is a bijection of bases. The fundamental theorem follows from $V \cong P(V) \oplus \ker P \cong \ker \phi \oplus \phi(V)$. □

The origin of linear algebra is the study of the properties of systems of linear equations. Let A be the $m \times n$ matrix of coefficients of m linear equations in n unknowns. A solution exists when the vector of constants is in the image of a linear transformation:

$$
\begin{aligned}
a_{11}x_1 + a_{12}x_2 + \cdots + a_{1n}x_n &= b_1 \\
a_{21}x_1 + a_{22}x_2 + \cdots + a_{2n}x_n &= b_2 \\
&\;\;\vdots \\
a_{m1}x_1 + a_{m2}x_2 + \cdots + a_{mn}x_n &= b_m
\end{aligned}
$$

$$
\Longleftrightarrow
\begin{bmatrix}
a_{11} & a_{12} & \cdots & a_{1n} \\
a_{21} & a_{22} & \cdots & a_{2n} \\
\vdots & \vdots & & \vdots \\
a_{m1} & a_{m2} & \cdots & a_{mn}
\end{bmatrix}
\begin{bmatrix}
x_1 \\ x_2 \\ \vdots \\ x_n
\end{bmatrix}
=
\begin{bmatrix}
b_1 \\ b_2 \\ \vdots \\ b_m
\end{bmatrix};
$$

the matrix formulation can be written $A\mathbf{x} = \mathbf{b}$ with $\mathbf{x} \in \mathbb{F}^{\times n}$ and $\mathbf{b} \in \mathbb{F}^{\times m}$. View $A\mathbf{x} = T_A(\mathbf{x}) = \mathbf{b}$, where $T_A\colon \mathbb{F}^{\times n} \to \mathbb{F}^{\times m}$ is the linear transformation given by multiplication by the matrix A. The Fundamental Theorem of Linear Algebra tells us the following:

Proposition 3.12 *For an $m \times n$ matrix A with entries in $\mathbb{F}$, the corresponding system of linear equations $A\mathbf{x} = \mathbf{b}$ has a solution if and only if $\mathbf{b}$ is in the image of $T_A\colon \mathbb{F}^{\times n} \to \mathbb{F}^{\times m}$. Solutions to the homogeneous system $A\mathbf{x} = \mathbb{O}$ are the elements of $\ker T_A$. If $n > m$, then there is always a nonzero solution to the homogeneous system.*

Proof The linear system $A\mathbf{x} = \mathbf{b}$ has a solution if and only if there is a vector $\mathbf{w} \in \mathbb{F}^{\times n}$ such that $T_A(\mathbf{w}) = \mathbf{b}$, that is, $\mathbf{b}$ is in the image of T_A. If $n > m$, then because $\dim \mathbb{F}^{\times n} = n = \dim \ker T_A + \dim T_A(\mathbb{F}^{\times n})$ and $\dim T_A(\mathbb{F}^{\times n}) \le m$, we have $\dim \ker T_A \ge n - m > 0$, and so there are nonzero vectors in $\ker T_A$. □

A sequence $W \xrightarrow{T} V \xrightarrow{L} U$ of linear transformations is *exact* at V if $\ker L = T(W)$. A *short exact sequence* of linear transformations,

$$
\{\mathbb{O}\} \to W \xrightarrow{T} V \xrightarrow{L} U \to \{\mathbb{O}\},
$$

is exact at W, V, and U, which implies that T is injective; the image of $\{\mathbb{O}\} \to W$ is the kernel of T, and so $\ker T = \{\mathbb{O}\}$; L is surjective because the kernel of

$U \to \{\mathbb{O}\}$ is all of U, which is the image of L. In our discussion of short exact sequences of groups, the right side of a short exact sequence was isomorphic to a quotient of the other two groups. For vector spaces, there is also a quotient construction that leads to the same conclusion.

Definition 3.13 Suppose $W \subset V$ is a subspace of a vector space V over a field $\mathbb{F}$. The *quotient* of V by W, denoted V/W, is the set of cosets of W in V as an abelian group with multiplication by a scalar given by $a(v + W) = av + W$.

Since every subgroup of an abelian group is normal in the group, the quotient of V/W is an abelian group. To see that V/W is a vector space, it suffices to show that multiplication by a scalar is well-defined. Suppose $v + W = v' + W$ and $a \in \mathbb{F}$. Then $a(v - v') = aw$ for some $w \in W$, and so $a(v - v') + W = av - av' + W = (av + W) - (av' + W) = W$ and $av + W = av' + W$ in V/W. The rest of the vector space axioms hold as they do in V.

Proposition 3.14 *If V is a finite-dimensional vector space and W is a subspace of V, then* $\dim V/W = \dim V - \dim W$. *If* $\{\mathbb{O}\} \to W \xrightarrow{T} V \xrightarrow{L} U \to \{\mathbb{O}\}$ *is a short exact sequence of vector spaces, then* $U \cong V/W$, *and so* $\dim V = \dim W + \dim U$.

Proof The surjection $\pi : V \to V/W$ given by $\pi(v) = v + W$ has kernel equal to W, and so the Fundamental Theorem of Linear Algebra implies $V \cong W \oplus V/W$. The dimension of V/W is $\dim V - \dim W$ from this isomorphism. For the short exact sequence, we take the injection $T : W \to V$ as an isomorphism between W and $T(W) \subset V$. By exactness, $T(W) \cong \ker L$, and since $V \cong \ker L \oplus L(V) \cong T(W) \oplus U \cong W \oplus U$, the dimension relation follows. $\square$

3.2 Tensor Products

There is also the notion of a product of vector spaces, which plays an important role in extending linear properties over many variables.

Definition 3.15 The *tensor product* of vector spaces V and W is the quotient vector space $\mathbb{F}(V \times W)/R$, where R is the subspace generated by the following expressions for all $v, v_1, v_2 \in V$, $w, w_1, w_2 \in W$, and $a \in \mathbb{F}$:

$$(v_1 + v_2, w) - (v_1, w) - (v_2, w); \quad (v, w_1 + w_2) - (v, w_1) - (v, w_2);$$
$$(av, w) - a(v, w); \quad (v, aw) - a(v, w).$$

The tensor product is denoted by $V \otimes W$. A coset of the form $(v, w) + R$ is denoted by $v \otimes w$ and is called a *simple tensor*. A typical element of $V \otimes W$ is a finite sum of simple tensors.

Suppose V and W are finite-dimensional vector spaces with bases $B = \{v_1, \ldots, v_n\}$ for V and $B' = \{w_1, \ldots, w_m\}$ for W. Any simple tensor $v \otimes w$ has $v = a_1 v_1 + \cdots + a_n v_n$ and $w = b_1 w_1 + \cdots + b_m w_m$. By applying the relations in R to $(a_1 v_1 + \cdots + a_n v_n) \otimes (b_1 w_1 + \cdots + b_m w_m)$ we get $\sum_{i,j} a_i b_j (v_i \otimes w_j)$. Elements of $\mathbb{F}(V \times W)$ are finite linear combinations of expressions of the form (v, w), and so $V \otimes W$ is spanned by the simple tensors and hence spanned by the set $\{v_i \otimes w_j\}$.

The raison d'être of tensor products is to give a simplifying framework in which to consider bilinear mappings. Suppose $T : V \times W \to U$ is a function. We say that T is a *bilinear mapping* if, for all $v, v' \in V$, $w, w' \in W$, and $a \in \mathbb{F}$,

$$T(v + v', w) = T(v, w) + T(v', w), \; T(v, w + w') = T(v, w) + T(v, w'),$$
$$\text{and } T(av, w) = T(v, aw) = aT(v, w).$$

The basic example is the canonical mapping $i : V \times W \to V \otimes W$ given by $i(v, w) = v \otimes w = (v, w) + R$.

Proposition 3.16 *Suppose V, W, and U are vector spaces over a field $\mathbb{F}$ and $T : V \times W \to U$ is a bilinear mapping. Then there exists a linear mapping $\widehat{T} : V \otimes W \to U$ for which $T \circ i = \widehat{T}$.*

Proof Define $\widehat{T}(v_i \otimes w_j) = T(v_i, w_j)$ and extend to all of $V \otimes W$ linearly. This gives a linear transformation from $V \otimes W$ to U. To see that $\widehat{T} \circ i = T$, suppose $(v, w) \in V \times W$. If $v = a_1 v_1 + \cdots + a_n v_n$ and $w = b_1 w_1 + \cdots + b_m w_m$, then the bilinear property of T implies that

$$T(v, w) = T(a_1 v_1 + \cdots + a_n v_n, b_1 w_1 + \cdots + b_m w_m) = \sum_{i,j} a_i b_j T(v_i, w_j).$$

We can write $i(v, w) = v \otimes w = \sum_{i,j} a_i b_j (v_i \otimes w_j)$. Then $\widehat{T} \circ i(v, w) = \sum_{i,j} a_i b_j T(v_i, w_j) = T(v, w)$. $\qquad\square$

The process can be reversed: Given a linear mapping $F : V \otimes W \to U$, consider the composite $F \circ i : V \times W \to V \otimes W \to U$. The definition of the tensor product implies that $F \circ i$ is a bilinear mapping. If we denote the set of all bilinear mappings from $V \times W$ to U by $\mathrm{Bil}(V \times W, U)$, then we obtain a bijection between $\mathrm{Bil}(V \times W, U)$ and $\mathrm{Hom}_{\mathbb{F}}(V \otimes W, U)$ by sending T to $\widehat{T}$, and in the other direction, we take F to $F \circ i$. The proposition implies that these are inverse functions giving the bijection.

Another important example of bilinear mappings arises when we consider the vector space $\text{Hom}_{\mathbb{F}}(V, \text{Hom}_{\mathbb{F}}(W, U))$. We have discussed the vector space structure on the set of linear mappings $W \to U$, $\text{Hom}_{\mathbb{F}}(W, U)$, so we can talk about the linear mappings from V to $\text{Hom}_{\mathbb{F}}(W, U)$. Any linear mapping $f : V \to \text{Hom}_{\mathbb{F}}(W, U)$ satisfies $f(av + v') = af(v) + f(v')$ inside the vector space of mappings from W to U. Since $f(v) : W \to U$ is a linear transformation, we also have $f(v)(bw + w') = bf(v)(w) + f(v)(w')$. This determines the bilinear transformation $\text{ad } f : V \times W \to U$ by $\text{ad } f(v, w) = f(v)(w)$ (the *adjoint mapping*). To see that $\text{ad } f$ is bilinear, we check

$$\begin{aligned}
\text{ad } f(av + v', w) &= f(av + v')(w) = af(v)(w) + f(v')(w) \\
&= a\big(\text{ad } f(v, w)\big) + \text{ad } f(v', w), \\
\text{ad } f(v, bw + w') &= f(v)(bw + w') = bf(v)(w) + f(v)(w') \\
&= b\big(\text{ad}(f)(v, w)\big) + \text{ad } f(v, w').
\end{aligned}$$

Since every bilinear transformation determines a linear mapping $V \otimes W \to U$, we obtain a transformation $\Psi : \text{Hom}_{\mathbb{F}}(V, \text{Hom}_{\mathbb{F}}(W, U)) \to \text{Hom}_{\mathbb{F}}(V \otimes W, U)$ defined $\Psi(f) = \widehat{\text{ad } f}$.

Proposition 3.17 *The mapping Ψ is a linear isomorphism.*

Proof We first establish that Ψ is a linear mapping. Suppose f_1 and f_2 are in $\text{Hom}_{\mathbb{F}}(V, \text{Hom}_{\mathbb{F}}(W, U))$. Then

$$\begin{aligned}
\Psi(af_1 + f_2)(v \otimes w) &= (af_1 + f_2)(v)(w) = af_1(v)(w) + f_2(v)(w) \\
&= \big(a\Psi(f_1) + \Psi(f_2)\big)(v \otimes w).
\end{aligned}$$

To see that Ψ is a bijection, we present an inverse. Suppose $F : V \otimes W \to U$ is a linear transformation. Then $F \circ i : V \times W \to U$ is a bilinear mapping. Such a mapping has an adjoint $\text{ad}(F \circ i) : V \to \text{Hom}_{\mathbb{F}}(W, U)$ given by $\text{ad}(F \circ i)(v)(w) = (F \circ i)(v, w) = F(v \otimes w)$. This determines a mapping $\Theta : \text{Hom}_{\mathbb{F}}(V \otimes W, U) \to \text{Hom}_{\mathbb{F}}(V, \text{Hom}_{\mathbb{F}}(W, U))$ by $\Theta(F) = \text{ad}(F \circ i)$. The composite $\Theta(\Psi(f))(v)(w) = \Theta(\widehat{\text{ad } f})(v)(w) = f(v)(w)$, that is, $\Theta \circ \Psi = \text{Id}$. Going the other way, $\Psi(\Theta(F))(v \otimes w) = \Psi(\text{ad}(F \circ i))(v, w) = F(v \otimes w)$; $\Psi \circ \Theta = \text{Id}$. $\square$

Corollary 3.18 *If $\dim V = n$ and $\dim W = m$, then $\dim V \otimes W = mn$.*

Proof Let $U = \mathbb{F}$. By Proposition 3.6, we have $\dim \text{Hom}_{\mathbb{F}}(V, \text{Hom}_{\mathbb{F}}(W, \mathbb{F})) = (\dim V)(\dim \text{Hom}_{\mathbb{F}}(W, \mathbb{F})) = (\dim V)(\dim W)$, and $\dim \text{Hom}_{\mathbb{F}}(V \otimes W, \mathbb{F}) = \dim V \otimes W$. $\square$

The isomorphism Ψ is called the *Hom-tensor interchange*. For linear transformations, there is also a tensor product.

Definition 3.19 Given two linear transformations $T: V \to W$ and $T': V' \to W'$, there is a linear transformation $T \otimes T': V \otimes V' \to W \otimes W'$ given by

$$T \otimes T'(v \otimes v') = T(v) \otimes T(v')$$

on simple tensors $v \otimes v'$.

The mapping $T \otimes T'$ may also be taken to be induced by the bilinear mapping

$$V \times V \xrightarrow{T \times T'} W \times W' \xrightarrow{i} W \otimes W'$$

given by $(i \circ T \times T')(v, v') = T(v) \otimes T'(v')$.

Let us work out $T \otimes T'$ when T and T' are multiplication by matrices. Suppose V has a basis $B = \{v_1, \ldots, v_n\}$, V' has a basis $B' = \{v'_1, \ldots, v'_{n'}\}$, W has a basis $C = \{w_1, \ldots, w_m\}$, and W' has a basis $C' = \{w'_1, \ldots, w'_{m'}\}$. If the matrix representation in these bases is given by the matrices $M = (a_{ij})$ for T and $N = (b_{kl})$ for T', then we can write

$$T(v_i) \otimes T'(v'_j) = (a_{1i}w_1 + \cdots + a_{mi}w_m) \otimes (b_{1j}w'_1 + \cdots + b_{m'j}w'_{m'})$$

$$= \sum_k \sum_l a_{ki} b_{lj} w_k \otimes w'_l.$$

The lexicographic orderings on $B \otimes B'$ and $C \otimes C'$ determine the entries of the matrix representing $T \otimes T'$: the $w_k \otimes w'_l$ coefficient of the matrix in the $v_i \otimes v'_j$ column is $a_{ki} b_{lj}$. If we hold i and k fixed and vary j and l, then we get an $m' \times n'$ block that corresponds to $a_{ki} N$ as all entries in N will appear in $a_{ki} b_{lj}$. Thus the matrix determined by $T \otimes T'$ can be presented in $m' \times n'$ blocks:

$$\begin{bmatrix} a_{11}N & a_{12}N & \cdots & a_{1n}N \\ a_{21}N & a_{22}N & \cdots & a_{2n}N \\ \vdots & \vdots & \ddots & \vdots \\ a_{m1}N & a_{m2}N & \cdots & a_{mn}N \end{bmatrix}$$

Such a product $M \otimes N$ is called the *tensor product* or *Kronecker product* of matrices $M = (a_{ij})$ and $N = (b_{kl})$. It can be defined for arbitrary pairs of linear transformations.

3.3 When $\mathbb{F} = \mathbb{C}$: Hermitian Inner Products

The vector space $\mathbb{R}^n$ is equipped with a positive definite inner product, the *dot product*: For $\mathbf{v} = (v_1, \ldots, v_n)$ and $\mathbf{w} = (w_1, \ldots, w_n)$, let $\langle \mathbf{v}, \mathbf{w} \rangle = \mathbf{v} \cdot \mathbf{w} = v_1 w_1 + \cdots + v_n w_n$. The geometry of $\mathbb{R}^n$ is based on the dot product.

For an arbitrary field $\mathbb{F}$, this formula could lead to undesirable properties. For example, if $\mathbb{F} = \mathbb{Z}/3\mathbb{Z}$, then $(1, 1, 1) \cdot (1, 1, 1) \equiv 0 \,(\mathrm{mod}\,3)$. Geometry over $\mathbb{Z}/3\mathbb{Z}$ is expected to be different, but the naive definition would give us nonzero vectors of zero length. For complex vector spaces, the naive definition gives complex numbers that we would like to interpret as lengths. To remedy this shortfall, we introduce *Hermitian inner products*. A Hermitian inner product assigns to a pair of vectors v and w in a complex vector space V a complex number denoted $\langle v, w \rangle \in \mathbb{C}$. This pairing satisfies the properties:

(1) for all $v, w \in V$, $\langle v, w \rangle = \overline{\langle w, v \rangle}$.
(2) for $c \in \mathbb{C}$, $\langle cv, w \rangle = c \langle v, w \rangle$.
(3) for all $v_1, v_2, w \in V$, $\langle v_1 + v_2, w \rangle = \langle v_1, w \rangle + \langle v_2, w \rangle$.
(4) For all $v \in V$, $\langle v, v \rangle$ is a real number, and $\langle v, v \rangle \geq 0$. Furthermore, $\langle v, v \rangle = 0$ if and only if $v = \mathbb{O}$.

Here $\overline{z} = \overline{a + bi} = a - bi$ denotes *complex conjugation*. Recall that $z = \overline{z}$ if and only if z is a real number. An inner product with property (4) is called a *positive definite* or *nondegenerate* inner product.

On $\mathbb{C}^n$ the standard Hermitian inner product is defined for $\mathbf{v} = (v_1, \ldots, v_n)^t$ and $\mathbf{w} = (w_1, \ldots, w_n)^t$ by

$$\langle \mathbf{v}, \mathbf{w} \rangle = v_1 \overline{w_1} + v_2 \overline{w_2} + \cdots + v_n \overline{w_n}.$$

For a finite-dimensional complex vector space V with a Hermitian inner product, by choosing a basis $B = \{v_1, \ldots, v_n\}$ for V we can present the inner product by a matrix $A = (c_{ij})$ where $c_{ij} = \langle v_i, v_j \rangle$. If we denote the coordinate mapping with respect to the basis B, $c_B \colon V \to \mathbb{C}^n$, then for $v = a_1 v_1 + \cdots + a_n v_n$ and $w = b_1 v_1 + \cdots b_n v_n$,

$$\langle v, w \rangle = c_B(v)^t A \,\overline{c_B(w)} = \sum_{i=1}^{n} a_i \overline{b_j} c_{ij}.$$

This follows because $\langle v, w \rangle = \langle a_1 v_1 + \cdots + a_n v_n, b_1 v_1 + \cdots + b_n v_n \rangle = a_1 \overline{b_1} \langle v_1, v_1 \rangle + \cdots + a_i \overline{b_j} \langle v_i, v_j \rangle + \cdots + a_n \overline{b_n} \langle v_n, v_n \rangle$.

If $\langle v, w \rangle$ is a Hermitian inner product, then the matrix A has certain properties. Since $\langle v_i, v_j \rangle = \overline{\langle v_j, v_i \rangle}$, the diagonal entries c_{ii} are real numbers. Furthermore, $c_{ji} = \overline{c_{ij}}$, and so the matrix A satisfies $A^t = \overline{A}$. A positive definite inner product requires that the matrix A is invertible. To see this, suppose $A = [\mathbf{c}_1 \ \mathbf{c}_2 \ \cdots \ \mathbf{c}_n]$ where $\mathbf{c}_i$ are the columns of A. Suppose w is a vector in V with $\langle v, w \rangle = 0$ for all $v \in V$. Then $c_B(v)^t A \,\overline{c_B(w)} = 0$ for all n vectors $c_B(v)$ in $\mathbb{C}^n$. If the inner product is nondegenerate, then $w = \mathbb{O}$, and the only solution to $A \,\overline{c_B(w)} = \mathbb{O}$ is the zero vector $w = \mathbb{O}$, and A is invertible.

Conversely, if A satisfies $A^t = \overline{A}$ and A is invertible, then the associated inner product $\langle v, w \rangle = c_B(v)^t A \overline{c_B(w)}$ is nondegenerate.

If V is a Hermitian inner product space and W is a linear subspace of V, then define the *orthogonal complement* of W (called "W perp") by

$$W^\perp = \{ v \in V \mid \langle v, w \rangle = 0 \text{ for all } w \in W \}.$$

Proposition 3.20 *Suppose V is a finite-dimensional Hermitian inner product space and W is a linear subspace of V. The subset $W^\perp$ is a linear subspace of V, and if the inner product is positive definite on W and V, then $W \cap W^\perp = \{\mathbb{O}\}$ and $V \cong W \oplus W^\perp$.*

Proof Suppose $w \in W$, v and u are in $W^\perp$, and $a \in \mathbb{C}$. Then $\langle av + u, w \rangle = a\langle v, w \rangle + \langle u, w \rangle = 0$ and $av + u \in W^\perp$, so $W^\perp$ is a linear subspace in V. Suppose $w \in W \cap W^\perp$. Then $\langle w, w \rangle = 0$, and so $w = \mathbb{O}$ by the positive definite condition.

To show $V \cong W \oplus W^\perp$, we need to show that every $v \in V$ can be written as $v = w + w'$ with $w \in W$ and $w' \in W^\perp$. Suppose $\{z_1, \ldots, z_k\}$ is any basis for W. The *Gram–Schmidt process* applied to $\{z_1, \ldots, z_k\}$ determines an orthonormal basis $\{w_1, \ldots, w_k\}$ for W. The algorithm proceeds as follows: to start, $w_1 = \dfrac{z_1}{\langle z_1, z_1 \rangle^{1/2}}$. Let $y_2 = z_2 - \langle z_2, w_1 \rangle w_1$ and $w_2 = \dfrac{y_2}{\langle y_2, y_2 \rangle^{1/2}}$. If $\{w_1, \ldots, w_j\}$ is an orthonormal set, then let

$$y_{j+1} = z_{j+1} - \langle z_j, w_j \rangle w_j - \langle z_j, w_{j-1} \rangle w_{j-1} - \cdots - \langle z_j, w_1 \rangle w_1,$$

and let $w_{j+1} = \dfrac{y_{j+1}}{\langle y_{j+1}, y_{j+1} \rangle^{1/2}}$. The reader will want to check that $\langle w_i, w_i \rangle = 1$ and $\langle w_i, w_j \rangle = 0$ when $i \neq j$. Also, $\text{Span}(\{w_1, \ldots, w_k\}) = W$.

For $v \in V$, let $v_W = \langle w_1, v \rangle w_1 + \langle w_2, v \rangle w_2 + \cdots + \langle w_k, v \rangle w_k$. Consider $v - v_W$. Notice that $v_W \in W$. Let us compute $\langle v - v_W, w_i \rangle$:

$$\begin{aligned}
\langle v - v_W, w_i \rangle &= \langle v, w_i \rangle - \langle v_W, w_i \rangle \\
&= \langle v, w_i \rangle - \langle \langle v, w_1 \rangle w_1 + \cdots + \langle v, w_k \rangle w_k, w_i \rangle \\
&= \langle v, w_i \rangle - \langle v, w_i \rangle \langle w_i, w_i \rangle = 0,
\end{aligned}$$

and so $w' = v - v_W \in W^\perp$ and $v = v_W + w'$. $\square$

Let us review eigenvalues and eigenvectors

If A is an $n \times n$ matrix with coefficients in $\mathbb{F}$, then we say that $\lambda \in \mathbb{F}$ is an *eigenvalue* of A if there is a nonzero vector $\mathbf{v} \in \mathbb{F}^{\times n}$ such that $A\mathbf{v} = \lambda \mathbf{v}$. In this case, we call $\mathbf{v}$ an *eigenvector* of A associated with λ.

1. Show that λ is an eigenvalue of A if and only if $\dim \ker(\lambda \operatorname{Id} - A) > 0$ if and only if $\det(\lambda \operatorname{Id} - A) = 0$.

2. Let $p(x) = \det(x \operatorname{Id} - A)$. Show that $p(x)$ is a monic ($p(x) = x^n + \cdots$) polynomial of degree n with coefficients in $\mathbb{F}$. The eigenvalues of A are the roots of $p(x)$. Deduce that $p(0) =$ the product of the eigenvalues of A, and if $p(x) = x^n - a_{n-1}x^{n-1} + \cdots$, then $a_{n-1} =$ the sum of the eigenvalues of A. We call $p(x)$ the *characteristic polynomial* of A.

3. Suppose that A and B are *similar* matrices, that is, $B = PAP^{-1}$ with an invertible matrix P. Then show that A and B have the same eigenvalues. (Hint: show that their characteristic polynomials are the same.)

4. If A is a diagonal matrix, that is, $a_{ij} = 0$ for $i \neq j$, then the eigenvalues are the diagonal entries. Suppose B is a matrix whose n eigenvectors form a basis for $\mathbb{F}^{\times n}$. Then show that B is similar to a diagonal matrix. Show that if the characteristic polynomial $p(x)$ for an $n \times n$ matrix C has n distinct roots, then C is similar to a diagonal matrix.

5. If A is an invertible matrix, then show that if λ is an eigenvalue of A, then $1/\lambda$ is an eigenvalue of A^{-1}.

While we are discussing eigenvalues and characteristic polynomials, we can ask if any monic polynomial over a field $\mathbb{F}$ is the characteristic polynomial of a matrix with coefficients in $\mathbb{F}$? The answer is yes, and we introduce a particular matrix to fill this need.

Definition 3.21 Given a monic polynomial $p(x) = a_0 + a_1 x + \cdots + a_{n-1}x^{n-1} + x^n$ in $\mathbb{F}[x]$, the *companion matrix* associated with $p(x)$ is the $n \times n$ matrix given by

$$
C_{p(x)} = \begin{bmatrix} 0 & 0 & \cdots & & -a_0 \\ 1 & 0 & \cdots & & -a_1 \\ & 1 & \ddots & & \vdots \\ & & \ddots & 0 & -a_{n-2} \\ & & & 1 & -a_{n-1} \end{bmatrix}.
$$

Proposition 3.22 *Given a monic polynomial $p(x)$ with its companion matrix $C = C_{p(x)}$, the characteristic polynomial $\det(x \operatorname{Id} - C)$ is $p(x)$.*

Proof We proceed by induction on the degree of the monic polynomial. To begin, let us take a degree-two polynomial $p(x) = a_0 + a_1 x + x^2$. The companion matrix is $C = \begin{bmatrix} 0 & -a_0 \\ 1 & -a_1 \end{bmatrix}$. The characteristic polynomial is given by

$$\det \begin{bmatrix} x & a_0 \\ -1 & x+a_1 \end{bmatrix} = x(x+a_1) + a_0 = a_0 + a_1 x + x^2.$$

Assume that the proposition holds for polynomials of degree less than n and suppose $p(x) = a_0 + a_1 x + \cdots + a_{n-1}x^{n-1} + x^n$. The determinant, $\det(x \,\mathrm{Id} - C)$, is given by

$$x \det \begin{bmatrix} x & \cdots & & & a_1 \\ -1 & \ddots & & & a_2 \\ & \ddots & & & \vdots \\ & & \ddots & x & a_{n-2} \\ & & & -1 & x+a_{n-1} \end{bmatrix} + (-1)^{n+1} a_0 \det \begin{bmatrix} -1 & x & \cdots & \\ & -1 & \ddots & \\ & & \ddots & x \\ & & & -1 \end{bmatrix},$$

which has been expanded by minors along the first row. By induction the first summand is $x(a_1 + a_2 x + \cdots + a_{n-1}x^{n-2} + x^{n-1})$, and the second summand is $(-1)^{n+1} a_0 (-1)^{n-1} = a_0$ because the matrix is upper triangular. Thus $\det(x\,\mathrm{Id} - C) = a_0 + a_1 x + \cdots + a_{n-1}x^{n-1} + x^n = p(x)$. $\qquad\square$

When we discuss the roots of monic polynomials later, this proposition affords us the opportunity to use matrices to combine various roots easily.

3.4 Matrix Groups

For a field $\mathbb{F}$, the general linear group $\mathrm{Gl}_n(\mathbb{F})$ is the group of invertible $n \times n$ matrices with entries in $\mathbb{F}$. The definition of the determinant of a square matrix extends to matrices over any field with the usual properties:

(1) $\det(AB) = \det(A)\det(B)$ for $n \times n$ matrices A and B;
(2) $A \in \mathrm{Gl}_n(\mathbb{F})$ if and only if $\det(A) \neq 0$.

Most proofs of these properties follow directly from the field axioms as in the case of matrices with entries in $\mathbb{R}$. Thus the properties (and formulas) hold for $n \times n$ matrices over all fields.

The determinant is a homomorphism $\det \colon \mathrm{Gl}_n(\mathbb{F}) \to \mathbb{F}^\times$ (recall that $\mathbb{F}^\times = \mathbb{F} - \{0\}$, the multiplicative group of units in $\mathbb{F}$). The kernel of det is the *special linear group*

$$\mathrm{Sl}_n(\mathbb{F}) = \big\{ A \mid A, \text{ an } n \times n \text{ matrix over } \mathbb{F} \text{ with } \det(A) = 1 \big\}.$$

To study this group, we apply Gaussian elimination from basic linear algebra. The equivalence between uniquely solvable systems of n linear equations in n unknowns and the invertibility of the matrix of coefficients is the basis for arguments about $Gl_n(\mathbb{F})$ and $Sl_n(\mathbb{F})$. Unique solvability translates to the property that an $n \times n$ matrix can be row-reduced to the identity matrix Id. The elementary row operations are:

(1) Interchange of row i and row j;
(2) Scaling row i by a nonzero scalar a;
(3) Adding a nonzero scalar multiple of row j to row i.

The bridge between these operations and the ring of all $n \times n$ matrices is the observation that *an elementary row operation on a matrix A is the result of matrix multiplication by a matrix P that does the row operation.* In fact, to construct the matrix P, simply apply the elementary row operation to the identity matrix. For example, for 3×3 matrices,

$$(1)\ \rho_1 \leftrightarrow \rho_2 \Leftrightarrow \begin{bmatrix} 0 & 1 & 0 \\ 1 & 0 & 0 \\ 0 & 0 & 1 \end{bmatrix}, \qquad (2)\ \text{scaling } \rho_3 \text{ by } c \Leftrightarrow \begin{bmatrix} 1 & 0 & 0 \\ 0 & 1 & 0 \\ 0 & 0 & c \end{bmatrix}.$$

Notice that these operation matrices have determinants that are not equal to one.

The third operation is best expressed by introducing a basis for all $n \times n$ matrices as a vector space over $\mathbb{F}$. Let B_{ij} denote the matrix with the ijth entry 1 and the rest of the entries zero. The collection $\{B_{ij} \mid 1 \le i, j \le n\}$ is a basis for $M_{n \times n}$, the vector space of all $n \times n$ matrices over the field $\mathbb{F}$. Furthermore, let $\mathbb{O}$ denote the matrix with all entries zero: then simple computations give

$$B_{ij} B_{kl} = \mathbb{O} \text{ for } j \ne k, \text{ and } B_{ij} B_{jl} = B_{il}. \tag{$\spadesuit$}$$

The elementary row operation "add a scalar multiple of row j to row i (of course, $i \ne j$) can be accomplished by multiplying (on the left) by the matrix $\text{Id} + b B_{ij}$ with the scalar $b \ne 0$. This matrix has determinant one for $i \ne j$. The matrices $\text{Id} + b B_{ij}$ are referred to as *transvections* [69].

For matrices of determinant one, the transvections play an important group-theoretic role:

Lemma 3.23 *For $n > 1$, the transvections generate $Sl_n(\mathbb{F})$.*

Proof Suppose $A = (a_{ij})$ is an $n \times n$ matrix with coefficients in $\mathbb{F}$ of determinant one. Then we can assume that $a_{21} \ne 0$. If not, we can add a row with a nonzero leading entry to row 2. If we now multiply A on the left by $\text{Id} + a_{21}^{-1}(1 - a_{11})B_{12}$, which is equivalent to adding $a_{21}^{-1}(1 - a_{11})$ times row 2 to row 1, then we get 1 in the $(1, 1)$ entry. By subtracting a_{k1} times row 1 from the other rows, we clear the first column of nonzero entries except the first entry, now 1.

Repeat this procedure to the $(1, 1)$ minor to make the $(2, 2)$ entry equal to 1 and the second column to have zero entries below the diagonal. Continuing, we get an upper triangular matrix with a diagonal of 1s. Finally, it is an easy step to use the 1s along the diagonal to clear the matrix of nonzero entries above the diagonal, giving the identity matrix. Each step is carried out by multiplying the matrix by a transvection. Thus we can write $P_N P_{N-1} \cdots P_1 A = \mathrm{Id}$.

A transvection $P = \mathrm{Id} + a B_{ij}$ has the inverse $P^{-1} = \mathrm{Id} - a B_{ij}$, and so the inverse of a transvection is a transvection. Therefore $A = P_1^{-1} \cdots P_{N-1}^{-1} P_N^{-1}$, and A is a product of transvections. $\qquad\square$

Corollary 3.24 *The center of* $\mathrm{Sl}_n(\mathbb{F})$ *is given by* $Z(\mathrm{Sl}_n(\mathbb{F})) = \{\lambda\, \mathrm{Id} \mid \lambda \in \mathbb{F}, \lambda^n = 1\}$.

Proof Because $\mathrm{Sl}_n(\mathbb{F})$ is generated by transvections, it suffices to consider the matrices that commute with all transvections $\mathrm{Id} + a B_{ij}$ for $i \neq j$. If A is in the center of $\mathrm{Sl}_n(\mathbb{F})$, then $A(\mathrm{Id} + a B_{ij}) = (\mathrm{Id} + a B_{ij})A$ for all $i \neq j$ and $a \in \mathbb{F}$. This condition implies $A B_{ij} = B_{ij} A$.

$$
B_{ij} A = \begin{bmatrix} \\ \\ \end{bmatrix} \quad A = \begin{bmatrix} -0- & | & 0 \\ a_{j1}\,a_{j2} \;\cdots\; a_{jj} \cdots a_{jn} \\ -0- & | & 0 \end{bmatrix}
$$

$$
A B_{ij} = A \begin{bmatrix} \\ \\ \end{bmatrix} = \begin{bmatrix} -0- & a_{1i} & 0 \\ & a_{ii} & \\ -0- & \vdots & 0 \\ & a_{ni} & \end{bmatrix}
$$

Because A is central, the rightmost matrices are equal. This implies that $a_{ij} = 0$ for $i \neq j$ and $a_{ii} = a_{jj}$. Varying i and j, it follows that A is a diagonal matrix with all entries equal. Because the determinant of A is one, $A = \lambda\, \mathrm{Id}$ with $\lambda^n = 1$. $\qquad\square$

For example, $Z(\mathrm{Sl}_2(\mathbb{F}_2)) = \{\pm\, \mathrm{Id}\}$, where $\mathbb{F}_2 = \mathbb{Z}/2\mathbb{Z}$.

The proof that A_n is simple when $n \geq 5$ uses the fact that A_n is generated by 3-cycles, each conjugate to the other. In $\mathrm{Sl}_n(\mathbb{F})$ the transvections generate the group. In fact, we can make the next step by analogy.

Lemma 3.25 *For $n > 2$, all transvections are conjugate in $\mathrm{Sl}_n(\mathbb{F})$.*

Proof Let $a \neq b$ be nonzero elements of $\mathbb{F}$. We first show that the transvections $\mathrm{Id} + aB_{ij}$ and $\mathrm{Id} + bB_{ij}$ are conjugate. Let P denote the diagonal matrix with 1s along the diagonal except for ab^{-1} in the jth entry and $a^{-1}b$ in the kth entry, where $k \neq i$, $k \neq j$ ($n > 2$). Then $\det P = 1$ and $P \in \mathrm{Sl}_n(\mathbb{F})$. We can present P as $\mathrm{Id} + (ab^{-1} - 1)B_{jj} + (a^{-1}b - 1)B_{kk}$ (not a transvection). The inverse of P switches the entries in the (j, j)- and (k, k)-places, giving $P^{-1} = \mathrm{Id} + (a^{-1}b - 1)B_{jj} + (ab^{-1} - 1)B_{kk}$. Then a straightforward computation using ($\spadesuit$) shows that

$$P(\mathrm{Id} + aB_{ij})P^{-1} = \mathrm{Id} + bB_{ij},$$

and these two transvections are conjugate.

Next, let Q denote the matrix with 1 in the (i, l) entry, -1 in the (l, i) entry, and zeroes in the (i, i) and (l, l) places. In terms of row operations, Q switches rows i and l and multiplies row l by -1, hence the determinant of Q is one, and Q is in $\mathrm{Sl}_n(\mathbb{F})$. We can present Q as $\mathrm{Id} - B_{il} + B_{li} - B_{ii} - B_{ll}$. The inverse Q^{-1} has the 1 and -1 switched. Another straightforward computation shows $Q(\mathrm{id} + aB_{ij})Q^{-1} = \mathrm{Id} + aB_{lj}$. Finally, let R switch columns j and k and negate the column k. Then R can be expressed as $\mathrm{Id} - B_{jk} + B_{kj} - B_{jj} - B_{kk}$. A calculation gives $R(\mathrm{Id} + aB_{ij})R^{-1} = \mathrm{Id} + aB_{ik}$. Thus, for any pair of transvections $\mathrm{Id} + aB_{ij}$ and $\mathrm{Id} + bB_{kl}$, $\mathrm{Id} + aB_{ij}$ is conjugate to $\mathrm{Id} + aB_{kj}$, which is conjugate to $\mathrm{Id} + aB_{kl}$, which is conjugate to $\mathrm{Id} + bB_{kl}$. Hence all transvections are conjugate in $\mathrm{Sl}_n(\mathbb{F})$ for $n > 2$. □

It follows that a normal subgroup N of $\mathrm{Sl}_n(\mathbb{F})$ that contains a transvection is all of $\mathrm{Sl}_n(\mathbb{F})$.

Get to know $\mathrm{Sl}_2(\mathbb{F})$

Our goal in this interlude is to explore the 2×2 case and to prove the following:

Proposition 3.26 *If N is a normal subgroup of $\mathrm{Sl}_2(\mathbb{F})$ and N contains a transvection, then $N = \mathrm{Sl}_2(\mathbb{F})$.*

1. Suppose $\left[\begin{smallmatrix} 1 & a \\ 0 & 1 \end{smallmatrix}\right] \in N$. It follows that $\left[\begin{smallmatrix} 1 & a \\ 0 & 1 \end{smallmatrix}\right]^{-1} = \left[\begin{smallmatrix} 1 & -a \\ 0 & 1 \end{smallmatrix}\right] \in N$. Show that

$$\begin{bmatrix} 0 & -1 \\ 1 & 0 \end{bmatrix} \begin{bmatrix} 1 & x \\ 0 & 1 \end{bmatrix} \begin{bmatrix} 0 & 1 \\ -1 & 0 \end{bmatrix} = \begin{bmatrix} 1 & 0 \\ -x & 1 \end{bmatrix}.$$

Thus, if a transvection with the $(1, 2)$ entry $x \neq 0$ is in N, then so is the transvection with $-x$ in the $(2, 1)$ entry. We can concentrate on the upper triangular case.

2. Show that
$$\begin{bmatrix} x & 0 \\ 0 & x^{-1} \end{bmatrix} \begin{bmatrix} 1 & a \\ 0 & 1 \end{bmatrix} \begin{bmatrix} x^{-1} & 0 \\ 0 & x \end{bmatrix} = \begin{bmatrix} 1 & ax^2 \\ 0 & 1 \end{bmatrix},$$
and so, for all $x \in \mathbb{F}$, $\begin{bmatrix} 1 & ax^2 \\ 0 & 1 \end{bmatrix}$ and $\begin{bmatrix} 1 & 0 \\ -ax^2 & 1 \end{bmatrix}$ are in N.

3. Show that for $x, y \in \mathbb{F}$,
$$\begin{bmatrix} 1 & ax^2 \\ 0 & 1 \end{bmatrix} \begin{bmatrix} 1 & -ay^2 \\ 0 & 1 \end{bmatrix} = \begin{bmatrix} 1 & a(x^2 - y^2) \\ 0 & 1 \end{bmatrix}.$$

Prove that if $2 \neq 0$ in $\mathbb{F}$, then any value of $z \in \mathbb{F}$ can be written $z = [2^{-1}(z + 1)]^2 - [2^{-1}(z - 1)]^2$. Hence, if $\begin{bmatrix} 1 & a \\ 0 & 1 \end{bmatrix}$ is in N, then, for all $x \in \mathbb{F}$, $\begin{bmatrix} 1 & x \\ 0 & 1 \end{bmatrix}$ is in N.

4. Show that $\mathrm{Sl}_2(\mathbb{F}_2)$ is a group of order 6. Here $\mathbb{F}_2 = \mathbb{Z}/2\mathbb{Z}$, the field of two elements. This group acts on $\mathbb{F}_2^{\times 2}$, the two-dimensional vector space over $\mathbb{F}_2$, as linear transformations. Since $(0, 0)$ is fixed, the group permutes the three points $(1, 0)$, $(0, 1)$, and $(1, 1)$. Use this to show that $\mathrm{Sl}_2(\mathbb{F}_2) \cong \Sigma_3$.

5. Show that $\mathrm{Sl}_2(\mathbb{F}_3)$ is a group of order 24. Here $\mathbb{F}_3 = \mathbb{Z}/3\mathbb{Z}$, the field of three elements $\{-1, 0, 1\}$. This group can be seen to act on the four lines through the origin in $\mathbb{F}_3^{\times 2}$ as linear transformations. Since $(0, 0)$ is fixed, the group permutes this set of four lines. Use this to show that $\mathrm{Sl}_2(\mathbb{F}_3)$ is isomorphic to Σ_4.

It follows that $\mathrm{Sl}_2(\mathbb{F}_2)$ and $\mathrm{Sl}_2(\mathbb{F}_3)$ are not simple groups.

Definition 3.27 Let $\mathrm{PSl}_n(\mathbb{F})$ denote the quotient group $\mathrm{Sl}_n(\mathbb{F})/Z(\mathrm{Sl}_n(\mathbb{F})) = \mathrm{Sl}_n(\mathbb{F})/\{\lambda \, \mathrm{Id}_n \mid \lambda \in \mathbb{F}, \lambda^n = 1\}$. This group is called the *projective special linear group* over the field $\mathbb{F}$.

The group $\mathrm{Sl}_n(\mathbb{F})$ acts on the set $\mathbb{F}P^{n-1} = \{\text{lines through the origin in } \mathbb{F}^{\times n}\}$. When $\mathbb{F} = \mathbb{R}$ or $\mathbb{C}$, $\mathbb{F}P^{n-1}$ is called the (real or complex) *projective space*, spaces that play an important role in geometry and topology. The action is given by $\rho \colon \mathrm{Sl}_n(\mathbb{F}) \to \Sigma(\mathbb{F}P^{\times(n-1)})$, where $\rho(A)(\mathrm{Span}\{\mathbf{v}\}) = \mathrm{Span}\{A\mathbf{v}\}$ for $\mathbf{v}$ a nonzero vector in $\mathbb{F}^{\times n}$.

The kernel of this action is determined by the fact that $\mathrm{Span}\{\mathbf{v}\} = \mathrm{Span}\{\mathbf{w}\}$ if and only if $\mathbf{w} = \lambda \mathbf{v}$ for some nonzero scalar λ in $\mathbb{F}$. For the canonical basis $\{e_1, \ldots, e_n\}$, the relation $Ae_i = \lambda_i e_i$ implies that A is a diagonal matrix. For any other basis $\{v_1, \ldots, v_n\}$, the same relations hold for $A \in \ker \rho$, and the entries along the diagonal must all be the same. Finally, being in $\mathrm{Sl}_n(\mathbb{F})$, $A = \lambda \,\mathrm{Id}$ with $\lambda^n = 1$, that is, $A \in Z(\mathrm{Sl}_n(\mathbb{F}))$. Taking a quotient, we get an associated faithful action

$$\bar{\rho} \colon \mathrm{PSl}_n(\mathbb{F}) \to \Sigma\left(\mathbb{F}P^{n-1}\right).$$

Lemma 3.28 *The group* $\mathrm{PSl}_n(\mathbb{F})$ *acts doubly transitively on the set* $\mathbb{F}P^{n-1}$.

Proof The action of the center of $\mathrm{Sl}_n(\mathbb{F})$ is the identity on $\mathbb{F}P^{n-1}$, so we can work in $\mathrm{Sl}_n(\mathbb{F})$. Suppose $\ell_1 = \mathrm{Span}\{v_1\}$ and $\ell_2 = \mathrm{Span}\{v_2\}$ are distinct lines in $\mathbb{F}^{\times n}$. Then $\{v_1, v_2\}$ is a linearly independent set. Extend to a basis $B = \{v_1, v_2, \ldots, v_n\}$ for $\mathbb{F}^{\times n}$. Consider the change-of-basis matrix c_B and let $A = c_B^{-1}$. Then $Ae_i = v_i$ for all $i = 1, \ldots, n$. Suppose $\lambda = \det A$. Define $C = AD$ where D is a diagonal matrix with entries $(1/\lambda, 1, \ldots, 1)$. Then $\det C = 1$ and $C \in \mathrm{Sl}_n(\mathbb{F})$. Furthermore, $C(\mathrm{Span}\{e_i\}) = \mathrm{Span}\{v_i\}$, and so the pair $(\mathrm{Span}\{e_1\}, \mathrm{Span}\{e_2\})$ maps to (ℓ_1, ℓ_2). For any other pair of distinct lines (ℓ_1', ℓ_2'), the same argument gives us the matrices C and C' in $\mathrm{Sl}_n(\mathbb{F})$ with

$$(\ell_1', \ell_2') \xleftarrow{C'} \left(\mathrm{Span}\{e_1\}, \mathrm{Span}\{e_2\}\right) \xrightarrow{C} (\ell_1, \ell_2),$$

and the composite $C'C^{-1}$ maps (ℓ_1, ℓ_2) to (ℓ_1', ℓ_2'). $\qquad\qquad\square$

With these results in hand, we can prove the following.

Theorem 3.29 *For $n > 2$, the projective special linear groups* $\mathrm{PSl}_n(\mathbb{F})$ *are simple.*

Proof Because $\mathrm{PSl}_n(\mathbb{F})$ acts doubly transitively on the set $\mathbb{F}P^{n-1}$, we can apply Iwasawa's Lemma from Chapter 2 to show that $\mathrm{PSl}_n(\mathbb{F})$ is simple if we can find a point in $\mathbb{F}P^{n-1}$ whose stabilizer contains an abelian subgroup whose conjugates generate $\mathrm{PSl}_n(\mathbb{F})$.

Let $\ell_1 = \mathrm{Span}\{e_1\}$ denote the line through the origin in $\mathbb{F}^{\times n}$ of multiples of e_1. Denote the stabilizer subgroup of ℓ_1 in $\mathrm{Sl}_n(\mathbb{F})$ by St_1 for the action $\rho \colon \mathrm{Sl}_n(\mathbb{F}) \to \Sigma(\mathbb{F}P^{n-1})$. By Proposition 2.21 this subgroup is maximal in $\mathrm{Sl}_n(\mathbb{F})$, that is, if H is a subgroup of $\mathrm{Sl}_n(\mathbb{F})$ with $St_1 \subset H \subset \mathrm{Sl}_n(\mathbb{F})$, then either $H = St_1$ or $H = \mathrm{Sl}_n(\mathbb{F})$.

Matrices in St_1 can be written uniquely as follows: Let $\mathbb{O}$ denote the zero vector in $\mathbb{F}^{\times(n-1)}$. Then

$$A = \begin{bmatrix} a & v^t \\ \mathbb{O} & B \end{bmatrix} \text{ with } a \in \mathbb{F}^{\times}, v \in \mathbb{F}^{\times(n-1)}, \text{ and } B \in \mathrm{Gl}_{n-1}(\mathbb{F}),$$

with the determinant of B equal to a^{-1}. This form follows because $\mathbf{e}_1$ is an eigenvector of any matrix in St_1.

Lemma 2.10 implies that stabilizers conjugate in an explicit manner: if $A \in Sl_n(\mathbb{F})$, then $ASt_1A^{-1} = $ the stabilizer subgroup of $\text{Span}\{A\mathbf{e}_1\}$.

The stabilizer subgroup St_1 contains a relevant subgroup. Let

$$K = \left\{ \begin{bmatrix} 1 & \mathbf{v}^t \\ \mathbb{O} & \text{Id}_{n-1} \end{bmatrix} \middle| \mathbf{v} \in \mathbb{F}^{\times(n-1)} \right\}.$$

In fact, K is an abelian subgroup of St_1 that is normal in St_1. To see this, note that

$$\begin{bmatrix} 1 & \mathbf{v}^t \\ \mathbb{O} & \text{Id}_{n-1} \end{bmatrix} \begin{bmatrix} 1 & \mathbf{w}^t \\ \mathbb{O} & \text{Id}_{n-1} \end{bmatrix} = \begin{bmatrix} 1 & (\mathbf{v}+\mathbf{w})^t \\ \mathbb{O} & \text{Id}_{n-1} \end{bmatrix}.$$

Thus K is isomorphic to $\mathbb{F}^{\times(n-1)}$. The following computation shows that K is normal in St_1:

$$\begin{bmatrix} a & \mathbf{v}^t \\ \mathbb{O} & B \end{bmatrix} \begin{bmatrix} 1 & \mathbf{w}^t \\ \mathbb{O} & \text{Id}_{n-1} \end{bmatrix} = \begin{bmatrix} a & \mathbf{v}^t + a\mathbf{w}^t \\ \mathbb{O} & B \end{bmatrix} = \begin{bmatrix} 1 & a\mathbf{w}^t B^{-1} \\ \mathbb{O} & \text{Id}_{n-1} \end{bmatrix} \begin{bmatrix} a & \mathbf{v}^t \\ \mathbb{O} & B \end{bmatrix}.$$

Finally, notice that K contains a transvection (take $\mathbf{v} = \mathbf{e}_2$), and by Lemma 3.23 the conjugates of K in $Sl_n(\mathbb{F})$ contain all transvections. Since $Sl_n(\mathbb{F})$ is generated by transvections, we have found a stabilizer that contains an abelian normal subgroup whose conjugates generate the whole group. The final condition we need for Iwasawa's Lemma is the following result.

Lemma 3.30 *For* $n > 2$, $[Sl_n(\mathbb{F}), Sl_n(\mathbb{F})] = Sl_n(\mathbb{F})$.

Proof It suffices to show that transvections can be expressed as commutators. Using relation ($\spadesuit$), it is straightforward to show that, for i, j, k distinct ($n > 2$),

$$\text{Id} + aB_{ij} = (\text{Id} + aB_{ik})(\text{Id} + B_{kj})(\text{Id} - aB_{ik})(\text{Id} - B_{kj}). \qquad \square$$

We have shown that $Sl_n(\mathbb{F})$ satisfies the conditions of Iwasawa's Lemma save for the fact that $Z(Sl_n(\mathbb{F})) \triangleleft Sl_n(\mathbb{F})$. All the relations hold for cosets of $Z(Sl_n(\mathbb{F}))$, and so $PSl_n(\mathbb{F})$ is simple for $n > 2$. $\qquad \square$

The case of $Sl_2(\mathbb{F})$ requires some number-theoretic ideas that you can polish off in the exercises at the end of the chapter. Familiar fields $\mathbb{F} = \mathbb{Q}, \mathbb{R}$, and $\mathbb{C}$ give simple groups $PSl_n(\mathbb{F})$ for $n \geq 2$. The case of finite fields such as $\mathbb{F}_p = \mathbb{Z}/p\mathbb{Z}$ for a prime p is interesting because $PSl_n(\mathbb{F}_p)$ is a finite group. We have shown that $PSl_n(\mathbb{F}_p)$ is a finite simple group for $n > 3$, and for $n = 2$ when $p > 3$. To put these groups into the set of examples of finite simple groups, let us determine their orders.

Let p be a prime. Beginning with the general linear group, it suffices to determine the number of linearly independent sets of n elements in $\mathbb{F}_p^{\times n}$. Such a set determines the columns of an invertible matrix, and vice versa. There are p^n vectors in $\mathbb{F}_p^{\times n}$. Leaving out the zero vector, there are $p^n - 1$ choices of the first column vector. Removing the multiples of that choice leaves $p^n - p$ choices for the second choice. The third choice is from the set of vectors in $\mathbb{F}_p^{\times n}$ that are not linear combinations of the first two choices. That gives $p^n - p^2$ possibilities. Continuing in this manner, we find

$$\# \mathrm{Gl}_n(\mathbb{F}_p) = (p^n - 1)(p^n - p)(p^n - p^2) \cdots (p^n - p^{n-1}).$$

Proposition 3.31 *For the finite field $\mathbb{F}_p$, $\# \mathrm{Sl}_n(\mathbb{F}_p) = \# \mathrm{Gl}_n(\mathbb{F}_p)/(p-1)$ and $\# \mathrm{PSl}_n(\mathbb{F}_p) = \# \mathrm{Gl}_n(\mathbb{F}_p)/(p-1)\gcd(p-1, n)$.*

Proof The determinant homomorphism $\det\colon \mathrm{Gl}_n(\mathbb{F}_p) \to \mathbb{F}_p^{\times}$ is a surjective homomorphism, so we get $\mathbb{F}_p^{\times} \cong \mathrm{Gl}_n(\mathbb{F})/\mathrm{Sl}_n(\mathbb{F}_p)$, and so $p - 1 = \# \mathrm{Gl}_n(\mathbb{F}_p)/\# \mathrm{Sl}_n(\mathbb{F}_p)$. Furthermore, $\mathrm{PSl}_n(\mathbb{F}_p) = \mathrm{Sl}_n(\mathbb{F}_p)/Z(\mathrm{Sl}_n(\mathbb{F}_p))$ and $\#Z(\mathrm{Sl}_n(\mathbb{F}_p)) = \#\{\lambda \in \mathbb{F}_p \mid \lambda^n = 1\}$. An elementary number theory argument shows that $\#Z(\mathrm{Sl}_n(\mathbb{F}_p)) = \gcd(p-1, n)$. □

For example, $\# \mathrm{PSl}_2(\mathbb{F}_5) = \dfrac{(5^2 - 1)(5^2 - 5)}{(5 - 1)\gcd(5 - 1, 2)} = \dfrac{24 \cdot 20}{4 \cdot 2} = 60$. In the exercises, you will prove that $\mathrm{PSl}_2(\mathbb{F}_5)$ is simple. However, there is only one simple group of order 60 (see exercises after Chapter 2), and so $\mathrm{PSl}_2(\mathbb{F}_5) \cong A_5$.

Exercises

3.1 Suppose W and U are subspaces of a finite-dimensional vector space V. Show that $W \cap U$ is also a subspace of V. Furthermore, prove that $\dim(W \oplus U) + \dim(W \cap U) = \dim W + \dim U$.

3.2 Let $\mathbb{F}$ be a field, and let $\mathbb{F}[x]$ be the ring of polynomials with coefficients in $\mathbb{F}$. Show that $\mathbb{F}[x]$ is an infinite-dimensional vector space over $\mathbb{F}$. Suppose k is another field that contains the field $\mathbb{F}$ (think $\mathbb{R} \subset \mathbb{C}$ or $\mathbb{Q} \subset \mathbb{R}$). If $a \in k$, then let $\mathrm{ev}_a\colon \mathbb{F}[x] \to k$ denote the mapping $\mathrm{ev}_a(p(x)) = p(a)$. Show that ev_a is a linear mapping. How would you interpret its kernel if nontrivial? How would you interpret $\ker \mathrm{ev}_a = \{0\}$? What is the image of ev_a in k?

3.3 When V has two bases B and B', then there is a *change-of-basis matrix* $C\colon \mathbb{F}^{\times n} \to \mathbb{F}^{\times n}$ that fits into the diagram

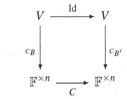

Here $c_B : V \to \mathbb{F}^{\times n}$ denotes the coordinate transformation expressing vectors in V in terms of the basis B, and $c_{B'} : V \to \mathbb{F}^{\times n}$ is the corresponding linear transformation for the basis B'. Determine an expression for C in terms of c_B and $c_{B'}$. When $V = \mathbb{F}^{\times 2}$, let $B = \{\mathbf{e}_1, \mathbf{e}_2\}$ and $B' = \{\mathbf{e}_1 - \mathbf{e}_2, \mathbf{e}_1 + \mathbf{e}_2\}$. What matrix is C in this case?

3.4 Consider the mapping $\varphi : \mathbb{F}^{\times n} \times \mathbb{F}^{\times m} \to \mathbb{F}^{\times (nm)}$ given by $\phi(\mathbf{v}, \mathbf{w}) = \mathbf{v}\mathbf{w}^t$ (matrix multiplication). This gives an $n \times m$ matrix that determines an nm-vector by ordering the entries lexicographically by index. Show that φ is bilinear and induces an isomorphism $\mathbb{F}^{\times n} \otimes \mathbb{F}^{\times m} \cong \mathbb{F}^{\times (nm)}$. What is φ when $n = m = 1$?

3.5 For vector spaces V, V_1, V_2 and W, W_1, W_2, prove that $(V_1 \oplus V_2) \otimes W \cong (V_1 \otimes W) \oplus (V_2 \otimes W)$ and $V \otimes (W_1 \oplus W_2) \cong (V \otimes W_1) \oplus (V \otimes W_2)$. Prove that $V \otimes \mathbb{F} \cong V$ for any vector space V over $\mathbb{F}$.

3.6 Recover the usual facts about Gauss–Jordan elimination for solving systems of linear equations $A\mathbf{x} = \mathbf{b}$ for an $m \times n$ matrix A and $\mathbf{b} \in \mathbb{F}^{\times m}$ from the fact that elementary row operations are given by multiplication by certain matrices. What goes wrong when there is no solution? What happens when the solution is not unique?

3.7 Given a 2×2 matrix $A = \left[\begin{smallmatrix} a & b \\ c & d \end{smallmatrix}\right]$ over a field $\mathbb{F}$ with determinant 1, show that A is similar to a matrix of the form $\left[\begin{smallmatrix} \lambda & 0 \\ 0 & \lambda^{-1} \end{smallmatrix}\right]$ or $\left[\begin{smallmatrix} 0 & -1 \\ 1 & a+d \end{smallmatrix}\right]$. (Hint: If A has eigenvalues in $\mathbb{F}$, then they are λ, λ^{-1} since $\det A = 1$. If A has no eigenvalues in $\mathbb{F}$, then $\{\mathbf{x}, A\mathbf{x}\}$ is a basis for $\mathbb{F}^{\times 2}$ for any nonzero vector $\mathbf{x} \in \mathbb{F}^{\times 2}$. In this basis the Cayley–Hamilton Theorem implies that A is similar to the second form, which is the companion matrix of the characteristic polynomial of A.)

3.8 Suppose $A \in \mathrm{Sl}_2(\mathbb{F})$ is similar to $A' = \left[\begin{smallmatrix} \lambda & 0 \\ 0 & \lambda^{-1} \end{smallmatrix}\right]$, and suppose that $A \in N$, a normal subgroup of $\mathrm{Sl}_2(\mathbb{F})$. Suppose $\lambda \neq \lambda^{-1}$ (that is, A is not in the center of $\mathrm{Sl}_2(\mathbb{F})$). Let $B = \left[\begin{smallmatrix} 1 & 1 \\ 0 & 1 \end{smallmatrix}\right]$ and form $C = A'B(A')^{-1}B^{-1}$. Show that C is in N and therefore N contains a transvection. Hence $N = \mathrm{Sl}_2(\mathbb{F})$.

3.9 Suppose $A \in \mathrm{Sl}_2(\mathbb{F})$ is similar to $A' = \left[\begin{smallmatrix} 0 & -1 \\ 1 & r \end{smallmatrix}\right]$, and suppose that $A \in N$, a normal subgroup of $\mathrm{Sl}_2(\mathbb{F})$. Let $B = \left[\begin{smallmatrix} 1 & x^2 \\ 0 & 1 \end{smallmatrix}\right]$ and $C = A'B(A')^{-1}B^{-1}$.

Show that C is in N and depends only on the choice of x, taking the form $\begin{bmatrix} 1 & -x^2 \\ -x^2 & 1+x^4 \end{bmatrix}$.

3.10 When you conjugate the matrix $\begin{bmatrix} 1 & -x^2 \\ -x^2 & 1+x^4 \end{bmatrix}$ with the matrix $\begin{bmatrix} x & x^{-1} \\ 0 & x^{-1} \end{bmatrix}$ you should get $\begin{bmatrix} 0 & 1 \\ -1 & 2+x^4 \end{bmatrix}$. Multiply the inverse of $\begin{bmatrix} 0 & 1 \\ -1 & 2+x^4 \end{bmatrix}$ by $\begin{bmatrix} 0 & 1 \\ -1 & 2+y^4 \end{bmatrix}$ to obtain a transvection in N as long as $x^4 - y^4 \neq 0$. We know that $t^4 - 1$ has at most four roots in $\mathbb{F}_p$ for $p > 3$. Show that this implies $N = \mathrm{Sl}_2(\mathbb{F})$, and so, for 2×2 matrices, we have shown that $\mathrm{PSl}_2(\mathbb{F}_p)$ is simple when $p \geq 5$.

3.11 Prove the stated properties of the matrices B_{ij}: $B_{ij}B_{kl} = \mathbb{O}$ if $j \neq k$, and $B_{ij}B_{jl} = B_{il}$. Also, $(\mathrm{Id} + bB_{ij})M = M'$ where M' is the result of adding b times row j to row i in M. Finally, show that $\det(\mathrm{Id} + bB_{ij}) = 1$ and $(\mathrm{Id} + bB_{ij})^{-1} = \mathrm{Id} - bB_{ij}$.

4

Representation Theory

...while, in the present state of our knowledge, many results in the pure theory are arrived at most readily by dealing with properties of substitution groups, it would be difficult to find a result that could be most readily obtained by consideration of groups of linear substitutions.

WILLIAM BURNSIDE, *Theory of Groups of Finite Order*
1st edition, 1897

Very considerable advances in the theory of groups of finite order have been made since the appearance of the first edition of this book. In particular, the theory of groups of linear substitutions has been the subject of numerous and important investigations by several writers; and the reason given in the original preface for omitting any account of it no longer holds good.

WILLIAM BURNSIDE, *Theory of Groups of Finite Order*
2nd edition, 1911

4.1 Definitions and Examples

A group G acting on a set X allows us to think of the elements of G as permutations of the set X – a concrete representation of the group elements of an abstract group. There is another representation of the elements of a group as a familiar object where we present the elements of the group as matrices.

For example, the group D_8 of symmetries of a square determines a collection of matrices by expressing each symmetry of a square as a symmetry of the plane that takes the square to itself. For the square with vertices $(1, 0)$, $(0, 1)$,

$(-1, 0)$, and $(0, -1)$, the generators of D_8 correspond to the matrices

$$r \mapsto \begin{bmatrix} 0 & -1 \\ 1 & 0 \end{bmatrix} \text{ and } f \mapsto \begin{bmatrix} 0 & -1 \\ -1 & 0 \end{bmatrix}.$$

The group D_8 has a presentation $\langle r, f \mid r^4 = e = f^2, fr = r^3 f \rangle$. We have identified r with the rotation of $\mathbb{R}^2$ through $\pi/2$ radians and f with reflection across the line $y = -x$. Extending this identification to the rest of D_8 by multiplying various powers of r and f gives a homomorphism $\rho \colon D_8 \to \mathrm{Gl}_2(\mathbb{R})$ and a group action of D_8 on $\mathbb{R}^2$ that consists of the isometries of the square. (You should check that $\rho(r)^4 = \mathrm{Id} = \rho(f)^2$ and $\rho(f) \circ \rho(r) = \rho(r^3) \circ \rho(f)$.)

The set **Doc**$^+$ of rotations of the regular dodecahedron gives a collection of orthonormal 3×3 matrices. To get some idea of which matrices, place the center of the dodecahedron at the origin. We can give coordinates for the vertices of the dodecahedron by noticing that there is a cube among the vertices of the dodecahedron. Let those vertices be $(\pm 1, \pm 1, \pm 1)$ for all possible choices of signs. The rest of the coordinates can now be determined by using the structure of a dodecahedron and the geometry of a regular pentagon. After a recommended romp with geometry, you will find that the coordinates are

$$(\pm 1, \pm 1, \pm 1), (\pm \phi, 0, \pm 1/\phi), (0, \pm 1/\phi, \pm \phi), (\pm 1/\phi, \pm \phi, 0),$$

where $\phi = \dfrac{\sqrt{5} + 1}{2} = 2 \cos(\pi/5)$ is the golden ratio, and all possible choices of signs are taken. The golden ratio appears because it is the ratio between the diagonal and side of a regular pentagon.

A rotation of $\mathbb{R}^3$ is determined by its fixed axis and the angle that the rest of $\mathbb{R}^3$ rotates around the axis. For the x-, y-, and z-axes, the rotations are given by

$$R_x(\theta) = \begin{bmatrix} 1 & 0 & 0 \\ 0 & \cos\theta & -\sin\theta \\ 0 & \sin\theta & \cos\theta \end{bmatrix}, \; R_y(\theta) = \begin{bmatrix} \cos\theta & 0 & -\sin\theta \\ 0 & 1 & 0 \\ \sin\theta & 0 & \cos\theta \end{bmatrix},$$

$$R_z(\theta) = \begin{bmatrix} \cos\theta & -\sin\theta & 0 \\ \sin\theta & \cos\theta & 0 \\ 0 & 0 & 1 \end{bmatrix}.$$

A famous theorem of Euler states that any rotation of space has an axis (equivalently, has at least one eigenvalue) and may be expressed as a product $R = R_z(\alpha) R_y(\beta) R_z(\gamma)$, where α, β, and γ satisfy $-\pi < \gamma \leq \pi$, $0 \leq \beta \leq \pi$, $-\pi < \alpha \leq \pi$. In aeronautical parlance, α is called yaw (a bearing), β pitch (elevation), and γ roll (bank angle).

For our purposes, we will not try to give a matrix for every element of **Doc**$^+$. Let us consider the rotations σ, τ, and ρ from Chapter 2. The axis ar around which ρ rotates the dodecahedron is the axis joining $(1, -1, -1)$ to $(-1, 1, 1)$. Because these points are vertices of the cube, it is easy to follow the standard basis $\{\mathbf{e}_1, \mathbf{e}_2, \mathbf{e}_3\}$, which lies in the centers of three faces of the cube, to their final positions after ρ, giving us

$$\rho \leftrightarrow \begin{bmatrix} 0 & 0 & -1 \\ -1 & 0 & 0 \\ 0 & 1 & 0 \end{bmatrix}.$$

The rotation τ may be taken around the axis joining the midpoints of edges $(\bot\phi, 0, 1/\phi)$ and $(\pm\phi, 0, -1/\phi)$. This axis lies along the x-axis, and once again, we can focus on the cube to see where the standard basis goes:

$$\tau \leftrightarrow \begin{bmatrix} 1 & 0 & 0 \\ 0 & -1 & 0 \\ 0 & 0 & -1 \end{bmatrix}.$$

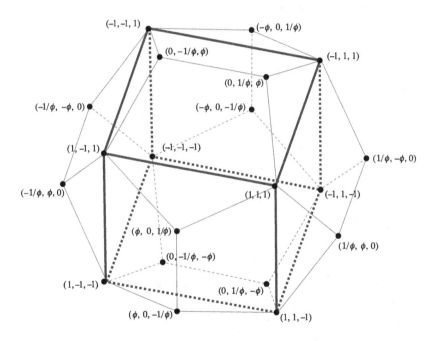

Figure 4.1

Finally, to determine the rotation matrix for σ, we need to find the center of the pentagon given by the points $(1, -1, 1)$, $(-1/\phi, \phi, 0)$, $(1, -1, -1)$, $(\phi, 0, -1/\phi)$, and $(\phi, 0, 1/\phi)$. The center is obtained by averaging the coordinates, giving $Z = \frac{1}{5}(2 + 2\phi - 1/\phi, \phi - 2, 0)$, a point in the x-y-plane.

Using the relation $\phi^2 - \phi - 1 = 0$ satisfied by the golden ratio, we can write $Z = (1/5)(\phi + 3, \phi - 2, 0)$. Let P denote the rotation $R_z(\theta)$ where $\theta = \arctan\left(\dfrac{\phi - 2}{\phi + 3}\right)$, which brings the ray $\mathbb{O}Z$ to $\mathbb{O}\mathbf{e}_1$, the x-axis. Rotate the y-z-plane by $R_x(2\pi/5)$, then apply $P^{-1} = R_z(-\theta)$. Thus

$$\sigma \leftrightarrow R_z(-\theta)R_x(2\pi/5)R_z(\theta).$$

With lots more geometry we could assign a matrix to every element of **Doc**$^+$ making it a finite subgroup of $\mathrm{Sl}_3(\mathbb{R})$ with the operation matrix multiplication. We generalize these representations of D_8 and **Doc**$^+$ as matrix groups and use linear algebra to obtain new opportunities to explore groups more deeply.

The focus of this chapter is on the linear actions of G on a vector space V.

Definition 4.1 A *representation* of a group G on a vector space V over a field $\mathbb{F}$ is a group homomorphism

$$\rho\colon G \twoheadrightarrow \mathrm{Gl}(V).$$

When the dimension of V over $\mathbb{F}$ is finite, it is called the *degree* of the representation.

The *trivial representation* is given by $\rho_0\colon G \to \mathrm{Gl}(V)$, where $\rho_0(g) = \mathrm{Id}_V\colon V \to V$ for all g in G. For the cyclic group of order n, $G = C_n \cong \mathbb{Z}/n\mathbb{Z}$, there is a complex representation $\rho\colon C_n \to \mathrm{Gl}_1(\mathbb{C}) = \mathbb{C}^\times$ given by $\rho(g) = e^{2\pi i/n} = \cos(2\pi/n) + i\sin(2\pi/n)$ where $g \in C_n$ is a generator. Extend ρ to a homomorphism by defining $\rho(g^k) = \rho(g)^k = e^{2\pi i k/n}$.

The action of a group G on a finite set X determines a representation of G on the vector space over $\mathbb{F}$ with basis X, defined in Chapter 3 by

$$\mathbb{F}X = \{a_1x_1 + a_2x_2 + \cdots + a_nx_n \text{ with } a_i \in \mathbb{F}, x_i \in X\}.$$

The groups $\Sigma(X)$ and $\mathrm{Gl}(\mathbb{F}X)$ are related by a homomorphism: Define the mapping $\Xi\colon \Sigma(X) \to \mathrm{Gl}(\mathbb{F}X)$ by $\Xi(\sigma) = T_\sigma$ where $T_\sigma(a_1x_1 + \cdots + a_nx_n) = a_1\sigma(x_1) + \cdots + a_n\sigma(x_n)$. The mapping T_σ is linear:

$$
\begin{aligned}
T_\sigma(av + v') &= T_\sigma\big(a(a_1x_1 + \cdots + a_nx_n) + (b_1x_1 + \cdots + b_nx_n)\big) \\
&= T_\sigma\big((aa_1 + b_1)x_1 + \cdots + (aa_n + b_n)x_n\big) \\
&= (aa_1 + b_1)\sigma(x_1) + \cdots + (aa_n + b_n)\sigma(x_n) \\
&= a\big(a_1\sigma(x_1) + \cdots + a_n\sigma(x_n)\big) + \big(b_1\sigma(x_1) + \cdots + b_n\sigma(x_n)\big) \\
&= aT_\sigma(v) + T_\sigma(v').
\end{aligned}
$$

Also, Ξ is a homomorphism:

$$\Xi(\sigma\tau)(v) = T_{\sigma\tau}(v) = a_1(\sigma\tau)(x_1) + \cdots + a_n(\sigma\tau)(x_n)$$
$$= T_\sigma\big(a_1\tau(x_1) + \cdots + a_n\tau(x_n)\big) = T_\sigma \circ T_\tau(v)$$
$$= \Xi(\sigma) \circ \Xi(\tau)(v).$$

A group action of G on X, $\rho\colon G \to \Sigma(X)$, determines a representation as the composite homomorphism $\hat{\rho} = \Xi \circ \rho\colon G \to \Sigma(X) \to \mathrm{Gl}(\mathbb{F}X)$:

$$\hat{\rho}(g)(a_1x_1 + a_2x_2 + \cdots + a_nx_n) = a_1\rho(g)(x_1) + a_2\rho(g)(x_2) + \cdots + a_n\rho(g)(x_n).$$

This construction is called the *permutation representation* associated with the action of G on the set X.

This construction applies to the actions of a group on itself. For example, G acts on itself by left multiplication $\rho(h)(g) = hg$. This action gives the *regular representation* $\rho\colon G \to \mathrm{Gl}(\mathbb{F}G)$, where, for $\sum_{g \in G} a_g g \in \mathbb{F}G$,

$$\rho(h)\left(\sum_{g \in G} a_g g\right) = \sum_{g \in G} a_g hg.$$

Suppose V is a finite-dimensional vector space with basis $B = \{\mathbf{b}_1, \ldots, \mathbf{b}_n\}$. The coordinate mapping $c_B\colon V \to \mathbb{F}^{\times n}$ given by $c_B(a_1\mathbf{b}_1 + \cdots + a_n\mathbf{b}_n) = [a_1, \ldots, a_n]^t \in \mathbb{F}^{\times n}$ is a linear isomorphism. If we have a representation $\rho\colon G \to \mathrm{Gl}(V)$, then we obtain an associated representation $\rho_B\colon G \to \mathrm{Gl}_n(\mathbb{F})$ given by $\rho_B(g) = c_B \circ \rho(g) \circ c_B^{-1}\colon \mathbb{F}^{\times n} \to \mathbb{F}^{\times n}$,

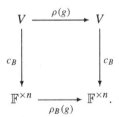

The representation ρ_B associates an invertible $n \times n$ matrix with each $g \in G$.

Definition 4.2 The *kernel of a representation* is $\ker \rho$, the subgroup of G of elements g with $\rho(g) = \mathrm{Id}_V$. We say that ρ is a *faithful representation* if ρ is an injective group homomorphism, that is, $\ker \phi = \{e\}$.

The regular representation has degree #G, the order of G, and it is faithful: To see this, suppose $\rho(g) = \mathrm{Id}$. Then $\rho(g)(h) = gh = h$ for any $h \in G$, and so $g = e$. Hence $\ker \rho = \{e\}$.

Get to know some representations

1. Suppose $\rho: G \to Gl(\mathbb{C}) = \mathbb{C}^\times$ is a representation of G of degree 1. Prove that $G/\ker \rho$ is an abelian group.
2. Suppose $\rho: G \to Gl_n(\mathbb{F})$ is a group representation. Show that the mapping given by the determinant $\det: Gl_n(\mathbb{F}) \to \mathbb{F}^\times$ is also a homomorphism. Show that composition with ρ gives $\det \circ \rho: G \to \mathbb{F}^\times$, a degree 1 representation.

Definition 4.3 Suppose $\rho: G \to Gl(V)$ is a representation. Define the *fixed subspace* of V by

$$V^G = \{v \in V \mid \rho(g)(v) = v \text{ for all } g \in G\}.$$

3. Show that V^G is a vector subspace of V. Show that $\rho(g)|_{V^G} = \text{Id}$, and so we get a trivial representation $\hat{\rho}: G \to Gl(V^G)$.
4. Develop the following examples:

   A. Let $G = C_3 = \langle g \mid g^3 = e \rangle$ be the cyclic group of order 3. Consider the mapping

   $$\rho(g) = \begin{bmatrix} 0 & 1 \\ -1 & -1 \end{bmatrix}.$$

   Show that this mapping leads to a representation of $\rho: C_3 \to Gl_2(\mathbb{R})$.

   B. Let $D_8 = \langle r, f \mid r^4 = f^2 = e, rf = fr^3 \rangle$ be the dihedral group of order 8. Consider the mapping

   $$\rho(r) = \begin{bmatrix} 0 & 1 \\ -1 & 0 \end{bmatrix}, \quad \rho(f) = \begin{bmatrix} 0 & -1 \\ -1 & 0 \end{bmatrix}.$$

   Show that this mapping gives a representation $\rho: D_8 \to Gl_2(\mathbb{R})$.

   C. Another representation of the dihedral group D_8 is given by $\rho: D_8 \to Gl_2(\mathbb{C})$,

   $$\rho(r) = \begin{bmatrix} i & 0 \\ 0 & -i \end{bmatrix}, \quad \rho(f) = \begin{bmatrix} 0 & 1 \\ 1 & 0 \end{bmatrix}.$$

   Show that this is a representation of D_8.

   D. Let $G = C_n = \langle a \mid a^n = e \rangle$ be the cyclic group of order n. Consider the mapping $\rho: C_n \to Gl_2(\mathbb{R})$,

   $$\rho(a) = \begin{bmatrix} \cos(2\pi/n) & -\sin(2\pi/n) \\ \sin(2\pi/n) & \cos(2\pi/n) \end{bmatrix}.$$

   Show that this mapping gives a representation of C_n of degree two.

E. Let $\rho\colon \Sigma_3 \to Gl_2(\mathbb{C})$ be defined on the generators $(1, 2)$ and $(1, 2, 3)$ by

$$\rho\big((1, 2)\big) = \begin{bmatrix} -1 & -1 \\ 0 & 1 \end{bmatrix}, \quad \rho\big((1, 2, 3)\big) = \begin{bmatrix} -1 & -1 \\ 1 & 0 \end{bmatrix}.$$

Show that this mapping leads to a representation of Σ_3.

F. Let $G = \{1, -1, i, -i\}$ denote the units in the ring of Gaussian integers $\mathbb{Z}[i]$. Let $\mathbb{C}$ denote the complex numbers as a vector space over $\mathbb{R}$. The complex representation $\rho\colon G \to \mathbb{C}^{\times}$, $\rho(g)(z) = gz$, determines a real representation $\rho'\colon G \to Gl_2(\mathbb{R})$. What is that representation? (Hint: Let $u = 1$ and $v = i$. Then $\{u, v\}$ is a basis for $\mathbb{C}$ over $\mathbb{R}$.)

G. Let $Q = \{\pm 1, \pm i, \pm j, \pm k\}$ be the group of *integral unit quaternions*. The elements satisfy $i^2 = j^2 = k^2 = -1$, $ij = k$, $jk = i$, $ki = j$, $ji = -k$, $kj = -i$, and $ik = -j$. Let $\rho\colon Q \to Gl_2(\mathbb{C})$ be given by

$$\rho(i) = \begin{bmatrix} i & 0 \\ 0 & -i \end{bmatrix}, \quad \rho(j) = \begin{bmatrix} 0 & 1 \\ -1 & 0 \end{bmatrix}.$$

Show that ρ extends to the rest of Q giving a representation.

As discussed in Chapter 3, we can add vector spaces by direct sum. We can extend this sum to add representations. Let $\rho_1\colon G \to Gl(V)$ and $\rho_2\colon G \to Gl(V')$ be representations of G, and form $\rho_1 \oplus \rho_2\colon G \to Gl(V \oplus V')$ where $\rho_1 \oplus \rho_2(g) = \rho_1(g) \oplus \rho_2(g)\colon V \oplus V' \to V \oplus V'$. We get a representation because

$$\rho_1 \oplus \rho_2(gh) = \rho_1(gh) \oplus \rho_2(gh) = \rho_1(g) \circ \rho_1(h) \oplus \rho_2(g) \circ \rho_2(h)$$
$$= \big(\rho_1(g) \oplus \rho_2(g)\big) \circ \big(\rho_1(h) \oplus \rho_2(h)\big),$$

where the last equality is a property of the composition of linear transformations between direct sums.

In the case of matrix representations $\rho_1\colon G \to Gl_n(\mathbb{F})$ and $\rho_2\colon G \to Gl_m(\mathbb{F})$, the sum $\rho_1 \oplus \rho_2$ has degree $n + m$, and, as matrices, we can write $\rho_1 \oplus \rho_2(g)$ in block form:

$$\rho_1 \oplus \rho_2(g) = \begin{bmatrix} \rho_1(g) & 0 \\ 0 & \rho_2(g) \end{bmatrix}.$$

This addition of representations for external direct products can be applied to internal direct products when the vector space decomposes into a direct sum of subspaces on which the representation acts separately. For example, suppose $G = C_2 \cong \{1, -1\}$, the multiplicative group, and let $\rho: G \to \mathrm{Gl}(V)$ be any representation. Suppose V is a vector space over a field $\mathbb{F}$ in which 2 is invertible. Let

$$V^{C_2} = \{v \in V \mid \rho(-1)(v) = v\}, \quad V^- = \{w \in V \mid \rho(-1)(w) = -w\}.$$

Define the linear transformation $P: V \to V$ by $P(v) = 2^{-1}(v + \rho(-1)(v))$. We will use the identity $\rho(-1)\rho(-1)(v) = \rho(-1 \cdot -1)(v) = \rho(1)(v) = v$. Then

$$
\begin{aligned}
P \circ P(v) &= 2^{-1}\big(P(v) + \rho(-1)\big(P(v)\big)\big) \\
&= 4^{-1}\big(v + \rho(-1)(v) + \rho(-1)(v) + \rho(-1)\rho(-1)(v)\big) \\
&= 4^{-1}\big(2v + 2\rho(-1)(v)\big) = 2^{-1}\big(v + \rho(-1)(v)\big) = P(v).
\end{aligned}
$$

By Lemma 3.11, $V \cong \ker P \oplus P(V)$. Notice that $w \in \ker P$ if and only if $\rho(-1)(w) = -w$, and so $\ker P = V^-$. Also, $\rho(-1)(P(v)) = 2^{-1}\rho(-1)(v + \rho(-1)(v)) = 2^{-1}(\rho(-1)(v) + \rho(-1)\rho(-1)(v)) = 2^{-1}(v + \rho(-1)(v)) = P(v)$, and so $P(V) \subset V^{C_2}$. If $v \in V^{C_2}$, then $P(v) = 2^{-1}(v + \rho(-1)(v)) = 2^{-1}(v + v) = v$, and $V^{C_2} = P(V)$. Since the representation on V^{C_2} is trivial, and $\rho(-1)$ is multiplication by -1 on V^-, the representation ρ can be expressed as $\rho = \rho_0 \oplus \sigma$, where $\sigma: C_2 \to \mathrm{Gl}(V^-)$ is determined by $\sigma(-1)(v) = -v$. This example can be generalized as we will see later.

A vector space is determined by its dimension, that is, if $\dim V = \dim W$, then there is a linear isomorphism $V \to W$. The notion of equivalence for representations must include the actions of the group.

Definition 4.4 Given two representations $\rho: G \to \mathrm{Gl}(V)$ and $\rho': G \to \mathrm{Gl}(W)$ and a linear mapping $T: V \to W$, we say that T is *G-equivariant* if for all $v \in V$ and $g \in G$, $\rho'(g)(T(v)) = T(\rho(g)(v))$, that is, the following diagram commutes:

$$
\begin{array}{ccc}
V & \xrightarrow{\rho(g)} & V \\
\Big\downarrow{\scriptstyle T} & & \Big\downarrow{\scriptstyle T} \\
W & \xrightarrow[\rho'(g)]{} & W
\end{array}
$$

The two representations are said to be *equivalent representations* if there is a G-equivariant isomorphism $T: V \to W$. We write $\rho \sim \rho'$ in this case.

If $\rho: G \to \mathrm{Gl}(V)$ is a representation of a group G, then we get the equivalent representation $\rho_B: G \to \mathrm{Gl}_n(\mathbb{F})$ from the coordinate isomorphism $c_B: V \to \mathbb{F}^{\times n}$. If we have two bases B and B', then ρ_B is equivalent to $\rho_{B'}$ by the change-of-basis isomorphism from the basis B to the basis B', $C_B^{B'} : \mathbb{F}^{\times n} \to$ $\mathbb{F}^{\times n}$. In fact, $\rho_{B'}(g) = C_B^{B'} \circ \rho_B(g) \circ (C_B^{B'})^{-1}$. A choice of basis leads to an equivalent matrix representation of G, and different choices of basis give equivalent representations.

Consider the example of the permutation representation $\rho_c : \Sigma_3 \to \text{Gl}(\mathbb{F}\Sigma_3)$ that comes from the conjugation action of Σ_3 on itself:

$$\rho_c(\sigma)\left(\sum_{g \in \Sigma_3} a_g g \right) = \sum_{g \in \Sigma_3} a_g \sigma g \sigma^{-1}.$$

By Cauchy's Formula the conjugacy classes in Σ_n are determined by their cycle presentation. The group is a basis for $\mathbb{F}\Sigma_3$, and this basis breaks into three disjoint pieces given by the conjugacy classes $A = \{(\,)\}$, $B = \{(1, 2), (1, 3),$ $(2, 3)\}$, and $C = \{(1, 2, 3), (1, 3, 2)\}$. Then $\mathbb{F}\Sigma_3 = \mathbb{F}A \oplus \mathbb{F}B \oplus \mathbb{F}C$, and Σ_3 acts on each subspace separately. We can write $\rho_c \sim \rho_0 \oplus \rho_B \oplus \rho_C$. If we order the basis as

$$b_1 = (\,), b_2 = (1, 2), b_3 = (1, 3), b_4 = (2, 3), b_5 = (1, 2, 3), b_6 = (1, 3, 2),$$

then we obtain a matrix representation. For example,

$$\rho_c\big((1, 2)\big) = \left[\begin{array}{cccc|cc} 1 & 0 & 0 & 0 & 0 & 0 \\ \hline 0 & 1 & 0 & 0 & 0 & 0 \\ 0 & 0 & 0 & 1 & 0 & 0 \\ 0 & 0 & 1 & 0 & 0 & 0 \\ \hline 0 & 0 & 0 & 0 & 0 & 1 \\ 0 & 0 & 0 & 0 & 1 & 0 \end{array}\right] = \left[\begin{array}{c|ccc|cc} 1 & & & & & \\ \hline & 1 & 0 & 0 & & \\ & 0 & 0 & 1 & & \\ & 0 & 1 & 0 & & \\ \hline & & & & 0 & 1 \\ & & & & 1 & 0 \end{array}\right].$$

More generally, if $\rho : G \to \Sigma(X)$ is an action of G on X, then the permutation representation associated with ρ splits into pieces that correspond to the orbits of the action.

There is another way to combine representations, in this case for possibly different groups. Suppose $\rho_1 : G_1 \to \text{Gl}(V)$ and $\rho_2 : G_2 \to \text{Gl}(W)$ are two representations. The *tensor product of representations* is the mapping

$$\rho_1 \otimes \rho_2 : G_1 \times G_2 \to \text{Gl}(V \otimes W), \quad (\rho_1 \otimes \rho_2)(g_1, g_2) = \rho_1(g_1) \otimes \rho_2(g_2).$$

To see that we get a representation, we compute, for $g, h \in G_1, k, l \in G_2$,

$$\rho_1 \otimes \rho_2(gh, kl) = \rho_1(gh) \otimes \rho_2(kl) = \big(\rho_1(g) \circ \rho_1(h)\big) \otimes \big(\rho_2(k) \circ \rho_2(l)\big)$$
$$= \big(\rho_1(g) \otimes \rho_2(k)\big) \circ \big(\rho_1(h) \otimes \rho_2(l)\big).$$

The tensor product of isomorphisms gives an isomorphism. For example, if $\rho_0 : C_2 \to \mathbb{C}^{\times}$ is the trivial representation and $\rho_{\pm} : C_2 \to \mathbb{C}^{\times}$ is the sign representation ($\rho_\pm(g) = -1$, $\rho_\pm(g^2) = 1$), then the tensor product of the representations is given by

$\rho_0 \otimes \rho_\pm$	(e, e)	(e, g)	(g, e)	(g, g)
	1	-1	1	-1

Given two representations of a single group G, we can combine them with the tensor product as the composite of homomorphisms

$$G \xrightarrow{\Delta} G \times G \xrightarrow{\rho_1 \otimes \rho_2} \mathrm{Gl}(V \otimes W),$$

where $\Delta(g) = (g, g)$ is the diagonal homomorphism. In the example above, we see that $(\rho_0 \otimes \rho_\pm) \circ \Delta = \rho_\pm$. This is a general fact: the one-dimensional trivial representation acts as an identity for the tensor product of representations of a single group G. This follows because $\mathbb{F} \otimes V \cong V \cong V \otimes \mathbb{F}$ for any vector space V over $\mathbb{F}$.

With a group action on a set X, we can talk of G-invariant subsets, for example, the orbits of the action. For representations, we have an analogous notion.

Definition 4.5 Given a representation $\rho \colon G \to \mathrm{Gl}(V)$ and a subspace $W \subset V$, we say that W is a G-*invariant subspace*, or W is *stable* under G, if $\rho(g)(w) \in W$ for all $g \in G$. The restriction of $\rho(g)$ to W determines a *subrepresentation*, denoted $\rho^W \colon G \to \mathrm{Gl}(W)$, for any G-invariant subspace $W \subset V$.

For example, if we have $\rho = \rho_1 \oplus \rho_2 \colon G \to \mathrm{Gl}(V_1 \oplus V_2)$, then $V_1 \cong V_1 \oplus \{\mathbb{O}\}$ and $V_2 \cong \{\mathbb{O}\} \oplus V_2$ are G-invariant subspaces of $V_1 \oplus V_2$ such that $\rho^{V_1} = \rho_1$ and $\rho^{V_2} = \rho_2$.

Definition 4.6 A representation $\rho \colon G \to \mathrm{Gl}(V)$ is *reducible* if there are G-invariant subspaces W_1 and W_2, each of nonzero dimension, such that $V = W_1 \oplus W_2$ and ρ is equivalent to $\rho^{W_1} \oplus \rho^{W_2}$. If V has no nontrivial G-invariant decompositions, we say that the representation ρ is *irreducible*.

Any degree one representation is irreducible; this follows because $1 = 1 + 0 = 0 + 1$, the only partitions of degree one into two parts. Suppose $G \cong \mathbb{Z}/n\mathbb{Z}$ is a cyclic group of order n. Let $\rho \colon G \to \mathbb{C}^\times = \mathbb{C} \setminus \{0\}$ be a representation. Then $\rho([0]) = 1$ and $\rho([n]) = \rho(n[1]) = (\rho([1]))^n = 1$, and so $\rho([1])$ is an nth root of unity, a complex number of the form $e^{2j\pi i/n}$ for some integer j. The multiplicative group $\mathbb{C}^\times$ has a representation $\bar{\rho} \colon \mathbb{C}^\times \to \mathrm{Gl}_2(\mathbb{R})$ given by

$$\bar{\rho}(a + ib) = \begin{bmatrix} a & -b \\ b & a \end{bmatrix}.$$

Notice that

$$\bar{\rho}\big((a + bi)(c + di)\big) = \bar{\rho}\big((ac - bd) + (ad + bc)i\big)$$

$$= \begin{bmatrix} ac - bd & -ad - bc \\ ad + bc & ac - bd \end{bmatrix} = \begin{bmatrix} a & -b \\ b & a \end{bmatrix} \begin{bmatrix} c & -d \\ d & c \end{bmatrix}$$

$$= \bar{\rho}(a + bi)\bar{\rho}(c + di).$$

In fact, $\bar{\rho}$ is an isomorphism. Thus, for any complex one-dimensional representation $\rho_1 \colon G \to \mathbb{C}^\times = \mathrm{Gl}_1(\mathbb{C})$, $\bar{\rho} \circ \rho_1 \colon G \to \mathrm{Gl}_2(\mathbb{R})$ determines a real two-dimensional real representation.

Given a representation $\rho \colon G \to \mathrm{Gl}(V)$, it may be possible to find a nontrivial G-invariant subspace $W \subset V$, but that does not guarantee that there is a complementary subspace W' that is also G-invariant for which $V \cong W \oplus W'$. The next theorem, proved by HEINRICH MASCHKE (1858–1908), clarifies when W' can also be found.

Maschke's Theorem *Suppose $\mathbb{F}$ is a field and G is a finite group for which the characteristic of the field $\mathbb{F}$ does not divide the order of G. Let $\rho \colon G \to \mathrm{Gl}(V)$ be a representation with V a finite-dimensional vector space over $\mathbb{F}$ and $W \subset V$ a G-invariant subspace. Then there is another subspace $W^0 \subset V$ such that*

(1) $V \cong W \oplus W^0$, (2) W^0 is a G-invariant subspace, and (3) $\rho \sim \rho^W \oplus \rho^{W^0}$.

Recall that the *characteristic of a field* $\mathbb{F}$ is the generator of the ideal that is the kernel of the ring homomorphism $\phi \colon \mathbb{Z} \to \mathbb{F}$ determined by $\phi(1) = 1$. For example, when $\mathbb{F} = \mathbb{Q}$, $\ker \phi = \{0\}$, and the characteristic of $\mathbb{Q}$ is zero. The characteristic of $\mathbb{F}_p = \mathbb{Z}/p\mathbb{Z}$ is p, a prime.

Proof As a subspace W has finite dimension. Let $\{w_1, \ldots, w_k\}$ be a basis for W. We can extend this basis for W to a basis $\{w_1, \ldots, w_k, w'_1, \ldots, w'_l\}$ for V. Let $W' = \mathrm{Span}\{w'_1, \ldots, w'_l\}$. Define $P \colon V \to V$ by $P(w + w') = w$ for $w \in W$ and $w' \in W'$, that is, $P(a_1 w_1 + \cdots + a_k w_k + b_1 w'_1 + \cdots + b_l w'_l) = a_1 w_1 + \cdots + a_k w_k$. Notice that $P(w) = w$ for all $w \in W$. For all $v \in V$, we can write $v = w + w'$, and $P(P(v)) = P(P(w + w')) = P(w) = w = P(v)$; hence P is a projection. By Lemma 3.11 $V \cong P(V) \oplus \ker P = W \oplus W'$. However, we do not know if W' is a G-invariant subspace.

To overcome this shortcoming, we form an *average* over G of the conjugates of P: Define

$$P^0 = \frac{1}{\#G} \sum_{g \in G} \rho(g) \circ P \circ \rho(g)^{-1} \colon V \to V.$$

Because the characteristic of $\mathbb{F}$ does not divide $\#G$, $1/\#G$ makes sense and is nonzero. Since P and $\rho(g)$ are linear, so is P^0. If $w \in W$ and $g \in G$, then

$\rho(g) \circ P \circ \rho(g^{-1})(w) = \rho(g)(P(\rho(g^{-1})(w)))$. Because W is G-invariant, $\rho(g^{-1})(w)$ is in W, and $P(\rho(g^{-1})(w)) = \rho(g^{-1})(w)$. Therefore $\rho(g) \circ P \circ \rho(g^{-1})(w) = w$ is in W. Thus

$$P^0(w) = \frac{1}{\#G} \sum_{g \in G} \rho(g) \circ P \circ \rho(g)^{-1}(w) = \frac{1}{\#G} \sum_{g \in G} w = w.$$

Since $P(V) \subset W$, and W is G-invariant, $P^0(V) \subset W$. Because P^0 is the identity on W, $P^0 \circ P^0 = P^0$. Let $W^0 = \ker P^0$. By Lemma 3.11, $V \cong P^0(V) \oplus \ker P^0 = W \oplus W^0$. To complete the proof of the theorem, we show that W^0 is a G-invariant subspace. First, we show that, for all $h \in G$, $\rho(h) \circ P^0 = P^0 \circ \rho(h)$:

$$\rho(h) \circ P^0 \circ \rho(h)^{-1} = \rho(h) \circ \left(\frac{1}{\#G} \sum_{g \in G} \rho(g) \circ P \circ \rho(g)^{-1} \right) \circ \rho(h^{-1})$$

$$= \frac{1}{\#G} \sum_{g \in G} \rho(h) \circ \rho(g) \circ P \circ \rho(g^{-1}) \circ \rho(h^{-1})$$

$$= \frac{1}{\#G} \sum_{g \in G} \rho(hg) \circ P \circ \rho(hg)^{-1} = P^0.$$

Thus $\rho(h) \circ P^0 \circ \rho(h)^{-1} = P^0$, and so $\rho(h) \circ P^0 = P^0 \circ \rho(h)$ for any $h \in G$. If $w^0 \in W^0$, then $P^0(\rho(h)(w^0)) = \rho(h)(P^0(w^0)) = \rho(h)(\mathbb{O}) = \mathbb{O}$, and so $\rho(h)(w^0)$ is in W^0.

Let $T: V \to W \oplus W^0$ be the isomorphism defined as in Lemma 3.11, $T(v) = (P^0(v), v - P^0(v))$. Since P^0 is G-equivariant, T is G-equivariant. Thus ρ is equivalent to $\rho^W \oplus \rho^{W^0}$. □

Corollary 4.7 *When the characteristic of the field $\mathbb{F}$ does not divide the order of G, a representation $\rho: G \to \mathrm{Gl}(V)$ is irreducible if and only if the only G-invariant subspaces of V are V and $\{\mathbb{O}\}$.*

For a finite group G, consider the regular representation $\rho: G \to \mathrm{Gl}(\mathbb{F}G)$. In the vector space $\mathbb{F}G$, we have the nonzero element $s = \sum_{g \in G} g$. Let W denote the subspace of $\mathbb{F}G$ spanned by $\{s\}$. If $h \in G$, then $\rho(h)(s) = \rho(h)(\sum_{g \in G} g) = \sum_{g \in G} hg = s$, and so W is a G-invariant subspace of $\mathbb{F}G$. The regular representation is not irreducible.

We say that a representation $\rho: G \to \mathrm{Gl}(V)$ is *completely reducible* if for any G-invariant subspace $W \subset V$ and $\{\mathbb{O}\} \neq W \neq V$, there is another G-invariant subspace W^0 such that $V \cong W \oplus W^0$ and $\rho = \rho^W \oplus \rho^{W^0}$. Maschke's Theorem tells us that as long as the characteristic of $\mathbb{F}$ does not divide the order of G, a representation $\rho: G \to \mathrm{Gl}(V)$ is completely reducible.

For an example of a representation that is not completely reducible, choose a prime p and let $C_p = \langle g \rangle$ be the cyclic group of order p. The representation $\rho \colon C_p \to \mathrm{Gl}_2(\mathbb{F}_p)$ where $\rho(g) = \begin{bmatrix} 1 & 1 \\ 0 & 1 \end{bmatrix}$, a shear, has $\rho(g^k) = \begin{bmatrix} 1 & k \\ 0 & 1 \end{bmatrix}$, with a single common eigenvector $\begin{bmatrix} 1 \\ 0 \end{bmatrix}$. If ρ were completely reducible, there would exist a second common eigenvector for the shear. (Can you say why?) But the characteristic polynomial of a shear is $(x-1)^2$, and there is only one eigenvector. Of course, the characteristic of $\mathbb{F}_p$ divides $\#C_p$, so the proof of Maschke's Theorem does not apply.

By induction on the dimension of V we can prove the following.

Proposition 4.8 *If* $\rho \colon G \to \mathrm{Gl}(V)$ *is completely reducible, then* ρ *is equivalent to* $\rho_1^{W_1} \oplus \rho_2^{W_2} \oplus \cdots \oplus \rho_k^{W_k}$ *with* $V \cong W_1 \oplus W_2 \oplus \cdots \oplus W_k$. *and each* $\rho_i^{W_i} \colon G \to \mathrm{Gl}(W_i)$ *irreducible.*

Get to know some irreducible representations

Let $G = \Sigma_3$, the group of permutations of $\{1, 2, 3\}$. Show that the following complex representations are irreducible representations of G.

1. There is the trivial representation $\rho_0 \colon \Sigma_3 \to \mathrm{Gl}(\mathbb{C}) = \mathbb{C}^\times$ given by $\rho_0(\sigma) = \mathrm{Id} = 1$. Why is the trivial representation $\rho_0 \colon \Sigma_3 \to \mathrm{Gl}_n(\mathbb{C})$ for $n > 1$ not irreducible?

2. Define the sign representation by $\rho_\pm \colon \Sigma_3 \to \mathbb{C}^\times$ given by $\rho_\pm(\sigma) = \mathrm{sgn}(\sigma)$, where the sign of a permutation σ is 1 if σ is even and -1 if σ is odd.

3. Let us construct a two-dimensional representation of Σ_3. The group Σ_3 acts on $\mathbb{C}^3$ by permuting entries, that is, $\rho(\sigma)(z_1, z_2, z_3) = (z_{\sigma(1)}, z_{\sigma(2)}, z_{\sigma(3)})$. Let $W = \{(a, b, c) \in \mathbb{C}^3 \mid a + b + c = 0\}$. Show that W has dimension 2 with a basis consisting of $\mathbf{b}_1 = \mathbf{e}_3 - \mathbf{e}_1$ and $\mathbf{b}_2 = \mathbf{e}_2 - \mathbf{e}_1$. Show that W is a Σ_3-invariant subspace of $\mathbb{C}^3$. Let $W^\perp = \{(a, a, a) \mid a \in \mathbb{C}\}$. Show that $\mathbb{C}^3 \cong W^\perp \oplus W$ and that Σ_3 acts trivially on $W^\perp$. Prove that $\rho^W \colon \Sigma_3 \to \mathrm{Gl}(W)$ induced by permuting indices is an irreducible representation. (Hint: To be reducible, a degree two representation must have a degree one G-invariant subspace.)

4. More generally, suppose $\rho \colon G \to \mathrm{Gl}_2(\mathbb{C})$ is a matrix representation for which there exist elements g and h in G for which $\rho(g) \circ \rho(h) \neq \rho(h) \circ \rho(g)$. Prove that ρ is irreducible.

5. Show that if ρ and ρ' are equivalent representations of a finite group G, then ρ is irreducible if and only if ρ' is irreducible.

Maschke's Theorem tells us that, under mild conditions, every representation is a sum of irreducible representations. In this way the irreducible representations are analogous to the prime numbers for factorizations of the integers. An indispensable property of irreducible representations is the following result of ISSAI SCHUR (1894–1939) [72], who established the foundations of representation theory.

Schur's Lemma *Suppose $\rho_1: G \to \mathrm{Gl}(V)$ and $\rho_2: G \to \mathrm{Gl}(W)$ are irreducible representations of a finite group G. Suppose $T: V \to W$ is a G-equivariant mapping. Then either T is an isomorphism, or T is the zero mapping. If V is a vector space over $\mathbb{C}$ and $T: V \to V$ is a G-equivariant isomorphism, then $T = \lambda\,\mathrm{Id}$ for some complex number λ.*

Proof Suppose $T(v) \neq \mathbb{O}$ for some $v \in V$. Let $T(V) = \{w \in W \mid w = T(v)$ for some $v \in V\}$ denote the image of T. The key point of the proof is that $T(V)$ and $\ker T$ are G-invariant subspaces of W and V, respectively. To see that $T(V)$ is a G-invariant subspace of W, let $w \in T(V)$. Since T is G-equivariant, $\rho_2(g) \circ T = T \circ \rho_1(g)$ for all $g \in G$. Therefore $\rho_2(g)(w) = \rho_2(g)(T(v)) = T(\rho_1(g)(v))$, and so $\rho_2(g)(w)$ is also in $T(V)$.

If $v \in \ker T$ and $g \in G$, then $T(\rho_1(g)(v)) = \rho_2(g)(T(v)) = \rho_2(g)(\mathbb{O}) = \mathbb{O}$, and $\ker T$ is a G-invariant subspace.

If ρ_2 is an irreducible representation, then $T(V) \subset W$ is a G-invariant subspace, and hence either $T(V) = W$ or $T(V) = \{\mathbb{O}\}$. Assuming that there is $v \in V$ such that $T(v) \neq \mathbb{O}$, we have that $T(V) = W$ and T is surjective.

If ρ_1 is irreducible, then $\ker T \subset V$ is a G-invariant subspace, and hence either $\ker T = V$ or $\ker T = \{\mathbb{O}\}$. Assuming that $T(v) \neq \mathbb{O}$ for some $v \in V$, we have that $\ker T = \{\mathbb{O}\}$ and T is injective. Thus, under the assumptions, T is an isomorphism.

Suppose $\mathbb{F} = \mathbb{C}$ and $T: V \to V$ is a G-equivariant isomorphism. Choose a basis $B = \{b_1, b_2, \ldots, b_n\}$ for V and let $c_B: V \to \mathbb{C}^n$ be the coordinate mapping. Then there is a matrix M_T making the following diagram commute:

$$
\begin{array}{ccc}
V & \xrightarrow{\ T\ } & V \\
{\scriptstyle c_B}\downarrow & & \downarrow{\scriptstyle c_B} \\
\mathbb{C}^n & \xrightarrow[M_T]{} & \mathbb{C}^n
\end{array}
$$

Let z be a complex variable. Let $\det(z\,\mathrm{Id} - M_T) = p(z)$ be the *characteristic polynomial* of the matrix. By the Fundamental Theorem of Algebra there is $\lambda \in \mathbb{C}$ for which $\det(\lambda\,\mathrm{Id} - M_T) = 0$. Thus $\ker(\lambda\,\mathrm{Id} - M_T)$ is a nontrivial subspace of V. We show that $\lambda\,\mathrm{Id} - M_T$ is G-equivariant. The coordinate isomorphism c_B determines an equivalent representation $\bar{\rho}_1 : G \to \mathrm{Gl}_n(\mathbb{C})$ by

$$\bar{\rho}_1(g) \colon \mathbb{C}^n \xrightarrow{c_B^{-1}} V \xrightarrow{\rho_1(g)} V \xrightarrow{c_B} \mathbb{C}^n.$$

To see that $\lambda \, \mathrm{Id} - M_T$ is G-equivariant, consider

$$\begin{aligned}
\bar{\rho}_1&(g) \circ (\lambda \, \mathrm{Id} - M_T) \\
&= c_B \circ \rho_1(g) \circ c_B^{-1} \circ \lambda \, \mathrm{Id} - c_B \circ \rho_1(g) \circ c_B^{-1} \circ M_T \\
&= \lambda \, \mathrm{Id} \circ c_B \circ \rho_1(g) \circ c_B^{-1} - c_B \circ \rho_1(g) \circ c_B^{-1} \circ c_B \circ T \circ c_B^{-1} \\
&= \lambda \, \mathrm{Id} \circ \bar{\rho}_1(g) - c_B \circ T \circ \rho_1(g) \circ c_B^{-1} \\
&= \lambda \, \mathrm{Id} \circ \bar{\rho}_1(g) - M_T \circ c_B \circ \rho_1(g) \circ c_B^{-1} = (\lambda \, \mathrm{Id} - M_T) \circ \bar{\rho}_1(g).
\end{aligned}$$

When $(\lambda \, \mathrm{Id} - M_T)(\mathbf{v}) = \mathbb{O}$, $\mathbf{v}$ is in the kernel of $\lambda \, \mathrm{Id} - M_T$, which is a G-invariant subspace of $\mathbb{C}^n$. But if ρ_1 is irreducible, then so is $\bar{\rho}_1$, and $\mathbb{C}^n = \ker(\lambda \, \mathrm{Id} - M_T)$, that is, $M_T \mathbf{v} = \lambda \mathbf{v}$ for all $\mathbf{v} \in \mathbb{C}^n$. For $T \colon V \to V$, we see that

$$\begin{aligned}
T(v) &= c_B^{-1} \circ c_B \circ T \circ c_B^{-1} \circ c_B(v) = c_B^{-1} \circ M_T \circ c_B(v) \\
&= c_B^{-1} \circ \lambda \, \mathrm{Id} \circ c_B(v) = \lambda c_B^{-1} \circ c_B(v) = \lambda v,
\end{aligned}$$

and it follows that $T(v) = \lambda v$ for all $v \in V$. $\qquad\qquad\square$

Although it is simple to state and prove, Schur's Lemma has some deep consequences.

Corollary 4.9 *Let G be a finite abelian group, and let V be a finite-dimensional vector space over $\mathbb{C}$. If $\rho \colon G \to \mathrm{Gl}(V)$ is an irreducible representation, then $\dim V = 1$.*

Proof Suppose g and h are in G. Then for all $v \in V$,

$$\rho(g) \circ \rho(h)(v) = \rho(gh)(v) = \rho(hg)(v) = \rho(h) \circ \rho(g)(v).$$

If we write $\rho(g) = T \colon V \to V$, then we have shown that $T \circ \rho(h) = \rho(h) \circ T$ for all $h \in G$. Hence $\rho(g)$ is G-equivariant. By Schur's Lemma, $\rho(g) = \lambda_g \, \mathrm{Id}$ for some $\lambda_g \in \mathbb{C}$. Because λ_g times any subspace of V is equal to the same subspace, every subspace of V is G-invariant. If ρ is irreducible, then V must have dimension one. $\qquad\qquad\square$

The Fundamental Theorem of finitely generated abelian groups [31] implies that a finite abelian group is isomorphic to a product $(\mathbb{Z}/n_1\mathbb{Z}) \times (\mathbb{Z}/n_2\mathbb{Z}) \times \cdots \times (\mathbb{Z}/n_t\mathbb{Z})$. On a single factor, let $[1] \in \mathbb{Z}/n_i\mathbb{Z}$ be the canonical generator of this cyclic group. Then $n_i[1] = [0]$ and $\rho(n_i[1]) = \rho([1])^{n_i} = \rho([0]) = 1 \in \mathbb{C}^\times$. So $\rho([1]) = \lambda_i$, where $\lambda_i = e^{2\pi i j/n_i}$ is an n_ith root of unity. If $\gcd(j, n_i) = 1$, then the representation is faithful on this factor.

In the general case, let λ_i be a choice of an n_ith root of unity for each i. Let $V_i = \mathbb{C}$. We define a representation $\rho\colon (\mathbb{Z}/n_1\mathbb{Z}) \times (\mathbb{Z}/n_2\mathbb{Z}) \times \cdots \times (\mathbb{Z}/n_t\mathbb{Z}) \to \mathrm{Gl}(V_1 \oplus V_2 \oplus \cdots \oplus V_t) \cong \mathrm{Gl}_t(\mathbb{C})$ by

$$
\rho\big(([a_1], [a_2], \ldots, [a_t])\big) = \begin{bmatrix} \lambda_1^{a_1} & & & \\ & \lambda_2^{a_2} & & \mathbb{O} \\ \mathbb{O} & & \ddots & \\ & & & \lambda_t^{a_t} \end{bmatrix}.
$$

The properties of product groups, roots of unity, and diagonal matrices imply that ρ is a representation.

Maschke's Theorem tells us that any completely reducible representation $\rho\colon G \to \mathrm{Gl}(V)$ decomposes as a sum $V = W_1 \oplus \cdots \oplus W_k$ with each W_i an irreducible representation of G. How many irreducible representations can a finite group have?

Lemma 4.10 *Suppose $\rho\colon G \to \mathrm{Gl}(V)$ is a representation and $V = W_1 \oplus \cdots \oplus W_k$ with each W_i irreducible. If $U \subset V$ is an irreducible subrepresentation of V, then $U \cong W_j$ for some j.*

Proof Let $u \in U$, $u \neq \mathbb{O}$. Because $V = W_1 \oplus \cdots \oplus W_k$, we can write $u = w_1 + \cdots + w_k$ for unique vectors $w_i \in W_i$, $1 \leq i \leq k$. There must be an index j with $w_j \neq \mathbb{O}$. Let $\mathrm{pr}_j\colon V \to W_j$ denote the projection of V onto W_j given by $\mathrm{pr}_j(w_1 + \cdots + w_k) = w_j$. Then $\mathrm{pr}_j(u) = w_j \neq \mathbb{O}$. Because $\mathrm{pr}_j\colon V \to W_j$ is G-equivariant and U is a G-invariant subspace of V, $\mathrm{pr}_j|_U\colon U \to W_j$ is G-equivariant and nonzero. Because both U and W_j are irreducible, by Schur's Lemma we know that $\mathrm{pr}_j|_U$ is an isomorphism. $\square$

This leads us to the following remarkable result.

Theorem 4.11 *For a finite group G and a field $\mathbb{F}$ for which the characteristic of $\mathbb{F}$ does not divide the order of G, let $\rho_{\mathrm{reg}}\colon G \to \mathrm{Gl}(\mathbb{F}G)$ be the regular representation and suppose $\mathbb{F}G = U_1 \oplus \cdots \oplus U_t$ with each U_i an irreducible representation of G. If $\rho\colon G \to \mathrm{Gl}(W)$ is any irreducible representation of G, then $W \cong U_j$ for some j.*

Proof Choose $w \in W$ with $w \neq \mathbb{O}$. Define the mapping $\phi\colon \mathbb{F}G \to W$ by

$$
\phi\bigg(\sum_{g \in G} \alpha_g g\bigg) = \sum_{g \in G} \alpha_g \rho(g)(w).
$$

The reader will want to check that ϕ is a linear mapping. Suppose $h \in G$. Then

$$\phi \circ \rho_{\text{reg}}(h)\left(\sum_{g \in G} \alpha_g g\right) = \phi\left(\sum_{g \in G} \alpha_g hg\right) = \sum_{g \in G} \alpha_g \rho(h) \circ \rho(g)(w)$$

$$= \rho(h)\left(\sum_{g \in G} \alpha_g \rho(g)(w)\right) = \rho(h) \circ \phi\left(\sum_{g \in G} \alpha_g g\right).$$

It follows that $\phi \circ \rho_{\text{reg}}(h) = \rho(h) \circ \phi$ for all $h \in G$ and ϕ is a G-equivariant mapping. Because W is an irreducible representation and $\phi(\mathbb{F}G)$ is a nontrivial ($w = 1\rho(e)(w)$) G-invariant subrepresentation, $W = \phi(\mathbb{F}G)$. Since ϕ is G-equivariant, $\ker \phi \subset \mathbb{F}G$ is a G-invariant subspace. By Maschke's Theorem we can write $\mathbb{F}G \cong \ker \phi \oplus U$ with U a subrepresentation of $\mathbb{F}G$. Furthermore, $U \cong \phi(\mathbb{F}G) = W$. Because W is irreducible, so is U, and, by the previous lemma, $W \cong U_j$ for some j. □

Corollary 4.12 *If G is a finite group, there are only finitely many nonisomorphic irreducible representations of G.*

Because $\dim \mathbb{F}G = \#G$, the order of G plays a role in controlling the number of irreducible representations of G. What kind of information can we determine about the collection of irreducible representations? For this, we explore some other aspects of linear algebra.

4.2 Characters

Each representation $\rho: G \to \text{Gl}(V)$ is equivalent to a matrix representation after choosing a basis B for V: $\rho: G \to \text{Gl}(V) \xrightarrow{c_B} \text{Gl}_n(\mathbb{F})$. For an $n \times n$ matrix $A = (a_{ij})$, we define its *trace* $\text{tr}(A) = \sum_{i=1}^{n} a_{ii}$, the sum of the diagonal entries of the matrix. The *character* of a representation ρ is the function

$$\chi_\rho: G \xrightarrow{\rho} \text{Gl}(V) \xrightarrow{c_B^*} \text{Gl}_n(\mathbb{F}) \xrightarrow{\text{tr}} \mathbb{F}.$$

To see that χ_ρ is independent of the choice of basis, we prove the following:

Proposition 4.13 *For $n \times n$ matrices A and B, $\text{tr}(AB) = \text{tr}(BA)$.*

Proof The diagonal entries of AB take the form $\sum_{j=1}^{n} a_{ij} b_{ji}$. Then

$$\text{tr}(AB) = \sum_{i=1}^{n} \sum_{j=1}^{n} a_{ij} b_{ji} = \sum_{j=1}^{n} \sum_{i=1}^{n} b_{ji} a_{ij} = \text{tr}(BA).$$

□

Corollary 4.14 *If P is an invertible $n \times n$ matrix, then $\text{tr}(PAP^{-1}) = \text{tr}(A)$.*

Proof $\text{tr}(PAP^{-1}) = \text{tr}(P(AP^{-1})) = \text{tr}((AP^{-1})P) = \text{tr}(A(P^{-1}P)) = \text{tr}(A)$.   □

If B' is another basis for V, let $M'_{\rho(g)} = c_{B'} \circ \rho(g) \circ c_{B'}^{-1}$. We obtain the diagram

$$
\begin{array}{ccccccc}
V & \xrightarrow{\ \text{Id}\ } & V & \xrightarrow{\ \rho(g)\ } & V & \xrightarrow{\ \text{Id}\ } & V \\
\downarrow{\scriptstyle c_{B'}} & & \downarrow{\scriptstyle c_B} & & \downarrow{\scriptstyle c_B} & & \downarrow{\scriptstyle c_{B'}} \\
\mathbb{F}^n & \xrightarrow{\ C_{B'}^B\ } & \mathbb{F}^n & \xrightarrow{\ M_{\rho(g)}\ } & \mathbb{F}^n & \xrightarrow{\ (C_{B'}^B)^{-1}\ } & \mathbb{F}^n
\end{array}
$$

Then $M'_{\rho(g)} = (C_{B'}^B)^{-1} \circ M_{\rho(g)} \circ C_{B'}^B$, and, by Corollary 4.14, $\text{tr}(M'_{\rho(g)}) = \text{tr}(M_{\rho(g)})$. For convenience, we write $\chi_\rho(g) = \text{tr}(\rho(g)) = \text{tr}(M_{\rho(g)})$ for any choice of basis giving the matrix.

As an example, consider the permutation representation of Σ_3 on $\mathbb{C}^3$. In the canonical basis $\{\mathbf{e}_1, \mathbf{e}_2, \mathbf{e}_3\}$, $\rho(\sigma)(\mathbf{e}_i) = \mathbf{e}_{\sigma(i)}$. We obtain the following chart:

$\sigma =$	Id	$(1,2)$	$(1,3)$	$(2,3)$	$(1,2,3)$	$(1,3,2)$
$\rho(\sigma) =$	$\begin{bmatrix}1&0&0\\0&1&0\\0&0&1\end{bmatrix}$	$\begin{bmatrix}0&1&0\\1&0&0\\0&0&1\end{bmatrix}$	$\begin{bmatrix}0&0&1\\0&1&0\\1&0&0\end{bmatrix}$	$\begin{bmatrix}1&0&0\\0&0&1\\0&1&0\end{bmatrix}$	$\begin{bmatrix}0&0&1\\1&0&0\\0&1&0\end{bmatrix}$	$\begin{bmatrix}0&1&0\\0&0&1\\1&0&0\end{bmatrix}$
$\text{tr}(\rho(g)) =$	3	1	1	1	0	0

Notice that the conjugacy classes in Σ_3 are $[(1,2)] = \{(1,2), (1,3), 2,3)\}$ and $[(1,2,3)] = \{(1,2,3), (1,3,2)\}$ and the character χ_ρ is constant on conjugacy classes. A tidier chart is given by

$[\sigma] =$	$[\text{Id}]$	$[(1,2)]$	$[(1,2,3)]$
$\chi_\rho(g)$	3	1	0

This reduction is a general fact.

Corollary 4.15 *Suppose $g, h \in G$ and $k = hgh^{-1}$. Then $\chi_\rho(k) = \chi_\rho(g)$, that is, χ_ρ is constant on members of a conjugacy class.*

Proof $\chi_\rho(k) = \chi_\rho(hgh^{-1}) = \text{tr}(\rho(h) \circ \rho(g) \circ \rho(h)^{-1}) = \text{tr}(\rho(g)) = \chi_\rho(g)$.
                                                                    □

The set of conjugacy classes of the group G is denoted by $\text{Cl}(G) = \{[g] \mid g \in G\}$, where $[g] = \{xgx^{-1} \mid x \in G\}$. When $\mathbb{F} = \mathbb{C}$, consider any function $f : G \to \mathbb{C}$ that satisfies $f(hgh^{-1}) = f(g)$ for all $g, h \in G$. Such a function is called a *class function*. By Corollary 4.15 a character induced by a complex representation is a class function. Denote the set of all class functions on the finite group G by

$$\text{Cf}(G) = \{f \colon G \to \mathbb{C} \mid f(h) = f(ghg^{-1}) \text{ for all } h, g \in G\}.$$

Proposition 4.16 *The set of class functions is a Hermitian inner product space over $\mathbb{C}$ of dimension $\#\text{Cl}(G)$, the number of conjugacy classes of G.*

Proof The set $\text{Cf}(G)$ is a complex vector space with pointwise addition and multiplication by scalars on complex-valued functions given by

$$(af_1 + f_2)(g) = af_1(g) + f_2(g).$$

The zero function $\mathbb{O}(g) = 0$ sends everything in G to zero and acts as the additive identity in $\text{Cf}(G)$. A Hermitian inner product on $\text{Cf}(G)$ is defined by

$$\langle f_1, f_2 \rangle = \frac{1}{\#G} \sum_{g \in G} f_1(g) \overline{f_2(g)}.$$

The inner product properties follow easily. Also,

$$\langle f, f \rangle = \frac{1}{\#G} \sum_{g \in G} f(g) \overline{f(g)} = \frac{1}{\#G} \sum_{g \in G} |f(g)|^2 \geq 0,$$

and $\langle f, f \rangle$ equals zero if and only if $f(g) = \mathbb{O}(g)$ for all $g \in G$.

A basis for $\text{Cf}(G)$ may be given by defining $\delta^{[g]}(h) = 1$ if $h \in [g]$ and 0 if $h \notin [g]$. This is the *indicator function* for the conjugacy class $[g]$. If $f \in \text{Cf}(G)$, then consider

$$F(h) = \frac{1}{\#[h]} \sum_{g \in G} f(g) \delta^{[g]}(h).$$

By its definition, $F(h) = f(h)$ for $h \in [g]$, and the set $\{\delta^{[g]}\}$ spans $\text{Cf}(G)$. If $a_1 \delta^{[g_1]}(x) + \cdots + a_s \delta^{[g_s]}(x) = 0$ for all $x \in G$, then evaluating at $x = g_i$ for each i implies that $a_i = 0$, and the set $\{\delta^{[g]} \mid [g] \in \text{Cl}(G)\}$ is a basis for $\text{Cf}(G)$. $\qquad\square$

The character of a complex representation is a class function.

Get to know characters

1. Let $\rho \colon D_8 \to \text{Gl}_2(\mathbb{R})$ denote the geometric representation of D_8 determined by

$$\rho(r) = \begin{bmatrix} \cos(\pi/2) & -\sin(\pi/2) \\ \sin(\pi/2) & \cos(\pi/2) \end{bmatrix} = \begin{bmatrix} 0 & -1 \\ 1 & 0 \end{bmatrix}, \quad \rho(f) = \begin{bmatrix} 0 & -1 \\ -1 & 0 \end{bmatrix}.$$

The group D_8 is given by $\{e, r, r^2, r^3, f, fr, fr^2, fr^3\}$. Work out the
rest of the matrices implied by the values of ρ on r and f. Then compute $\chi_\rho(g)$ for all $g \in D_8$.

2. Do the same for D_{10} where $\rho(r) = \begin{bmatrix} \cos(2\pi/5) & -\sin(2\pi/5) \\ \sin(2\pi/5) & \cos(2\pi/5) \end{bmatrix}$ and $\rho(f) = \begin{bmatrix} 1 & 0 \\ 0 & -1 \end{bmatrix}$.

3. Find all of the degree one representations of the cyclic group $C_4 \cong \mathbb{Z}/4\mathbb{Z}$ of order four. A degree one representation is its own character. How many distinct characters did you find?

4. Find all of the degree one representations of the cyclic group $C_5 \cong \mathbb{Z}/5\mathbb{Z}$ of order five.

5. Prove that $f: G \to \mathbb{C}$ is a class function if and only if $f(gh) = f(hg)$.

A character is a list of complex numbers (or elements of the base field) associated with a representation and indexed over the conjugacy classes of G. How well do such lists capture the inner workings of a group?

Proposition 4.17 *If $\rho_1: G \to \mathrm{Gl}(V)$ and $\rho_2: G \to \mathrm{Gl}(W)$ are equivalent representations, then $\chi_{\rho_1} = \chi_{\rho_2}$.*

Proof The G-equivariant isomorphism $T: V \to W$ satisfies $T \circ \rho_1(g) = \rho_2(g) \circ T$ for all $g \in G$. Then $\rho_2(g) = T \circ \rho_1(g) \circ T^{-1}$. The associated matrices are conjugates. $\square$

Proposition 4.18 *If $\rho = \rho_1 \oplus \rho_2: G \to \mathrm{Gl}(V_1 \oplus V_2)$ is a sum of representations, then $\chi_\rho = \chi_{\rho_1} + \chi_{\rho_2}$. If $\rho = \rho_1 \otimes \rho_2: G_1 \times G_2 \to \mathrm{Gl}(V \otimes W)$, then $\chi_\rho = \chi_{\rho_1} \chi_{\rho_2}$.*

Proof Choose bases for V_1 and V_2; their disjoint union is a basis for $V = V_1 \oplus V_2$, and the matrix representation for ρ takes the form

$$M_g^\rho = \begin{bmatrix} M_g^{\rho_1} & \mathbb{O} \\ \mathbb{O} & M_g^{\rho_2} \end{bmatrix}.$$

It follows that $\mathrm{tr}(M_g^\rho) = \mathrm{tr}(M_g^{\rho_1}) + \mathrm{tr}(M_g^{\rho_2}) = \chi_{\rho_1}(g) + \chi_{\rho_2}(g)$.

The statement for $\rho_1 \otimes \rho_2$ follows from the block structure of $A \otimes B$ for square matrices A and B. (See Chapter 3.) The diagonal blocks take the form $a_{ii} B$, and so the trace of $A \otimes B$ is $\sum_{i=1}^n a_{ii} \mathrm{tr}(B) = \mathrm{tr}(A) \mathrm{tr}(B)$. $\square$

When G acts on a finite set X, the permutation representation

$$\bar{\rho} = \Xi \circ \rho: G \xrightarrow{\rho} \Sigma(X) \xrightarrow{\Xi} \mathrm{Gl}(\mathbb{F}X)$$

has a character with a combinatorial description. The element $g \in G$ acts on a vector in $\mathbb{F}X$ by

$$\bar{\rho}(g)(a_1 x_1 + a_2 x_2 + \cdots + a_m x_m)$$
$$= a_1 \rho(g)(x_1) + a_2 \rho(g)(x_2) + \cdots + a_m \rho(g)(x_m).$$

The matrix determined by $\bar{\rho}(g)$ with X as a basis has 1 in a diagonal entry exactly when $\rho(g)(x_i) = x_i$, that is, x_i is an element fixed by $\rho(g)$. The trace of $\bar{\rho}(g)$ is the number of elements fixed by $\rho(g)$, $\#\text{Fix}(g)$.

Proposition 4.19 *If G acts on a finite set X, then the associated permutation representation $\bar{\rho} \colon G \to \text{Gl}(\mathbb{F}X)$ has character $\chi_{\bar{\rho}}(g) = \#\text{Fix}(g) = \#\{x \in X \mid \rho(g)(x) = x\}$.*

In the case of the natural action of Σ_n on $[n] = \{1, 2, \ldots, n\}$ and a permutation $\sigma \in \Sigma_n$, we express σ as a product of disjoint cycles. The fixed elements of σ are the elements of $[n]$ that do not appear in any of the cycles. For example, for $(12)(357) \in \Sigma_8$, $\text{Fix}((12)(357)) = \{4, 6, 8\}$ and $\chi_{\rho}((12)(357)) = 3$.

4.3 Values of Characters*

The values of characters contain information about the group and its representation. For example,

$$\chi_{\rho}(e) = \text{tr}(\text{Id}_V) = \dim V,$$

so $\chi_{\rho}(e)$ gives the degree of the representation. Suppose $\rho \colon G \to \text{Gl}_n(\mathbb{C})$ is a complex matrix representation of G. The trace of the $n \times n$ matrix $\rho(g)$ is related to the $(n-1)$th coefficient in the characteristic polynomial of $\rho(g)$. Let $p(z) = \det(z \, \text{Id} - \rho(g)) = z^n - a_{n-1} z^{n-1} + \cdots \pm a_0$. Letting $z = 0$, we see that $a_0 = (-1)^n \det(\rho(g))$. Over the complex numbers, we have that the matrix $\rho(g)$ is similar to an upper triangular matrix. Proceed by induction on n. When $n = 1$, a 1×1 matrix is already upper triangular. Recall that the Fundamental Theorem of Algebra tells us that $p(z)$ has a complex root λ_1 and hence an eigenvalue for which there is an eigenvector $\mathbf{v}_1$. Extend this vector to a basis $B = \{\mathbf{v}_1, \mathbf{v}_2, \ldots, \mathbf{v}_n\}$ for $\mathbb{C}^{\times n}$. In this basis, $c_B \circ \rho(g) \circ c_B^{-1} = \left[\begin{array}{c|c} \lambda_1 & * \\ \hline 0 & C \end{array} \right]$. By induction there is a basis $B' = \{\mathbf{v}_1, \mathbf{w}_2, \ldots, \mathbf{w}_n\}$ for which the $(n-1) \times (n-1)$ matrix C is conjugate to an upper triangular matrix and $M = c_{B'} \circ \rho(g) \circ c_{B'}^{-1} = \left[\begin{array}{c|c} \lambda_1 & * \\ \hline 0 & C' \end{array} \right]$. Over the basis B',

$$\det(z \, \text{Id} - \rho(g)) = \det(c_{B'}) \det(z \, \text{Id} - \rho(g)) \det(c_{B'})^{-1}$$
$$= \det(c_{B'}(z \, \text{Id} - \rho(g)) c_{B'}^{-1}) = \det(z \, \text{Id} - (c_{B'} \circ \rho(g) \circ c_{B'}^{-1}))$$
$$= (z - \lambda_1)(z - m_{22}) \cdots (z - m_{nn}).$$

Furthermore,

$$p(z) = (z - \lambda_1) \cdots (z - m_{nn}) = z^n - (m_{11} + \cdots + m_{nn})z^{n-1} + \cdots .$$

Since $\mathrm{tr}(\rho(g)) = \mathrm{tr}(c_{B'} \circ \rho(g) \circ c_{B'}^{-1}) = \lambda_1 + \cdots + m_{nn} = a_{n-1}$, $\mathrm{tr}(\rho(g))$ is the sum of the eigenvalues of $\rho(g)$.

The eigenvalues of a matrix representation are among a special class of complex numbers.

Definition 4.20 An *algebraic number* $\alpha \in \mathbb{C}$ is the root of a monic polynomial with rational coefficients. If α is the root of a monic polynomial with integer coefficients, we say that α is an *algebraic integer*.

For any complex number ζ, let $\mathrm{ev}_\zeta \colon \mathbb{Q}[x] \to \mathbb{C}$ be the ring homomorphism $\mathrm{ev}_\zeta(p(x)) = p(\zeta)$. The kernel of ev_ζ is an ideal in $\mathbb{Q}[x]$, and so $\ker \mathrm{ev}_\zeta = (m_\zeta(x))$ because $\mathbb{Q}[x]$ is a principal ideal domain. The polynomial $m_\zeta(x)$ may be zero, and then ev_ζ is injective, and ζ generates a subring of $\mathbb{C}$ isomorphic to $\mathbb{Q}[x]$. In this case, we say that ζ is a *transcendental* number. For example, the works of CHARLES HERMITE (1822–1901) and FERDINAND VON LINDEMANN (1852–1939) demonstrated that e and π are transcendental. This settled the problem from Ancient Greece of *squaring-the-circle*. When $m_\zeta(x)$ is nonzero, it may be taken to be monic, and we call this the *minimal polynomial* satisfied by ζ.

Theorem 4.21 *The sum of algebraic numbers (integers) is an algebraic number (integer). The product of algebraic numbers (integers) is an algebraic number (integer). If $\zeta \neq 0$ is an algebraic number, then so is $1/\zeta$.*

Proof (Following [49]) Suppose ζ and α are algebraic numbers (integers). Let $m_\zeta(x)$ and $m_\alpha(x)$ in $\mathbb{Q}[x]$ ($\mathbb{Z}[x]$) denote the minimal monic polynomials for which $m_\zeta(\zeta) = 0 = m_\alpha(\alpha)$. Denote the companion matrices for these polynomials by C_ζ and C_α, respectively. As we observed in Chapter 3, the characteristic polynomials of C_ζ and C_α are $m_\zeta(x)$ and $m_\alpha(x)$. Since ζ and α are eigenvalues, there are nonzero vectors $\mathbf{u}$ and $\mathbf{v}$ such that $C_\zeta \mathbf{u} = \zeta \mathbf{u}$ and $C_\alpha \mathbf{v} = \alpha \mathbf{v}$. The tensor product matrix $C_\zeta \otimes C_\alpha$ has rational (integral) entries and satisfies

$$C_\zeta \otimes C_\alpha(\mathbf{u} \otimes \mathbf{v}) = C_\zeta \mathbf{u} \otimes C_\alpha \mathbf{v} = \zeta \mathbf{u} \otimes \alpha \mathbf{v} = (\zeta \alpha)(\mathbf{u} \otimes \mathbf{v}).$$

Thus $\zeta \alpha$ is an eigenvalue of the matrix $C_\zeta \otimes C_\alpha$ and hence a root of a monic rational (integral) polynomial, and $\zeta \alpha$ is an algebraic number (integer).

For the sum $\zeta + \alpha$, consider the matrix $C_\zeta \otimes \mathrm{Id} + \mathrm{Id} \otimes C_\alpha$. Then

$$(C_\zeta \otimes \mathrm{Id} + \mathrm{Id} \otimes C_\alpha)(\mathbf{u} \otimes \mathbf{v}) = (\zeta \mathbf{u}) \otimes \mathbf{v} + \mathbf{u} \otimes (\alpha \mathbf{v}) = (\zeta + \alpha)\mathbf{u} \otimes \mathbf{v}.$$

In this case, $\zeta + \alpha$ is an eigenvalue of the matrix $C_\zeta \otimes \mathrm{Id} + \mathrm{Id} \otimes C_\alpha$ with rational (integral) entries. Thus $\zeta + \alpha$ is the root of a monic rational (integer) polynomial and an algebraic number (integer).

For the reciprocal of an algebraic number, suppose $m_\zeta(\zeta) = \zeta^s + a_{s-1}\zeta^{s-1} + \cdots + a_1\zeta + a_0 = 0$. Then dividing by ζ^s, we get $a_0(1/\zeta)^s + a_1(1/\zeta)^{s-1} + \cdots + a_{s-1}(1/\zeta) + 1 = 0$. Divide by a_0 to get a monic polynomial with rational coefficients with $1/\zeta$ as a root. $\qquad\Box$

We next prove an important property of algebraic integers that will be useful later.

Proposition 4.22 *If α is an algebraic integer, and α is a rational number, then α is an integer.*

Proof Suppose α is a root of a monic polynomial $p(x) = a_0 + a_1 x + \cdots + a_{n-1}x^{n-1} + x^n$ with integral coefficients. Suppose further that $\alpha = k/l$ with $k, l \in \mathbb{Z}, l \neq 0$, and $\gcd(k, l) = 1$. Then $l^n p(k/l) = 0$ gives the relation

$$k^n = -a_0 l^n - a_1 k l^{n-1} - \cdots - a_{n-1} l k^{n-1}$$
$$= l\left(-a_0 l^{n-1} - a_1 k l^{n-2} - \cdots - a_{n-1} k^{n-1}\right).$$

Since l divides the right side, l divides k^n. Any prime that divides l must divide k. However, $\gcd(k, l) = 1$, and so $l = 1$. Thus k/l is an integer. $\qquad\Box$

For a representation $\rho \colon G \to \mathrm{Gl}(V)$, let $A = c_B \circ \rho(g) \circ c_B^{-1} \in \mathrm{Gl}_n(\mathbb{C})$ for some choice of basis for V. Suppose λ is an eigenvalue of A and $q(x)$ is a complex polynomial. Then $q(A) = c_k A^k + c_{k-1} A^{k-1} + \cdots + c_1 A + c_0 \,\mathrm{Id}$ is an $n \times n$ complex matrix. If we multiply the matrix $q(A)$ by an eigenvector $\mathbf{v}$ associated to λ, then we get

$$q(A)\mathbf{v} = c_k A^k \mathbf{v} + \cdots + c_1 A \mathbf{v} + c_0 \mathbf{v} = c_k \lambda^k \mathbf{v} + \cdots + c_1 \lambda \mathbf{v} + c_0 \mathbf{v} = q(\lambda)\mathbf{v}.$$

Because an element g is in G, a finite group, g has an order m. Then $\rho(g)^m = \rho(g^m) = \rho(e) = \mathrm{Id}$. Hence $A^m - \mathrm{Id} = \mathbb{O}$, and the matrix A satisfies the polynomial $x^m - 1$.

Define the evaluation mapping $\mathrm{ev}_A \colon \mathbb{C}[x] \to M_{n \times n}$, where $M_{n \times n}$ is the ring of $n \times n$ matrices with entries in $\mathbb{C}$, and $\mathrm{ev}_A(q(x)) = q(A)$. Because powers of a matrix commute with one another, ev_A is a ring homomorphism. The kernel of ev_A is a principal ideal $(m_A(x))$ with $m_A(x)$ the monic polynomial of least degree satisfying $m_A(A) = \mathbb{O}$. Since $A^m - \mathrm{Id} = \mathbb{O}$, we have that $x^m - 1 \in \ker \mathrm{ev}_A$ and $m_A(x)$ divides $x^m - 1$. Furthermore, if $p(x) = \det(x\,\mathrm{Id} - A)$ is the characteristic polynomial of A, then the Cayley–Hamilton Theorem [4] states that $p(A) = \mathbb{O}$ and $m_A(x)$ also divides $p(x)$. However, for an eigenvalue λ of A with associated eigenvector $\mathbf{v}$, $A\mathbf{v} = \lambda\mathbf{v}$, we have $m_A(\lambda)\mathbf{v} = m_A(A)\mathbf{v} =$

$\mathbb{O}\mathbf{v} = \mathbb{O}$, and so $m_A(\lambda) = 0$. It follows that the roots of $p(x)$ are also roots of $m_A(x)$. Since $m_A(x)$ divides $p(x)$, the eigenvalues of A are the roots of both polynomials. Because $m_A(x)$ divides $x^m - 1$, the roots of $m_A(x)$ are among the mth roots of unity.

Theorem 4.23 *The values of the character associated with a complex representation $\rho \colon G \to \mathrm{Gl}(V)$ are algebraic integers.*

Corollary 4.24 *For a complex representation $\rho \colon G \to \mathrm{Gl}(V)$, $\chi_\rho(g^{-1}) = \overline{\chi_\rho(g)}$.*

Proof Because the eigenvalues of $\rho(g)$ lie among the mth roots of unity, where m is the order of g in G, the eigenvalues of $\rho(g^{-1}) = \rho(g)^{-1}$ are the inverses of the eigenvalues of $\rho(g)$. Since $(e^{i\theta})^{-1} = e^{-i\theta} = \overline{e^{i\theta}}$, the sum of eigenvalues that determines $\chi_\rho(g^{-1})$ is the sum of the complex conjugates of the summands giving $\chi_\rho(g)$. □

As an example, consider the degree two representation of the dihedral group D_{2n} given by $\rho \colon D_{2n} \to \mathrm{Gl}_2(\mathbb{R})$, $\rho(r) = \left[\begin{smallmatrix} \cos(2\pi/n) & -\sin(2\pi/n) \\ \sin(2\pi/n) & \cos(2\pi/n) \end{smallmatrix}\right]$, and $\rho(f) = \left[\begin{smallmatrix} 1 & 0 \\ 0 & -1 \end{smallmatrix}\right]$. Then $\chi_\rho(r^j) = 2\cos(2\pi j/n) = e^{2\pi i j/n} + e^{-2\pi i j/n}$, a sum of nth roots of unity and hence an algebraic integer.

This property of the values of characters will play a role in later chapters.

Get to know algebraic integers

1. Prove that the set of algebraic numbers is a countable subfield of $\mathbb{C}$.
2. Suppose p is a prime and $\alpha = 1/\sqrt{p}$. What is the minimal polynomial over $\mathbb{Q}$ of α? What is the minimal polynomial of $\sqrt{\alpha}$? Which are algebraic integers?
3. If α is an algebraic integer and $m_\alpha(x)$ is its minimal polynomial over $\mathbb{Q}$, show that $m_\alpha(x) \in \mathbb{Z}[x]$.
4. If $\alpha \in \mathbb{C}$ is an algebraic number, show that $\mathbb{Q}[\alpha]$ is a finite-dimensional vector space over $\mathbb{Q}$.
5. If $\alpha \in \mathbb{C}$ is an algebraic number, show that $\mathbb{Q}[\alpha] = \{p(\alpha) \mid p(x) \in \mathbb{Q}[x]\}$ is a field.

4.4 Irreducible Characters

Characters are class functions. Some sums of characters correspond to sums of representations. We know that there are only finitely many irreducible representations of a finite group G. The set of characters spans a linear subspace of $\mathrm{Cf}(G)$. We next determine the dimension of that subspace and find out how it is related to the number of irreducible representations.

Definition 4.25 A character χ_ρ is an *irreducible character* if $\rho\colon G \to \mathrm{Gl}(V)$ is an irreducible representation.

Theorem 4.26 *The set of irreducible complex representations of G forms an orthonormal basis for the Hermitian inner product space* $\mathrm{Cf}(G)$.

To prepare for the proof of this theorem, we introduce some properties of the trace. In Chapter 3, we introduced the vector space $\mathrm{Hom}_{\mathbb{F}}(V, W)$ where V and W are vector spaces over a field $\mathbb{F}$:

$$\mathrm{Hom}_{\mathbb{F}}(V, W) = \{T\colon V \to W \mid T \text{ is a linear transformation}\}.$$

When G acts linearly on V and W, there is an important subspace of $\mathrm{Hom}_{\mathbb{F}}(V, W)$:

Definition 4.27 If $\rho_1\colon G \to \mathrm{Gl}(V)$ and $\rho_2\colon G \to \mathrm{Gl}(W)$ are representations of a group G, then

$$\mathrm{Hom}_G(V, W) = \{T\colon V \to W \mid T \text{ is linear and } G\text{-equivariant}\},$$

that is, we have $\rho_2(g)(T(v)) = T(\rho_1(g)(v))$ for all $g \in G$ and $v \in V$.

When ρ_1 and ρ_2 are irreducible complex representations of G, Schur's Lemma tells us that $\dim \mathrm{Hom}_G(V, W) = 0$ when ρ_1 is not equivalent to ρ_2, and the vector space is one-dimensional, indexed on an eigenvalue, when ρ_1 is equivalent to ρ_2.

It turns out that the values of $\langle \chi_{\rho_1}, \chi_{\rho_2} \rangle$ are discernible from these dimensions. Fix representations $\rho_1\colon G \to \mathrm{Gl}(V)$ and $\rho_2\colon G \to \mathrm{Gl}(W)$ and consider the following sequence of linear mappings:

$$\mathrm{Hom}_G(V, W) \overset{i}{\hookrightarrow} \mathrm{Hom}_{\mathbb{C}}(V, W) \overset{\Phi}{\longrightarrow} \mathrm{Hom}_G(V, W) \overset{i}{\hookrightarrow} \mathrm{Hom}_{\mathbb{C}}(V, W),$$

where i denotes the inclusion mapping $\mathrm{Hom}_G(V, W) \hookrightarrow \mathrm{Hom}_{\mathbb{C}}(V, W)$, and the function $\Phi\colon \mathrm{Hom}_{\mathbb{C}}(V, W) \to \mathrm{Hom}_G(V, W)$ is defined by

$$\Phi(L) = \frac{1}{\#G} \sum_{g \in G} \rho_2(g^{-1}) \circ L \circ \rho_1(g).$$

Notice that $\Phi(L)$ is G-equivariant: Let $h \in G$. Then

$$\rho_2(h)^{-1} \circ \Phi(L) \circ \rho_1(h) = \frac{1}{\#G} \sum_{g \in G} \rho_2(h)^{-1} \circ \rho_2(g)^{-1} \circ L \circ \rho_1(g) \circ \rho_1(h)$$

$$= \frac{1}{\#G} \sum_{g \in G} \rho_2\big((gh)^{-1}\big) \circ L \circ \rho_1(gh) = \Phi(L).$$

Also, notice that if $T \in \mathrm{Hom}_G(V, W)$, then

$$\Phi(T) = \frac{1}{\#G} \sum_{g \in G} \rho_2(g)^{-1} \circ T \circ \rho_1(g) = \frac{1}{\#G} \sum_{g \in G} T = T.$$

The composite $\Phi \circ i$ is the identity on $\mathrm{Hom}_G(V, W)$.

To understand the other composite $i \circ \Phi \colon \mathrm{Hom}_{\mathbb{C}}(V, W) \to \mathrm{Hom}_{\mathbb{C}}(V, W)$, suppose dim $V = n$ and dim $W = m$. Make a choice of bases for V and W and identify $\mathrm{Hom}_{\mathbb{C}}(V, W)$ with the space $M_{m \times n}$ of $m \times n$ complex matrices. The composite determines a mapping $i \circ \Phi \colon M_{m \times n} \to M_{m \times n}$. Here is an interesting lemma from linear algebra.

Lemma 4.28 *Let A be an $m \times m$ matrix, and let C be an $n \times n$ matrix. Let $_A\Theta_C \colon M_{m \times n} \to M_{m \times n}$ be defined by*

$$_A\Theta_C(M) = AMC.$$

Then $_A\Theta_C$ is a linear transformation such that $\mathrm{tr}(_A\Theta_C) = \mathrm{tr}(A)\,\mathrm{tr}(C)$.

Proof To compute the trace in the general case of a linear transformation $T \colon V \to V$, let $\mathcal{B} = \{v_1, \ldots, v_n\}$ be a basis for V. Then $T(v_i) = a_{i1}v_1 + \cdots + a_{in}v_n$. If we denote $(T(v_i))_j = a_{ij}$, then $\mathrm{tr}(T) = \sum_{i=1}^{n}(T(v_i))_i$.

The dimension of $M_{m \times n}$ is mn. The canonical basis for $M_{m \times n}$ is given by $\{B_{ij}\}$ where $1 \leq i \leq m$, $1 \leq j \leq n$ and B_{ij} has 1 in the (i, j)th entry and zeroes elsewhere. To compute $\mathrm{tr}(_A\Theta_C)$, we denote the (i, j)th entry of $AB_{ij}C$ by $(AB_{ij}C)_{ij}$. Then $\mathrm{tr}(_A\Theta_C) = \sum_{i,j}(AB_{ij}C)_{ij}$. We can write

$$AB_{ij}C = A[\mathbb{O} \; \cdots \; \mathbb{O} \; \mathbf{e}_i \; \mathbb{O} \; \cdots \; \mathbb{O}]C, \text{ with } \mathbf{e}_i \text{ in the } j\text{th column of } B_{ij},$$

$$= [A\mathbb{O} \; \cdots \; A\mathbf{e}_i \; \cdots \; A\mathbb{O}]C$$

$$= \begin{bmatrix} 0 & \cdots & a_{1i} & \cdots & 0 \\ 0 & \cdots & a_{2i} & \cdots & 0 \\ \vdots & & \vdots & & \vdots \\ 0 & \cdots & a_{mi} & \cdots & 0 \end{bmatrix} \begin{bmatrix} b_{11} & b_{12} & \cdots & b_{1n} \\ b_{21} & b_{22} & \cdots & b_{2n} \\ \vdots & & & \vdots \\ b_{n1} & b_{n2} & \cdots & b_{nn} \end{bmatrix}$$

$$= \begin{bmatrix} a_{1i}b_{j1} & a_{1i}b_{j2} & \cdots & a_{1i}b_{jn} \\ \vdots & \vdots & & \vdots \\ i \rightarrow & & a_{ii}b_{jj} & \\ & & & \\ & & j \uparrow & \end{bmatrix}.$$

Thus $(AB_{ij}C)_{ij} = a_{ii}b_{jj}$, and

$$\mathrm{tr}(_A\Theta_C) = \sum_{i,j}(AB_{ij}C)_{ij} = \sum_{i,j}a_{ii}b_{jj} = \left(\sum_i a_{ii}\right)\left(\sum_j b_{jj}\right)$$

$$= \mathrm{tr}(A)\,\mathrm{tr}(C). \qquad \square$$

Proposition 4.29 $\mathrm{tr}(i \circ \Phi) = \langle \chi_{\rho_1}, \chi_{\rho_2} \rangle$.

Proof Using the notation of the previous lemma, we have

$$i \circ \Phi(T) = \frac{1}{\#G} \sum_{g \in G} \rho_2(g^{-1}) \circ T \circ \rho_1(g) = \frac{1}{\#G} \sum_{g \in G} \rho_{2(g^{-1})}\Theta_{\rho_1(g)}(T).$$

Taking the trace, we get

$$\mathrm{tr}(i \circ \Phi) = \frac{1}{\#G} \sum_{g \in G} \mathrm{tr}(_{\rho_2(g^{-1})}\Theta_{\rho_1(g)}) = \frac{1}{\#G} \sum_{g \in G} \mathrm{tr}(\rho_2(g)^{-1})\,\mathrm{tr}(\rho_1(g))$$

$$= \frac{1}{\#G} \sum_{g \in G} \overline{\chi_{\rho_2}(g)}\chi_{\rho_1}(g) = \langle \chi_{\rho_1}, \chi_{\rho_2} \rangle. \qquad \square$$

Proposition 4.30 $\mathrm{tr}(\Phi \circ i) = \dim \mathrm{Hom}_G(V, W)$.

Proof $\Phi \circ i \colon \mathrm{Hom}_G(V, W) \to \mathrm{Hom}_G(V, W)$ is the identity mapping. $\quad\square$

First Orthogonality Theorem *If $\rho_1 \colon G \to \mathrm{Gl}(V)$ and $\rho_2 \colon G \to \mathrm{Gl}(W)$ are representations of G, then $\langle \chi_{\rho_1}, \chi_{\rho_2} \rangle = \dim \mathrm{Hom}_G(V, W)$. If ρ_1 and ρ_2 are equivalent irreducible representations over $\mathbb{C}$, then $\langle \chi_{\rho_1}, \chi_{\rho_2} \rangle = 1$; if ρ_1 and ρ_2 are not equivalent, then $\langle \chi_{\rho_1}, \chi_{\rho_2} \rangle = 0$.*

Proof To relate this to the inner product and the dimension, consider the composite $i \circ \Phi$. Since i is injective, $\ker(i \circ \Phi) = \ker \Phi$. Also, Φ is surjective onto $\mathrm{Hom}_G(V, W)$ because $\Phi \circ i = \mathrm{Id}$. The Fundamental Theorem of Linear Algebra (Chapter 3) implies that $\mathrm{Hom}_{\mathbb{C}}(V, W) \cong \mathrm{Hom}_G(V, W) \oplus \ker \Phi$. Choose a basis for $\mathrm{Hom}_G(V, W)$ and a basis for $\ker \Phi$ giving a basis for $\mathrm{Hom}_{\mathbb{C}}(V, W)$. The matrix representing $i \circ \Phi$ with respect to such a basis takes the form $\left[\begin{smallmatrix} \mathrm{Id} & \mathbb{O} \\ \mathbb{O} & \mathbb{O} \end{smallmatrix}\right]$. The trace is equal to the dimension of $\mathrm{Hom}_G(V, W)$, and so $\langle \chi_{\rho_1}, \chi_{\rho_2} \rangle = \dim \mathrm{Hom}_G(V, W)$.

If ρ_1 and ρ_2 are irreducible but not equivalent, then by Schur's Lemma $\mathrm{Hom}_G(V, W) = \{\mathbb{O}\}$. The composition $i \circ \Phi = \mathbb{O}$, and the trace is also zero. When ρ_1 is equivalent to ρ_2, we can take $V = W$, and Schur's Lemma tells us that $\dim \mathrm{Hom}_G(V, V) = 1$. Therefore $\langle \chi_{\rho_1}, \chi_{\rho_2} \rangle = 1$.   □

For any representation over $\mathbb{C}$, Maschke's Theorem guarantees that it can be completely decomposed into irreducible subrepresentations. When $\rho = \rho_1 \oplus \rho_2 \oplus \cdots \oplus \rho_s$, the associated character is given by $\chi_\rho = \chi_{\rho_1} + \chi_{\rho_2} + \cdots + \chi_{\rho_s}$. The First Orthogonality Theorem implies that the set of irreducible characters forms an orthonormal set in the Hermitian inner product space $\mathrm{Cf}(G)$. Let

$$\mathrm{Irr}(G) = \{\chi_1, \ldots, \chi_t\}$$

denote the set of all irreducible characters of G. Let $W = \mathrm{Span}(\mathrm{Irr}(G))$ denote the set of all complex linear combinations of the irreducible characters, a subspace of $\mathrm{Cf}(G)$. As a Hermitian inner product space, $\mathrm{Cf}(G) \cong W \oplus W^\perp$, and every class function f can be written $f = w + w'$, where w is a linear combination of irreducible characters, and w' lies in the orthogonal complement of W. By the definition of $W^\perp$, $\langle w', \chi_i \rangle = 0$ for all $\chi_i \in \mathrm{Irr}(G)$.

Any class function $f \colon G \to \mathbb{C}$ can be combined with a representation $\rho \colon G \to \mathrm{Gl}(V)$ to make a linear transformation $T_f^\rho \colon V \to V$ defined by

$$T_f^\rho(v) = \sum_{g \in G} f(g)\rho(g)(v).$$

Because f is a class function, we see that T_f^ρ is G-equivariant: For $h \in G$,

$$\rho(h) \circ T_f^\rho \circ \rho(h)^{-1}(v)$$

$$= \sum_{g \in G} \rho(h) \circ f(g)\rho(g) \circ \rho(h^{-1})(v)$$

$$= \sum_{g \in G} f(g)\rho(hgh^{-1})(v) = \sum_{g \in G} f(hgh^{-1})\rho(hgh^{-1})(v)$$

$$= T_f^\rho(v).$$

When ρ is irreducible and w' is a class function in $W^\perp$, Schur's Lemma implies that $T_{w'}^\rho(v) = \lambda v$ for some $\lambda \in \mathbb{C}$. The trace of $T_{w'}^\rho$ is $(\dim V)\lambda$. Computing the trace directly, we get

$$\mathrm{tr}(T_{w'}^\rho) = \sum_{g \in G} w'(g)\,\mathrm{tr}(\rho(g)) = \sum_{g \in G} w'(g)\chi_\rho(g) = \#G\langle w', \overline{\chi_\rho} \rangle$$

$$= \#G\langle \chi_\rho, \overline{w'} \rangle.$$

Therefore $\lambda = \dfrac{\#G}{\dim V}\langle \chi_\rho, \overline{w'} \rangle$. Let us carry out this construction for the regular representation $\rho_{\mathrm{reg}} \colon G \to \mathrm{Gl}(\mathbb{C}G)$. In this case, $\lambda = 0$ because the regular representation is a sum of irreducible subrepresentations and $w' \in W^\perp$.

However,

$$T_{w'}^{\text{reg}}(e) = \sum_{g \in G} w'(g)\rho_{\text{reg}}(g)(e) = \sum_{g \in G} w'(g)g \in \mathbb{C}G.$$

Since this sum is $\mathbb{O}$ and the group elements form a basis for $\mathbb{C}G$, $w'(g) = 0$ for all $g \in G$, that is, $w' = \mathbb{O}$, the zero function. It follows that $f = w$ and $\text{Cf}(G) = \text{Span}(\text{Irr}(G))$.

This completes the proof of Theorem 4.26. We can now apply the familiar linear algebra of orthonormal bases. For example, suppose χ_ρ is a character of a representation $\rho \colon G \to \text{Gl}(V)$ and $V \cong W_1 \oplus \cdots \oplus W_k$ with each W_l irreducible. Let χ_i denote the irreducible character associated with W_i. Then $\chi_\rho = \chi_1 + \cdots + \chi_k$, and

$$\langle \chi_\rho, \chi_i \rangle = \langle \chi_1, \chi_i \rangle + \cdots + \langle \chi_k, \chi_i \rangle.$$

Since χ_i and χ_j are irreducible characters, the inner products on the right are either zero or one depending on when ρ^{W_j} is equivalent to the representation associated with χ_i. Thus $\langle \chi_\rho, \chi_i \rangle$ equals the number of times the character χ_i appears in the expression of χ_ρ as a sum of irreducible characters.

Applying this to the representation ρ, we can write

$$V \cong m_1 W_1 \oplus m_2 W_2 \oplus \cdots \oplus m_t W_t,$$

where $m_i W_i = W_i \oplus \cdots \oplus W_i$ (m_i times). Then $\chi_\rho = m_1 \chi_1 + \cdots + m_t \chi_t$, and

$$\langle \chi_\rho, \chi_\rho \rangle = \sum_{i=1}^{t} m_i \langle \chi_\rho, \chi_i \rangle = \sum_{i=1}^{t} m_i^2.$$

Theorem 4.31 *If χ_ρ is a character of a representation $\rho \colon G \to \text{Gl}(V)$, then $\langle \chi_\rho, \chi_\rho \rangle$ is a positive integer. Furthermore, $\langle \chi_\rho, \chi_\rho \rangle = 1$ if and only if ρ is irreducible.*

Corollary 4.32 *A class function $f \in \text{Cf}(G)$ is the character of a representation of G if and only if $\langle f, \chi_i \rangle$ is a nonnegative integer for all irreducible characters χ_i.*

For the regular representation, $\mathbb{C}G \cong m_1 W_1 \oplus \cdots \oplus m_t W_t$, where W_i is irreducible, and $m_i \geq 1$. By Theorem 4.11, $\{W_1, \ldots, W_t\}$ is the set of all distinct irreducible representations of G. Denote the dimension of W_i by d_i.

Proposition 4.33 *The character of the regular representation χ_{reg} satisfies $\chi_{\text{reg}} = d_1 \chi_1 + \cdots + d_t \chi_t$, that is, the number of times an irreducible summand appears in the regular representation is its dimension.*

Proof We know that $\langle \chi_{\text{reg}}, \chi_i \rangle$ gives the number of times that W_i appears in the decomposition of $\mathbb{C}G$ into a sum of irreducible representations. Suppose

$\{w_1, \ldots, w_d\}$ is a basis for W_i with $d = d_i$. Consider the linear transformation $\phi_j \colon \mathbb{C}G \to W_i$ given on the basis for $\mathbb{C}G$ by $\phi_j(h) = \rho_{\mathrm{reg}}(h)(w_j)$. Observe that ϕ_j is G-equivariant for $j = 1, \ldots, d$: For $x, h \in G$,

$$\begin{aligned}
\big[\phi_j \circ \rho_{\mathrm{reg}}(x)\big](h) = \phi_j\big(\rho_{\mathrm{reg}}(x)(h)\big) &= \phi_j(xh) = \rho_{\mathrm{reg}}(xh)(w_j) \\
&= \rho_{\mathrm{reg}}(x) \circ \rho_{\mathrm{reg}}(h)(w_j) = \rho_{\mathrm{reg}}(x)\big(\phi_j(h)\big) \\
&= \big[\rho_{\mathrm{reg}}(x) \circ \phi_j\big](h).
\end{aligned}$$

We claim that $\{\phi_1, \ldots, \phi_d\}$ is a basis for $\mathrm{Hom}_G(\mathbb{C}G, W_i)$. Suppose $L \in \mathrm{Hom}_G(\mathbb{C}G, W_i)$. Then, since $L(e) \in W_i$, $L(e) = \lambda_1 w_1 + \cdots + \lambda_d w_d$. Being G-equivariant, for any $x \in G$,

$$\begin{aligned}
L(x) = L(xe) = \big[L \circ \rho_{\mathrm{reg}}(x)\big](e) &= \big[\rho_{\mathrm{reg}}(x) \circ L\big](e) \\
&= \rho_{\mathrm{reg}}(x)(\lambda_1 w_1 + \cdots + \lambda_d w_d) \\
&= \lambda_1 \phi_1(x) + \cdots + \lambda_d \phi_d(x).
\end{aligned}$$

It follows that $\mathrm{Span}(\{\phi_1, \ldots, \phi_d\}) = \mathrm{Hom}_G(\mathbb{C}G, W_i)$. Suppose $c_1 \phi_1 + \cdots + c_d \phi_d = \mathbb{O}$ in $\mathrm{Hom}_G(\mathbb{C}G, W_i)$. Then

$$\mathbb{O} = (c_1 \phi_1 + \cdots + c_d \phi_d)(e) = c_1 \phi_1(e) + \cdots + c_d \phi_d(e) = c_1 w_1 + \cdots + c_d w_d.$$

Since $\{w_1, \ldots, w_d\}$ is linearly independent, $c_1 = \cdots = c_d = 0$, and the set $\{\phi_1, \ldots, \phi_d\}$ is linearly independent.

By the First Orthogonality Theorem, $\langle \chi_{\mathrm{reg}}, \chi_i \rangle = \dim \mathrm{Hom}_G(\mathbb{C}G, W_i) = d_i = m_i$. $\qquad\square$

Corollary 4.34 *If $\{W_1, \ldots, W_t\}$ is a complete set of distinct irreducible representations of G, then $\#G = \sum_{i=1}^{t} (\dim W_i)^2$.*

Proof $\#G = \dim \mathbb{C}G = \chi_{\mathrm{reg}}(e) = d_1 \chi_1(e) + \cdots + d_t \chi_t(e) = d_1^2 + \cdots + d_t^2.$ $\qquad\square$

For the group Σ_3, we know that Σ_3 falls into three conjugacy classes, $[e]$, $[(1, 2)]$, and $[(1, 2, 3)]$. Since this gives the dimension of $\mathrm{Cf}(\Sigma_3)$, there are three irreducible representations of the group Σ_3, namely, the trivial representation, the sign representation, and a third representation W of degree two. Since $\#\Sigma_3 = 6 = 1^2 + 1^2 + 2^2$, we deduce that $\mathbb{C}\Sigma_3 \cong \mathbb{C} \oplus \mathbb{C} \oplus W \oplus W$.

Corollary 4.35 *Two representations with the same characters are isomorphic.*

Proof For $i = 1, 2$, the expression of χ_{ρ_i} in terms of the basis $\mathrm{Irr}(G)$ determines the number of times each irreducible summand appears in each vector space. When the characters are equal, the decompositions into sums of irreducible representations will be the same. $\qquad\square$

While discussing the degrees of irreducible representations, we obtain a further restriction on their values.

Proposition 4.36 *For an irreducible representation* $\rho: G \to \text{Gl}(V)$, $\dim V = \chi_\rho(e)$ *is a divisor of* $\#G$.

Proof Consider a conjugacy class $[g_i]$ of G. Define the linear transformation $\rho([g_i]) = \sum_{x \in [g_i]} \rho(x): V \to V$. For any $h \in G$,

$$\rho(h) \circ \rho([g_i]) \circ \rho(h)^{-1} = \rho(h) \circ \sum_{x \in [g_i]} \rho(x) \circ \rho(h)^{-1}$$
$$= \sum_{x \in [g_i]} \rho(hxh^{-1}) = \rho([g_i]).$$

Therefore $\rho([g_i])$ is G-equivariant. Because ρ is irreducible, by Schur's Lemma, $\rho([g_i]) = \lambda_i \, \text{Id}$, for some $\lambda_i \in \mathbb{C}$.

Next, consider the subset of G given by $[g_i][g_j] = \{xy \mid x \in [g_i], y \in [g_j]\}$. If $h \in G$, then $h(xy)h^{-1} = (hxh^{-1})(hyh^{-1})$, and so the conjugates of elements of $[g_i][g_j]$ also lie in $[g_i][g_j]$. If $g_k \in [g_i][g_j]$, then it follows that $[g_k] \subset [g_i][g_j]$. Therefore $[g_i][g_j]$ is a union of conjugacy classes. Since

$$\rho([g_i]) \circ \rho([g_j]) = \sum_{x \in [g_i]} \sum_{y \in [g_j]} \rho(x) \circ \rho(y) = \sum_{x \in [g_i]} \sum_{y \in [g_j]} \rho(xy),$$

we can write $\rho([g_i]) \circ \rho([g_j])$ as a sum of $\rho([g_k])$:

$$\rho([g_i]) \circ \rho([g_j]) = \sum_k a_{ijk} \rho([g_k]),$$

where $a_{ijk} = \#\{(x, y) \mid x \in [g_i], y \in [g_j], xy \in [g_k]\}$.

When ρ is irreducible, $\rho([g_i]) \circ \rho([g_j]) = \lambda_i \lambda_j \, \text{Id}$, and so we deduce that

$$\lambda_i \lambda_j = \sum_k a_{ijk} \lambda_k.$$

This relation implies that the subring of $\mathbb{C}$ of integer linear combinations of the λ_i is finitely generated, and it follows that the numbers λ_i are algebraic integers (see Section 4.3 and the exercises at the end of this chapter).

We can compute the trace of $\rho([g_i])$ in two ways:

$$\text{tr}(\rho([g_i])) = \text{tr}(\lambda_i \, \text{Id}) = \lambda_i \chi_\rho(e) \text{ and}$$
$$\text{tr}(\rho([g_i])) = \sum_{x \in [g_i]} \text{tr}(\rho(x)) = \#[g_i] \chi_\rho(g_i).$$

Then

$$1 = \langle \chi_\rho, \chi_\rho \rangle = \frac{1}{\#G} \sum_{g \in G} \chi_\rho(g) \overline{\chi_\rho(g)} = \frac{1}{\#G} \sum_{i=1}^{t} \#[g_i] \chi_\rho(g_i) \overline{\chi_\rho(g_i)}$$

$$- \frac{1}{\#G} \sum_{i=1}^{t} \lambda_i \chi_\rho(e) \overline{\chi_\rho(g_i)} - \frac{\chi_\rho(e)}{\#G} \sum_{i=1}^{t} \lambda_i \chi_\rho(g_i^{-1}).$$

For each i, λ_i and $\chi_\rho(g_i^{-1})$ are algebraic integers. The sum of these products is also an algebraic integer. So $\#G/\chi_\rho(e)$ is an algebraic integer and a rational number. This implies $\#G/\chi_\rho(e)$ is an integer, that is, $\chi_\rho(e)$ divides $\#G$. $\qquad\square$

4.5 Character Tables

The set $\mathrm{Cl}(G)$ of conjugacy classes of G may be given as $\{[e], [g_2], \ldots, [g_t]\}$, where we have chosen $g_i \in [g_i]$, a representative, and the choices satisfy $[g_i] \cap [g_j] = \emptyset$ for $i \neq j$. The set of choices $S = \{e = g_1, g_2, \ldots, g_t\}$ is called a *transversal* by analogy with the transversal of the set of cosets of a subgroup. The Hermitian inner product space $\mathrm{Cf}(G)$ of class functions on G has dimension $\#\mathrm{Cl}(G)$ and an orthonormal basis $\mathrm{Irr}(G)$ of irreducible characters on G. The values of each character are determined by evaluating them on a representative of each conjugacy class.

The values determined by the irreducible characters can be summarized in a *character table* with entries $a_{ij} = \chi_{\rho_i}(g_j)$, where ρ_i is an irreducible representation, and g_j is the representative of a conjugacy class in a transversal. Each irreducible character ρ_i determines a row vector $(\chi_{\rho_i}(g_1), \chi_{\rho_i}(g_2), \ldots, \chi_{\rho_i}(g_t))$.

To compute the inner product of characters, we need to sum over all $g \in G$. We can multiply each $\chi_{\rho_i}(g_i)$ by the number of elements in $[g_i]$, which is $[G : C_G(g_i)]$, to obtain

$$\langle \chi_{\rho_i}, \chi_{\rho_j} \rangle = \frac{1}{\#G} \sum_{g \in G} \chi_{\rho_i}(g)\overline{\chi_{\rho_j}(g)} = \sum_{i=1}^{t} \frac{\#[g_i]}{\#G} \chi_{\rho_i}(g_i)\overline{\chi_{\rho_j}(g_i)}.$$

The usual inner product on class functions, when restricted to the transversal, can be expressed by

$$\begin{bmatrix} f(g_1), f(g_2), \ldots, f(g_t) \end{bmatrix} \begin{bmatrix} 1 & & & & & \mathbb{O} \\ & \ddots & & & & \\ & & \#[g_i] & & & \\ & & & \ddots & & \\ \mathbb{O} & & & & \#[g_t] \end{bmatrix} \begin{bmatrix} \overline{F(g_1)} \\ \overline{F(g_2)} \\ \vdots \\ \vdots \\ \overline{F(g_t)} \end{bmatrix}$$

$$= \#G\langle f, F \rangle.$$

The First Orthogonality Theorem tells us that the rows of a character table are orthonormal with respect to the inner product on $\mathrm{Cf}(G)$. This is a considerable restriction on the $t \times t$ matrix of entries in the character table. Furthermore, we know that the first column lists the degrees of the irreducible representations.

We can sometimes use this restriction to complete the values of a table. For example, let $G = \Sigma_3$. The three conjugacy classes are [Id], [(1, 2)], [(1, 2, 3)] with transversal $S = \{\text{Id}, (1, 2), (1, 2, 3)\}$. We know two degree one characters given by the trivial representation and the sign representation. Putting these data into a table gives

class	$[e]$	$[(1, 2)]$	$[(1, 2, 3)]$
# in class	1	3	2
χ_{triv}	1	1	1
χ_{sgn}	1	-1	1
χ_3	d_3	x	y

We leave the final row unknown to deduce the values from the preceding results. The degree of χ_3 is 2 according to Corollary 4.34, and $\#\Sigma_3 = 6 = 1^2 + 1^2 + 2^2$. The inner products of the third row with the first and second rows lead to the relations $2 + 3x + 2y = 0$ and $2 - 3x + 2y = 0$. It follows that $x = 0$ and $y = -1$, giving the final row of the table:

class	[Id]	$[(1, 2)]$	$[(1, 2, 3)]$
# class	1	3	2
χ_{triv}	1	1	1
χ_{sgn}	1	-1	1
χ_3	2	0	-1

For small groups, this approach can be sufficient to construct the characters of all irreducible representations, even if you do not know the matrices that constitute the representation. For large groups, the computations are more complicated.

There is another condition a character table must satisfy.

Second Orthogonality Theorem *The standard Hermitian inner product on the columns of a character table give an orthogonal set. Furthermore, the squared norm of a column gives the order of the centralizer of any element in the associated conjugacy class.*

Proof Another basis for the vector space of class functions $\mathrm{Cf}(G)$ is given by $\{\delta^{[g_1]}, \ldots, \delta^{[g_t]}\}$, where $\delta^{[g_i]}(x) = 1$ if $x \in [g_i]$ and $\delta^{[g_i]}(x) = 0$ otherwise. By construction the $\delta^{[g_i]}$ are class functions, and so they can be expressed as linear combinations of irreducible characters. Let $\delta^{[g_i]} = \alpha_1 \chi_1 + \cdots + \alpha_t \chi_t$.

Because the irreducible characters form an orthonormal basis, we can compute

$$\alpha_k = \left\langle \delta^{[g_i]}, \chi_k \right\rangle = \frac{1}{\#G} \sum_{x \in G} \delta^{[g_i]}(x) \overline{\chi_k(x)} = \frac{\#[g_i]}{\#G} \overline{\chi_k(g_i)}.$$

It follows that

$$\delta^{[g_i]}(x) = \sum_{k=1}^{t} \frac{\#[g_i]}{\#G} \overline{\chi_k(g_i)} \chi_k(x) = \#[g_i] \frac{1}{\#G} \sum_{k=1}^{t} \chi_k(x) \overline{\chi_k(g_i)}.$$

Because $\delta^{[g_i]}(x) = 0$ when $x \notin [g_i]$ and equals 1 when $x \in [g_i]$, the columns in the character table are orthogonal as t-vectors. For $x = g_i$,

$$\sum_{k=1}^{t} \chi_k(g_i) \overline{\chi_k(g_i)} = \sum_{k=1}^{t} |\chi_k(g_i)|^2 = \frac{\#G}{\#[g_i]} \delta^{[g_i]}(g_i) = \frac{\#G}{[G : C_G(g_i)]}$$

$$= \#C_G(g_i). \qquad \qquad \square$$

Get to know character tables

Suppose G is an abelian group. The conjugacy classes of an abelian group coincide with the elements of the group, and so there are as many irreducible characters of an abelian group as there are elements. This implies that all irreducible representations have degree one.

1. For an abelian group G, we can also see that an irreducible representation has degree one by showing that, for all $g \in G$, $\rho(g) \colon V \to V$ is G-equivariant. By Schur's Lemma, $\rho(g) = \lambda_g \, \mathrm{Id}$. Show that this implies degree one.

2. A degree one representation $\rho \colon G \to \mathrm{Gl}_1(\mathbb{C}) = \mathbb{C}^\times$ is a homomorphism of G into the multiplicative group of nonzero complex numbers. Since ρ must take its values in the roots of unity, use this fact to construct a character table for C_n, the cyclic group of order n, for $n = 2, 3, 4$. What is the general form for all n?

3. Suppose the abelian group G is isomorphic to a product $H_1 \times H_2$. Using the tensor product of representations and the fact that the tensor product of one-dimensional vector spaces is one-dimensional, deduce that the tensor products of the irreducible representations of H_1 and H_2 determine the irreducible representations of G. Work out the character table for $C_2 \times C_2$ and C_6 from this observation.

4. If p and q are primes and $\omega_p = e^{2\pi i/p}$ and $\omega_q = e^{2\pi i/q}$ are primitive roots of unity for p and q, describe a character table for C_{pq}.

5. The Fundamental Theorem of Finite Abelian Groups allows us to fac-
tor a finite abelian group G as a product of cyclic groups, $G: C_{n_1} \times C_{n_2} \times \cdots \times C_{n_k}$. Describe the subsequent character table for G in terms of these data.

4.6 More Fundamental Concepts

The orthogonality theorems offer considerable help to construct a character table. However, there are more aspects of structure to consider. How does the
character table reflect subgroup data, quotients, and other fundamental concepts? Let us consider each in turn.

I. Quotients. Suppose N is a normal subgroup of G. The canonical surjection
$\pi: G \to G/N$ is a homomorphism, and so representations of G/N become
representations of G by precomposition with π. Such a representation is called
a *lift* of $\rho: G/N \to \text{Gl}(V)$, denoted $\tilde{\rho} = \rho \circ \pi: G \to \text{Gl}(V)$. A key property
of lifts is the following:

Proposition 4.37 *If $\rho: G/N \to \text{Gl}(V)$ is an irreducible representation, then
the lift $\tilde{\rho}: G \to \text{Gl}(V)$ is also irreducible.*

Proof Suppose $W \subset V$ is a G-invariant subspace with respect to $\tilde{\rho}: G \to \text{Gl}(V)$. Then $\rho(gN)(w) = \tilde{\rho}(g)(w) \in W$ for all $w \in W$ and $g \in G$. This
implies W is also a G/N-invariant subspace of V. If ρ is irreducible, then
$W = \{\mathbb{O}\}$ or V, and so $\tilde{\rho}$ is irreducible. □

The values of the associated character satisfy $\tilde{\chi}(g) = \chi(gN)$. Since N is
normal in G, if $g \in N$, then the conjugacy class of g lies in N, $[g] \subset N$. For
example, consider the subgroup $V = \{e, (1, 2)(3, 4), (1, 3)(2, 4), (1, 4)(2, 3)\}$
of A_4; we know V is normal in A_4 by Cauchy's Formula. The quotient A_4/V
is isomorphic to C_3, whose character table is given by

	e	$(1, 2, 3)V$	$(1, 3, 2)V$
χ_1	1	1	1
χ_2	1	ω	ω^2
χ_3	1	ω^2	ω

,

where $\omega = e^{2\pi i/3}$. Recall that $1 + \omega + \omega^2 = 0$ because $x^3 - 1 = (x - 1)(x^2 + x + 1)$ and that $\overline{\omega} = \omega^2$.

The lifts to A_4, which has four conjugacy classes, are

$[g_i]$	$[e]$	$[(1,2)(3,4)]$	$[(1,2,3)]$	$[(1,3,2)]$
$\#[g_i]$	1	3	4	4
$\tilde{\chi}_1$	1	1	1	1
$\tilde{\chi}_2$	1	?	ω	ω^2
$\tilde{\chi}_3$	1	?	ω^2	ω
χ_4	?	?	?	?

Since $(1,2)(3,4)$ is in V, the value of the lifts $\tilde{\chi}_2$ and $\tilde{\chi}_3$ on the conjugacy class $[(1,2)(3,4)]$ is one. It suffices to fill in the final row. Corollary 4.34 requires $1^2 + 1^2 + 1^2 + ?^2 = 12$, and so the fourth representation has degree 3. The orthogonality of columns implies $\chi_4((1,2)(3,4)) = -1$ and $\chi_4((1,2,3)) = 0 = \chi_4((1,3,2))$. The complete character table is given by

$[g_i]$	$[e]$	$[(1,2)(3,4)]$	$[(1,2,3)]$	$[(1,3,2)]$
$\#[g_i]$	1	3	4	4
$\tilde{\chi}_1$	1	1	1	1
$\tilde{\chi}_2$	1	1	ω	ω^2
$\tilde{\chi}_3$	1	1	ω^2	ω
χ_4	3	-1	0	0

This is a lovely computation. From a much simpler quotient we were able to fill in the entire character table without having to find the missing representation. Of course, we needed to know the conjugacy classes of A_4. The order of A_4 is small enough that there was only one missing representation.

Before we treat more complicated computations, we can reverse the process and ask what quotients might emerge from looking at a character table. Every normal subgroup of a group G is the kernel of a homomorphism with domain G. Since a representation $\rho\colon G \to \mathrm{Gl}(V)$ is a homomorphism, $\ker \rho = \{g \in G \mid \rho(g) = \mathrm{Id}_V\}$ is a normal subgroup of G. Character tables consist of the values of traces. So $g \in \ker \rho$ means that $\mathrm{tr}(\rho(g)) = \mathrm{tr}(\mathrm{Id}_V) = \chi_\rho(e)$, the degree of ρ.

Definition 4.38 The *kernel* of a character χ is the subset of G given by $\ker \chi = \{g \in G \mid \chi(g) = \chi(e)\}$.

A character is constant on a conjugacy class, so $\ker \chi$ is a union of conjugacy classes. Because a character comes from a representation $\chi = \chi_\rho$, $\ker \chi$ is explicitly known.

Theorem 4.39 *For a representation* $\rho\colon G \to \mathrm{Gl}(V)$, $\ker \chi_\rho = \ker \rho$.

Proof For $g \in \ker \rho$, $\chi(g) = \mathrm{tr}(\rho(g)) = \chi(e)$, so $\ker \rho \subset \ker \chi_\rho$. Generally, we have that $\chi_\rho(g) = \omega_1 + \cdots + \omega_n$, where $\dim V = n$ and ω_i are Nth roots of unity for $N = \#G$. The norm $|\chi_\rho(g)| = |\omega_1 + \cdots + \omega_n| \le |\omega_1| + \cdots + |\omega_n| = n$. Suppose some $\omega_i \neq \omega_j$. Renumbering, say $\omega_1 \neq \omega_2$. If $\omega_1 = \cos\alpha + i\sin\alpha$ and $\omega_2 = \cos\beta + i\sin\beta$, then

$$
\begin{aligned}
|\omega_1 + \cdots + \omega_n| &= |\cos\alpha + i\sin\alpha + \cos\beta + i\sin\beta + \omega_3 + \cdots + \omega_n| \\
&\le |\cos\alpha + i\sin\alpha + \cos\beta + i\sin\beta| + n - 2 \\
&= \sqrt{(\cos\alpha + \cos\beta)^2 + (\sin\alpha + \sin\beta)^2} + n - 2 \\
&= \sqrt{2(1 + \cos\alpha\cos\beta + \sin\alpha\sin\beta)} + n - 2 \\
&= \sqrt{2\big(1 + \cos(\alpha - \beta)\big)} + n - 2.
\end{aligned}
$$

When $\alpha \not\equiv \beta \pmod{2\pi}$, $\cos(\alpha - \beta) < 1$ and $|\omega_1 + \cdots + \omega_n| < n$. Therefore $|\chi(g)| = \chi(e)$ implies $\omega_1 = \omega_2 = \cdots = \omega_n$ and $\chi_\rho(g) = n\omega_1$. If $\chi_\rho(g) = \chi_\rho(e)$, then $\omega_1 = 1$ and $g \in \ker\rho$. □

Corollary 4.40 *If an irreducible character χ has an element $g \in G$ such that $g \neq e$ and $\chi(g) = \chi(e)$, then the group G is not simple.*

The character table also allows us to determine the normal subgroups of G. Suppose N is a normal subgroup of G. Then we can associate with G/N its set of irreducible characters $\mathrm{Irr}(G/N) = \{\chi_1, \ldots, \chi_s\}$. If we consider the regular representation of G/N, $\rho_{\mathrm{reg}} \colon G/N \to \mathrm{Gl}(\mathbb{C}G/N)$, then $\rho_{\mathrm{reg}} \circ \pi \colon G \to \mathrm{Gl}(\mathbb{C}G/N)$ is the lift of ρ_{reg}. The kernel of $\rho_{\mathrm{reg}} \circ \pi$ is N, and $N = \bigcap \ker\tilde{\chi}_i$ for χ_i, where $\langle \tilde{\chi}_{\mathrm{reg}}, \tilde{\chi}_i \rangle \neq 0$.

Reversing this process, we can select indices $1 \le i_1, \ldots, i_k \le t$, take irreducible characters $\{\chi_{i_1}, \ldots, \chi_{i_k}\} \subset \mathrm{Irr}(G)$ and let $M = \bigcap_{j=1}^{k} \ker\chi_{i_j}$. Then M is normal in G, being an intersection of normal subgroups.

For example, recall the commutator subgroup of G, denoted G', which is generated by all products of the form $ghg^{-1}h^{-1}$. Since $k(ghg^{-1}h^{-1})k^{-1} = (kgk^{-1})(khk^{-1})(kgk^{-1})^{-1}(khk^{-1})^{-1}$, G' is a normal subgroup of G. The quotient G/G' is abelian, and so the irreducible representations of G/G' are all degree one. We refer to degree one representations as *linear representations* . Each linear representation lifts to a linear representation of G. On the other hand, a linear representation of G, $\rho \colon G \to \mathbb{C}^\times$ has the property that $\rho(ghg^{-1}h^{-1}) = \rho(g)\rho(h)\rho(g)^{-1}\rho(h)^{-1}$ in an abelian group, so $\rho(ghg^{-1}h^{-1}) = 1$; every commutator lies in $\ker\rho$. Let $p \colon G/G' \to G/\ker\rho$ denote the homomorphism $p(gG') = g\ker\rho$. There is the induced homomorphism $\hat{\rho} \colon G/\ker\rho \to \mathbb{C}^\times$ that is an isomorphism onto the image of ρ. We can factor ρ as $\hat{\rho} \circ p \circ \pi \colon G \to G/G' \to G/\ker\rho \to \mathbb{C}^\times$, which shows that ρ is a lift of a linear character on G/G'. Because G/G' is abelian, there are $[G : G']$ many linear representations of G.

Proposition 4.41 *The linear representations of G are in one-to-one correspondence with the irreducible representations of G/G'. Furthermore, the commutator subgroup in G can be determined from the character table of G as $G' = \bigcap_{\lambda_i} \ker \lambda_i$, where $\{\lambda_i\}$ is the set of linear representations of G.*

Let us look at a mystery character table [84] and see what we can learn from it:

	$[e]$	$[g_2]$	$[g_3]$	$[g_4]$	$[g_5]$	$[g_6]$	$[g_7]$
χ_1	1	1	1	1	1	1	1
χ_2	1	1	1	-1	1	-1	-1
χ_3	1	1	1	1	-1	-1	-1
χ_4	1	1	1	-1	-1	1	1
χ_5	2	-2	0	0	0	$\sqrt{2}$	$-\sqrt{2}$
χ_6	2	2	-2	0	0	0	0
χ_7	2	-2	0	0	0	$-\sqrt{2}$	$\sqrt{2}$

The group has order $4 \cdot 1^2 + 3 \cdot 2^2 = 16$, a 2-group. The cardinalities of the conjugacy classes can be determined from the inner product of the columns with themselves, so $\#[g_2] = 1$, $\#[g_3] = 2$, $\#[g_4] = \#[g_5] = 4$, and $\#[g_6] = \#[g_7] = 2$. There are four linear representations, and so $G' = \bigcap_{k=1}^{4} \ker \chi_k = [e] \cup [g_2] \cup [g_3]$, a subgroup of index 4. We do not see the value i among values of the linear representations, so $G/G' \cong C_2 \times C_2$. There are other normal subgroups, including $\ker \chi_3 = [e] \cup [g_2] \cup [g_3] \cup [g_4]$, which has order 8.

While discussing linear representations we can view a linear representation λ as a class function. Given an irreducible character χ, we get another class function by multiplication $\lambda\chi$. This is a character from the tensor product construction. Furthermore,

$$\langle \lambda\chi, \lambda\chi \rangle = \frac{1}{\#G} \sum_{g \in G} \lambda(g)\chi(g)\overline{\lambda(g)\chi(g)} = \frac{1}{\#G} \sum_{g \in G} |\lambda(g)|^2 \chi(g)\overline{\chi(g)}$$

$$= 1 \cdot \langle \chi, \chi \rangle = 1.$$

So $\lambda\chi$ is also irreducible. In the character table above, we see that $\chi_4 = \chi_2\chi_3$ and $\chi_7 = \chi_2\chi_5$.

II. Subgroups. The inclusion of a subgroup H in G is a homomorphism. Any representation $\rho: G \to \mathrm{Gl}(V)$ determines a representation $\mathrm{Res}_H \rho: H \hookrightarrow G \xrightarrow{\rho} \mathrm{Gl}(V)$, the *restriction* of ρ to H. Unlike the case of quotients, if ρ is an irreducible representation of G, then $\mathrm{Res}_H \rho$ need not be irreducible. For example, if G is nonabelian and H is an abelian subgroup of G, then any irreducible representation of G of degree greater than one implies that $\text{Res}_H \rho$ is a decomposable representation of H.

We can learn some useful things from restrictions. Suppose $g \in G$ and $H = \langle g \rangle$ is the cyclic subgroup generated by g in G. If $\rho \colon G \to \text{Gl}(V)$ is a representation of G, then $\text{Res}_H \rho \colon H \to \text{Gl}(V)$ decomposes into linear representations, $\text{Res}_H \rho = \theta_1 + \cdots + \theta_n$ with $n = \dim V$. If we let $V \cong U_1 \oplus \cdots \oplus U_n$ be the decomposition into H-invariant, dimension one subspaces, then $U_i = \text{Span}\{u_i\}$, and u_i is an eigenvector for $\text{Res}_H \rho(g)$. Because $\text{Res}_H \rho(g) = \rho(g)$, it follows that $\rho(g)$ is diagonalizable. The fact that eigenvalues of $\rho(g)$ are among the $\#G$th roots of unity follows as well: $\text{Res}_H \rho|_{U_i} \colon U_i \to U_i$ is multiplication by a complex number and $g^{\#G} = e$.

For any subgroup H of G and $h \in H$, let $\text{Res}_H \chi_\rho(h) = \text{tr}(\text{Res}_H \rho(h)) = \text{tr}(\rho(h))$ denote the restriction to a subgroup of a character χ_ρ. The restriction gives a character on H, which can be decomposed $\text{Res}_H \chi = m_1\theta_1 + \cdots + m_s\theta_s$, where m_i are nonnegative integers, and θ_i are irreducible characters on H. Then

$$\sum_{i=1}^{s} m_i^2 = \langle \text{Res}_H \chi, \text{Res}_H \chi \rangle_H = \frac{1}{\#H} \sum_{h \in H} \chi(h)\overline{\chi(h)}.$$

We can compare this sum with the related computation for $\langle \chi, \chi \rangle_G$: If χ is an irreducible character on G, then

$$1 = \langle \chi, \chi \rangle_G = \frac{1}{\#G} \sum_{g \in G} \chi(g)\overline{\chi(g)}$$

$$= \frac{1}{\#G} \left(\sum_{h \in H} \chi(h)\overline{\chi(h)} + \sum_{g \in G - H} \chi(g)\overline{\chi(g)} \right)$$

$$= \frac{\#H}{\#G} \left(\frac{1}{\#H} \sum_{h \in H} \chi(h)\overline{\chi(h)} + \frac{1}{\#H} \sum_{g \in G - H} |\chi(g)|^2 \right)$$

$$= \frac{1}{[G : H]} \langle \text{Res}_H \chi, \text{Res}_H \chi \rangle + r$$

with $r \geq 0$.

Proposition 4.42 *If H is a subgroup of G and χ is an irreducible character of G, then $\text{Res}_H \chi = m_1\theta_1 + \cdots + m_s\theta_s$, where θ_i are irreducible characters of H, and*

$$\sum_{i=1}^{s} m_i^2 \leq [G : H],$$

with equality if and only if $\chi(g) = 0$ for all $g \in G - H$.

		()	(1, 2, 3)	(1, 3, 2)			()	[(1, 2)]	[(1, 2, 3)]
C_3:	θ_1	1	1	1	Σ_3:	χ_1	1	1	1
	θ_2	1	ω	ω^2		χ_2	1	-1	1
	θ_3	1	ω^2	ω		χ_3	2	0	-1

For example, consider the character tables for Σ_3 and for C_3, the cyclic group of order 3, which is isomorphic to the subgroup $H = \langle (1, 2, 3) \rangle$. In this case, $\text{Res}_H \chi_3 = \theta_2 + \theta_3$. This follows because $1 + \omega + \omega^2 = 0$ and so $\omega + \omega^2 = -1$. We also have

$$\langle \text{Res}_H \chi_3, \text{Res}_H \chi_3 \rangle_H = \langle \theta_2 + \theta_3, \theta_2 + \theta_3 \rangle_H = 2 = \big[\Sigma_3 : \langle (1, 2, 3) \rangle \big].$$

Therefore $\chi_3((12)) = 0$.

Corollary 4.43 *If H is an abelian subgroup of G, then the degree of any irreducible representation of G is less than or equal to $[G : H]$.*

Proof Let χ be the character associated with an irreducible representation ρ of G. The degree of $\text{Res}_H \rho$ is the same as that of ρ. The character $\text{Res}_H \chi$ can be written as a sum of linear characters, and

$$\chi(e) = \text{Res}_H \chi(e) = m_1 \theta_1(e) + \cdots + m_s \theta_s(e) = m_1 + \cdots + m_s,$$

because abelian H implies $\theta_i(e) = 1$. Then $m_1 + \cdots + m_s \le m_1^2 + \cdots + m_s^2 \le [G : H]$, so the degree of ρ is bounded by $[G : H]$.   □

A generalization of the case of an abelian subgroup of G was proved in a 1937 paper of ALFRED H. CLIFFORD (1908–1992).

Theorem 4.44 *(Clifford) Suppose N is a normal subgroup of G and χ is an irreducible character of G. If ψ_1 is an irreducible character of N with $\langle \text{Res}_N \chi, \psi_1 \rangle_N = d$, then $\text{Res}_N \chi = \psi_1 + \cdots + \psi_m$ with each ψ_i irreducible and a conjugate of ψ_1.*

Proof The *conjugate of a class function* f in $\text{Cf}(N)$ by an element $g \in G$ is the function $f^g(n) = f(gng^{-1})$. When N is normal in G, we show that f^g is also in $\text{Cl}(N)$: Suppose $h, n \in N$ and there is $x \in N$ such that $h = xnx^{-1}$. Then

$$f^g(h) = f\big(ghg^{-1}\big) = f\big(g(xnx^{-1})g^{-1}\big)$$
$$= f\big((gxg^{-1})(gng^{-1})(gx^{-1}g^{-1})\big) = f\big(y(gng^{-1})y^{-1}\big),$$

because $N \triangleleft G$ and $x \in N$, we have $gxg^{-1} = y \in N$, and because $f \in \text{Cf}(N)$,

$$= f\big(gng^{-1}\big) = f^g(n).$$

Let $\rho: G \to \text{Gl}(V)$ be the irreducible representation for which $\chi = \chi_\rho$. Denote $\text{Res}_N \rho = \rho_N$. Suppose W is an N-invariant subspace in V. The image $\rho(g)(W) = \{\rho(g)(w) \mid w \in W\}$ is a subspace of V, and for $n \in N$,

$$\rho(n)\big(\rho(g)(w)\big) = \rho(g)\big(\rho(g^{-1}ng)(w)\big) = \rho(g)(w');$$

$g^{-1}ng \in N$, and W is N-invariant, so $\rho(g^{-1}ng)(w) = w' \in W$. Thus $\rho(g)(W)$ is also N-invariant for all $g \in G$.

If $U \subset \rho(g)(W)$ is an N-invariant subspace, then $\rho(g^{-1})(U) \subset W$ is N-invariant. If W is irreducible, then $\rho(g^{-1})(U) = \{\mathbb{O}\}$ or $U = W$. Then $U = \{\mathbb{O}\}$ or $U = \rho(g)(W)$, and $\rho(g)(W)$ is also irreducible. The direct sum $\bigoplus_{g \in G} \rho(g)(W) \subset V$ is G-invariant. If V is irreducible, then $\bigoplus_{g \in G} \rho(g)(W) = \{\mathbb{O}\}$ or V. Therefore the representation ρ_N of N decomposes as $V \cong W \oplus \rho(g_2)(W) \oplus \cdots \oplus \rho(g_l)(W)$ with all summands isomorphic, and the summands all of the same degree. Let $\theta_i : N \to \text{Gl}(\rho(g_i)(W))$ denote the irreducible representation of N corresponding to each summand.

Let ψ_1 denote the character for $\theta_1 : N \to \text{Gl}(W)$. On the summand given by $\rho(g_i)(W)$, we can write $\rho_N(n) = \rho(g_i) \circ \rho(g_i^{-1}ng_i) \circ \rho(g_i^{-1})$, and so $\text{tr}(\rho_N(n)) = \text{tr}(\rho(g_i^{-1}ng_i))$ and $\psi_i = \psi_1^{g_i}$. If W appears as a summand d times in the decomposition of V, then $\langle \text{Res}_N \chi, \psi_1 \rangle = d$. Because $\rho(g_i)$ is an isomorphism of N-invariant irreducible subspaces, $\rho(g_i)(W)$ appears d times as well in the decomposition of V. Thus V is decomposed into isomorphic summands of an irreducible N-invariant subspace, and the character for each summand is a conjugate of ψ_1. $\qquad\square$

Get to know lifts and restrictions

1. Based on the character table presented below, (1) what is the order of the group? (2) What is the cardinality of each conjugacy class of the group? (3) Is the group abelian? (4) Find all the normal subgroups of the group. (5) What is the cardinality of the commutator subgroup? (6) Is this the character table of a familiar group?

	$[e]$	$[g_2]$	$[g_3]$	$[g_4]$	$[g_5]$	$[g_6]$
χ_1	1	1	1	1	1	1
χ_2	1	−1	−1	1	1	−1
χ_3	1	−1	1	−1	1	−1
χ_4	1	1	−1	−1	1	1
χ_5	2	−2	0	0	−1	1
χ_6	2	2	0	0	−1	−1

2. Suppose $N \lhd G$ is a normal subgroup and $\chi \in \mathrm{Irr}(G)$ is an irreducible character. Suppose $\langle \mathrm{Res}_N \chi, \chi_1 \rangle_N$ is nonzero, where χ_1 is the trivial representation of N. Prove that N is contained in the kernel of χ.

3. Suppose N is a subgroup of G of index 2. Suppose χ is an irreducible character of G. Show that $\mathrm{Res}_N \chi$ is either irreducible or $\mathrm{Res}_N \chi$ is a sum of two distinct irreducible characters of N of the same degree.

4. Use the orthogonality relations to show that if N is a normal subgroup of G and $g \in G$, then $\#C_{G/N}(gN) \leq \#C_G(g)$.

5. Establish the following properties of conjugates of class functions: Suppose $N \lhd G$ is a normal subgroup and $f \in \mathrm{Cf}(N)$; then (1) $f^g(n) = f(g^{-1}ng)$ is also a class function on N; (2) $f^{(gh)} = (f^g)^h$ for all $g, h \in G$; (3) $\langle f^g, F^g \rangle_N = \langle f, F \rangle_N$ for $f, F \in \mathrm{Cf}(N)$; (4) for a class function θ on G, $\langle \mathrm{Res}_N \theta, f^g \rangle_N = \langle \mathrm{Res}_N \theta, f \rangle_N$ for all $f \in \mathrm{Cf}(N)$.

III. Induction from subgroups. Suppose H is a subgroup of G and $\rho \colon H \to \mathrm{Gl}(W)$ is a representation of H. Is it possible to construct a representation of G that restricts to ρ on H? Recall from Section 2.5 the basic ingredients of the construction of the transfer: For a subgroup H of G, $G = \bigcup gH$, the union of left cosets of H in G. Suppose $s = [G : H]$ and $S = \{e = g_1, g_2, \ldots, g_s\}$ is a *transversal* of coset representatives of H, that is, $G/H = \{g_1 H, g_2 H, \ldots, g_s H\}$. It follows that every element $x \in G$ can be written as $x = g_i h$ for some $g_i \in S$ and $h \in H$.

Let $V = W_1 \oplus W_2 \oplus \cdots \oplus W_s$, where each W_i is a copy of W. To keep track of the transversal, we decorate each copy W_i of W with $g_i \in S$, that is, denote by $g_i \cdot w$ an element $w \in W_i$. An element $v \in V$ takes the form $v = \sum_{i=1}^s g_i \cdot w_i$.

We define the *induced representation* from ρ on H to G, $\mathrm{Ind}_H^G \rho \colon G \to \mathrm{Gl}(V)$, as follows: Suppose $x \in G$. Then for each i, $xg_i = g_j h_i^x$. On an element $g_i \cdot w_i$, let $\mathrm{Ind}_H^G \rho(x)(g_i \cdot w_i) = g_j \cdot (\rho(h_i^x)(w_i))$, that is, let $\rho(h_i^x)$ act on $w_i \in W$ and place it in the g_j-indexed copy of W. To see that we get a representation, we need to check that we get a homomorphism from this construction. Let $x, y \in G$. Then

$$(yx)g_i = y(xg_i) = y\left(g_j h_i^x\right) = (yg_j)h_i^x = \left(g_k h_j^y\right)h_i^x = g_k\left(h_j^y h_i^x\right).$$

Writing $\rho_H^G = \mathrm{Ind}_H^G \rho$,

$$\rho_H^G(yx)(g_i \cdot w_i) = g_k \cdot \left(\rho\left(h_j^y h_i^x\right)(w_i)\right) = g_k \cdot \rho\left(h_j^y\right)\left(\rho(h_i^x)(w_i)\right)$$
$$= \rho_H^G(y)\left(g_j \cdot \rho(h_i^x)(w_i)\right) = \rho_H^G(y) \circ \rho_H^G(x)(g_i \cdot w_i).$$

For example, consider the subgroup $\langle i \rangle$ of Q, the unit quaternion group, generated by i. There is a representation of $\langle i \rangle$ given by identifying each power of i with the complex multiplier it gives, $\rho : \langle i \rangle \to \mathrm{Gl}(\mathbb{C})$. The quotient $Q/\langle i \rangle$ is the set $\{1\langle i \rangle, j\langle i \rangle\}$. The induced representation of Q from ρ has $V = 1 \cdot \mathbb{C} \oplus j \cdot \mathbb{C}$. Since $j1 = j1$ and $jj = -1$, $\mathrm{Ind}_{\langle i \rangle}^{Q} \rho(j)(1 \cdot z_1 + j \cdot z_2) = 1 \cdot (-z_2) + j \cdot z_1$. Carrying on to $-j$ and writing out the matrix forms of the representation gives us

$$\mathrm{Ind}_{\langle i \rangle}^{Q} \rho(\pm j) = \begin{bmatrix} 0 & \mp 1 \\ \pm 1 & 0 \end{bmatrix}.$$

Similar computations give us

$$\mathrm{Ind}_{\langle i \rangle}^{Q} \rho(\pm k) = \begin{bmatrix} 0 & \mp i \\ \mp i & 0 \end{bmatrix}, \quad \mathrm{Ind}_{\langle i \rangle}^{Q} \rho(\pm i) = \begin{bmatrix} \pm i & 0 \\ 0 & \mp i \end{bmatrix},$$

$$\mathrm{Ind}_{\langle i \rangle}^{Q} \rho(\pm 1) = \begin{bmatrix} \pm 1 & 0 \\ 0 & \pm 1 \end{bmatrix}.$$

The traces of these matrices are given in the table for $\bar{\rho} = \mathrm{Ind}_{\langle i \rangle}^{Q} \rho$:

	$[1]$	$[-1]$	$[\pm i]$	$[\pm j]$	$[\pm k]$
$\chi_{\bar{\rho}}$	2	-2	0	0	0

We compute $\langle \chi_{\bar{\rho}}, \chi_{\bar{\rho}} \rangle$ to get

$$\langle \chi_{\bar{\rho}}, \chi_{\bar{\rho}} \rangle = \frac{1}{8} \sum_{g \in Q} \mathrm{tr}\big(\bar{\rho}(g)\big) \cdot \overline{\mathrm{tr}\big(\bar{\rho}(g)\big)} = \frac{1}{8}\big(2 \cdot 2 + (-2) \cdot (-2)\big) = 1,$$

and so we have constructed an irreducible representation of Q.

It is not true that a representation induced from an irreducible representation on a subgroup is always irreducible. For example, if G is abelian, then the induced representation has degree $[G : H]$, which is greater than one. As we will see in the examples below, a careful examination of the induced representation may reveal an irreducible summand.

How do we relate the character of an induced representation on G to the character on H? Introduce the *indicator function* on H,

$$\delta_H(g) = \begin{cases} 1 & \text{if } g \in H, \\ 0 & \text{if } g \notin H. \end{cases}$$

The induced representation on G from $\rho : H \to \mathrm{Gl}(W)$ associates a linear isomorphism $\mathrm{Ind}_H^G \rho(x) : V \to V$ where $V = g_1 \cdot W \oplus g_2 \cdot W \oplus \cdots \oplus g_s \cdot W$ and $\{g_1 = e, g_2, \ldots, g_s\}$ is a set of coset representatives of H in G. The linear transformation $\mathrm{Ind}_H^G \rho(x)$ maps $g_i \cdot W$ to $g_l \cdot W$ when $x g_i = g_l h_i^x$ with $h_i^x \in H$. If $i \neq k$, then this part of the linear transformation contributes zero to the trace.

In the case $xg_i = g_i h_i^x$, we have $h_i^x = g_i^{-1} x g_i$, and the part of the trace coming from $g_i \cdot W$ is the trace of $\rho(h_i^x) = \rho(g_i^{-1} x g_i)$. We summarize this by

$$\text{tr}\big(\text{Ind}_H^G \rho(x)\big) = \sum_{i=1}^{s} \delta_H\big(g_i^{-1} x g_i\big) \, \text{tr}\big(\rho(g_i^{-1} x g_i)\big)$$

$$= \sum_{i=1}^{s} \delta_H\big(g_i^{-1} x g_i\big) \chi_\rho\big(g_i^{-1} x g_i\big).$$

Here the factor $\chi_\rho(g_i^{-1} x g_i)$ only makes sense if $g_i^{-1} x g_i \in H$. It if is not in H, then $\delta_H(g_i^{-1} x g_i) = 0$, and we ignore the lack of definition as a zero.

To see that this computation is independent of the choice of transversal, suppose $g_i H = l_i H$, that is, $l_i = g_i h$. Then, when $g_i^{-1} x g_i \in H$, $l_i^{-1} x l_i = h^{-1}(g_i^{-1} x g_i)h$, a conjugate in H of $g_i^{-1} x g_i$, and so $\chi_\rho(g_i^{-1} x g_i) = \chi_\rho(l_i^{-1} x l_i)$. To remove the dependence on a choice of transversal, we can vary each choice of g_i by any $h \in H$. Summing over all such choices for $i = 1, \ldots, s$, we get every element in G, each value repeated $\#H$ times. We can write

$$\chi_{\text{Ind}_H^G \rho}(x) = \text{tr}\big(\text{Ind}_H^G \rho(x)\big) = \frac{1}{\#H} \sum_{y \in G} \delta_H\big(y^{-1} x y\big) \chi_\rho\big(y^{-1} x y\big).$$

With this formula we prove the following:

Frobenius Reciprocity *For representations $\rho \colon H \to \text{Gl}(W)$ and $\theta \colon G \to \text{Gl}(V)$ with subgroup H of G and $V = W \oplus (g_2 \cdot W) \oplus \cdots \oplus (g_s \cdot W)$ for a set of coset representatives $\{e = g_1, g_2, \ldots, g_s\}$ for G/H, we have*

$$\langle \chi_{\text{Ind}_H^G \rho}, \chi_\theta \rangle_G = \langle \chi_\rho, \text{Res}_H \chi_\theta \rangle_H.$$

Proof Let us work more generally with all class functions. The induced character suggests a mapping $\text{Ind}_H^G \colon \text{Cf}(H) \to \text{Cf}(G)$ defined by

$$\text{Ind}_H^G f(x) = \frac{1}{\#H} \sum_{y \in G} \delta_H\big(y^{-1} x y\big) f\big(y^{-1} x y\big).$$

When $y^{-1} x y \notin H$, we multiply by zero, giving zero. The restriction of a class function on G, say F, to H is simply $F|_H$, written $\text{Res}_H F$. Then

$$\big\langle \text{Ind}_H^G f, F \big\rangle_G = \frac{1}{\#G} \sum_{g \in G} \text{Ind}_H^G f(g) \overline{F(g)}$$

$$= \frac{1}{\#G} \sum_{g \in G} \frac{1}{\#H} \sum_{x \in G} \delta_H\big(x^{-1} g x\big) f\big(x^{-1} g x\big) \overline{F(g)}$$

$$= \frac{1}{\#G} \frac{1}{\#H} \sum_{g \in G} \sum_{x \in G} \delta_H(x^{-1}gx) f(x^{-1}gx) \overline{F(x^{-1}gx)},$$

but $x^{-1}gx = h \in H$ if and only if $g = xhx^{-1}$, so

$$= \frac{1}{\#G} \frac{1}{\#H} \sum_{x \in G} \sum_{h \in H} f(h)\overline{F(h)} = \frac{1}{\#H} \sum_{h \in H} f(h)\overline{F(h)}$$

$$= \langle f, \operatorname{Res}_H F \rangle_H.$$

When applied to characters, $\langle \chi_{\operatorname{Ind}_H^G \rho}, \chi_\theta \rangle_G = \langle \operatorname{Ind}_H^G \chi_\rho, \chi_\theta \rangle_G = \langle \chi_\rho, \operatorname{Res}_H \chi_\theta \rangle_H$. $\square$

Let us put all this technology to work. The group A_4 has four irreducible characters, three of which come from the normal subgroup V. The fourth character has the entries

$$\theta_4 \ \Big| \quad 3 \qquad -1 \qquad 0 \qquad 0$$

The entries may be determined by the orthogonality theorems. However, it would be nice to identify the source of the representation. Taking V as a subgroup of A_4, we can use induction to construct a representation on A_4 from one on V. Using the formula $\operatorname{Ind}_V^{A_4} \chi(g) = \frac{1}{4} \sum_{x \in A_4} \delta_V(xgx^{-1}) \chi(xgx^{-1})$, we can compute $\operatorname{Ind}_V^{A_4} \chi_2 = \theta_4$. Constructing $\operatorname{Ind}_V^{A_4} \chi_2$ in this manner leads to the desired values for the character θ_4. Checking that $\langle \operatorname{Ind}_V^{A_4} \chi_2, \operatorname{Ind}_V^{A_4} \chi_2 \rangle = 1$ verifies that $\operatorname{Ind}_V^{A_4} \chi_2$ is irreducible.

Next consider the group A_5. We know quite a bit about it, but let us see how far we can go in constructing the character table, and from that table deduce the interesting properties of A_5.

The conjugacy classes of A_5 are

$$\operatorname{Cl}(A_5) = \{[( )], [(1,2)(3,4)], [(1,2,3)], [(1,2,3,4,5)], [(2,1,3,4,5)]\}.$$

By Cauchy's Formula the disjoint cycle type of a permutation is preserved under conjugation. In A_5, all 3-cycles are conjugate, but the 5-cycles split into two orbits. Using the same reasoning, observe that the commutator subgroup $(A_5)'$ is all of A_5: Consider $ghg^{-1}h^{-1}$ where $h = (x, y, z)$, a 3-cycle. Then ghg^{-1} can give any other 3-cycle. The product of 3-cycles $(ghg^{-1})h^{-1} = (a, b, c)(a, c, d) = (b, c, d)$, and so all 3-cycles appear in $(A_5)'$. However, the 3-cycles generate A_5, so $(A_5)' = A_5$. It follows that there is only one linear representation among the irreducible representations of A_5.

By Corollary 4.34 the degrees of the nontrivial irreducible representations satisfy $59 = d_2^2 + d_3^2 + d_4^2 + d_5^2$ with $d_i \mid 60$. Thus the d_i are among

$\{2, 3, 4, 5, 6\}$. The unique solution (with $d_2 \leq d_3 \leq d_4 \leq d_5$) is $d_2 = d_3 = 3$, $d_4 = 4$, and $d_5 = 5$.

To induce representations up to A_5, we begin with the subgroup A_4 of A_5 of index $[A_5 : A_4] = 5$. We first induce to A_5 the trivial representation of A_4. Choosing coset representatives for A_5/A_4, after some bookkeeping, let

$$g_1 = (\ ), \quad g_2 = (1, 2)(3, 5), \quad g_3 = (1, 2, 5),$$
$$g_4 = (1, 2, 3, 5, 4), \text{ and } g_5 = (1, 3, 4, 5, 2).$$

The values of $\mathrm{Ind}_{A_4}^{A_5} \theta_1$ are computed on the transversal of the conjugacy classes directly:

	$[(\ )]$	$[(1, 2)(3, 4)]$	$[(1, 2, 3)]$	$[(1, 2, 3, 4, 5)]$	$[(2, 1, 3, 4, 5)]$
$\mathrm{Ind}_{A_4}^{A_5} \theta_1(g_i)$	5	1	2	0	0

The value $\mathrm{Ind}_{A_4}^{A_5} \theta_1((1, 2, 3)) = 2$ is derived from the conjugate expression $g_1^{-1}(1, 2, 3)g_1^{-1} = (\ )(1, 2, 3)(\ ) = (1, 2, 3) \in A_4$ and $g_4^{-1}(1, 2, 3)g_4 = (4, 5, 3, 2, 1)(1, 2, 3)(1, 2, 3, 5, 4) = (4, 1, 2) \in A_4$.

We check if we have an irreducible representation by computing $\langle \mathrm{Ind}_{A_4}^{A_5} \theta_1,$ $\mathrm{Ind}_{A_4}^{A_5} \theta_1 \rangle = \frac{1}{60}(5^2 + 15 \cdot 1^2 + 20 \cdot 2^2) = 2$. It is not irreducible. However, it is a representation, so it is a sum of irreducibles. So far, among the irreducibles for A_5, we only know the trivial character ψ_1. Let us see if it is a summand: By Frobenius Reciprocity,

$$\langle \mathrm{Ind}_{A_4}^{A_5} \theta_1, \psi_1 \rangle_{A_5} = \langle \theta_1, \mathrm{Res}_{A_4} \psi_1 \rangle_{A_4} = \langle \theta_1, \theta_1 \rangle_{A_4} = 1.$$

So we can write $\mathrm{Ind}_{A_4}^{A_5} \theta_1 = \psi_1 +\ ?$. Consider the character $? = \mathrm{Ind}_{A_4}^{A_5} \theta_1 - \psi_1$ with values

	$[(\ )]$	$[(1, 2)(3, 4)]$	$[(1, 2, 3)]$	$[(1, 2, 3, 4, 5)]$	$[(2, 1, 3, 4, 5)]$
$\mathrm{Ind}_{A_4}^{A_5} \theta_1(g_i) - \psi_1(g_i)$	4	0	1	-1	-1

The inner product of this character with itself is one, and we have found an irreducible character. Label it $\psi_4 = \mathrm{Ind}_{A_4}^{A_5} \theta_1 - \psi_1$.

The same calculation can be used to compute $\mathrm{Ind}_{A_4}^{A_5} \theta_2$. In this case the values are

	$[(\ )]$	$[(1, 2)(3, 4)]$	$[(1, 2, 3)]$	$[(1, 2, 3, 4, 5)]$	$[(2, 1, 3, 4, 5)]$
$\mathrm{Ind}_{A_4}^{A_5} \theta_2(g_i)$	5	1	-1	0	0

The value $\mathrm{Ind}_{A_4}^{A_5} \theta_2((1, 2, 3)) = -1$ is derived from $\theta_2((1, 2, 3)) + \theta_2((4, 1, 2)) = \omega + \omega^2 = -1$. Checking for irreducibility, we get $\langle \mathrm{Ind}_{A_4}^{A_5} \theta_2, \mathrm{Ind}_{A_4}^{A_5} \theta_2 \rangle = \frac{1}{60}(5^2 + 15 \cdot 1^2 + 20 \cdot (-1)^2) = 1$, an irreducible character. Label $\mathrm{Ind}_{A_4}^{A_5} \theta_2$

by ψ_5. The character table for A_5 thus far takes the form

$[g_i]$	$[()]$	$[(1,2)(3,4)]$	$[(1,2,3)]$	$[(1,2,3,4,5)]$	$[(2,1,3,4,5)]$
$\#[g_i]$	1	15	20	12	12
ψ_1	1	1	1	1	1
ψ_2	3				
ψ_3	3				
ψ_4	4	0	1	-1	-1
ψ_5	5	1	-1	0	0

To get at the remaining rows, we consider another subgroup of A_5, a Sylow 5-subgroup $P_5 = \langle(1,2,3,4,5)\rangle$ of index $[A_5 : P_5] = 12$. More bookkeeping is required to choose coset representatives:

$$g_1 = (\ ), g_2 = (1,2,3), g_3 = (1,2,4), g_4 = (1,2,5), g_5 = (1,3,4),$$
$$g_6 = (1,3,5), g_7 = (1,4,5), g_8 = (2,3,4), g_9 = (2,3,5),$$
$$g_{10} = (2,4,5), g_{11} = (3,4,5), g_{12} = (1,2)(3,5).$$

Denote by $\bar{\theta}_i$ the irreducible representations of P_5, where $i = 1, \ldots, 5$ and $\bar{\theta}_i((1,2,3,4,5)) = \zeta^{i-1}$ with $\zeta = e^{2\pi i/5}$. Consider

$$\mathrm{Ind}_{A_4}^{A_5} \bar{\theta}_2(g) = \sum_{k=1}^{12} \delta_{P_5}(g_i^{-1} g g_i) \bar{\theta}_2(g_i^{-1} g g_i).$$

For $g = (\ )$, we get 12, and for $(1,2)(3,4)$ and $(1,2,3)$, we get zero. The conjugacy class of $(1,2,3,4,5)$ contains the generator, and we get a nonzero contribution to the sum from g_1 and g_{12} giving $\zeta + \zeta^4$. The conjugacy class of $(2,1,3,4,5)$ contains $(1,2,3,4,5)^2$, and the nonzero values come from g_5 and g_7 giving the sum $\zeta^2 + \zeta^3$.

	$[()]$	$[(1,2)(3,4)]$	$[(1,2,3)]$	$[(1,2,3,4,5)]$	$[(2,1,3,4,5)]$
$\mathrm{Ind}_{P_5}^{A_5} \bar{\theta}_2(g_i)$	12	0	0	$\zeta + \zeta^4$	$\zeta^2 + \zeta^3$

From the degree we know that $\mathrm{Ind}_{A_4}^{A_5} \bar{\theta}_2$ is not irreducible. Let us determine the irreducible summands of $\mathrm{Ind}_{A_4}^{A_5} \bar{\theta}_2$. So far we know ψ_1, ψ_4, and ψ_5:

$$\langle \mathrm{Ind}_{A_4}^{A_5} \bar{\theta}_2, \psi_1 \rangle = \frac{1}{60}\left(12 + 12(\zeta + \zeta^4) + 12(\zeta^2 + \zeta^3)\right) = 0,$$

$$\langle \mathrm{Ind}_{A_4}^{A_5} \bar{\theta}_2, \psi_4 \rangle = \frac{1}{60}\left(12 \cdot 4 + 12 \cdot (\zeta + \zeta^4)(-1) + 12 \cdot (\zeta^2 + \zeta^3)(-1)\right) = 1,$$

$$\langle \mathrm{Ind}_{A_4}^{A_5} \bar{\theta}_2, \psi_5 \rangle = \frac{1}{60}(12 \cdot 5) = 1.$$

This uses the relation $\zeta + \zeta^2 + \zeta^3 + \zeta^4 = -1$. The computations also give us the character

	[( )]	[(1, 2)(3, 4)]	[(1, 2, 3)]	[(1, 2, 3, 4, 5)]	[(2, 1, 3, 4, 5)]
$\operatorname{Ind}_{A_4}^{A_5} \bar{\theta}_2 - \psi_4 - \psi_5$	3	−1	0	$\zeta + \zeta^4 + 1$	$\zeta^2 + \zeta^3 + 1$

Let $\psi_2 = \operatorname{Ind}_{A_4}^{A_5} \bar{\theta}_2 - \psi_4 - \psi_5$. The inner product of ψ_2 with itself is one, and so we have found another irreducible character. The same details allow us to compute $\operatorname{Ind}_{P_5}^{A_5} \bar{\theta}_3$ leading to an irreducible character ψ_3 with the same entries as ψ_2 but with the last two entries reversed. An amusing computation with the 5th roots of unity shows that $\zeta + \zeta^4 + 1 = \dfrac{1 + \sqrt{5}}{2}$ and $\zeta^2 + \zeta^3 + 1 = \dfrac{1 - \sqrt{5}}{2}$. This completes the character table for A_5:

$[g_i]$	[( )]	[(1, 2)(3, 4)]	[(1, 2, 3)]	[(1, 2, 3, 4, 5)]	[(2, 1, 3, 4, 5)]
$\#[g_i]$	1	15	20	12	12
ψ_1	1	1	1	1	1
ψ_2	3	−1	0	$\dfrac{1 + \sqrt{5}}{2}$	$\dfrac{1 - \sqrt{5}}{2}$
ψ_3	3	−1	0	$\dfrac{1 - \sqrt{5}}{2}$	$\dfrac{1 + \sqrt{5}}{2}$
ψ_4	4	0	1	−1	−1
ψ_5	5	1	−1	0	0

From the table we see that the only character with a nontrivial kernel is the trivial representation, and so the only normal subgroups of A_5 are $\{(\,)\}$ and A_5.

The construction of the table from the properties of the group hides some of the more specific ways in which A_5 may be represented. For example, the degree 3 representations can be realized from the isomorphism between **Doc**$^+$, the group of rotations of a dodecahedron, and A_5. The rotation matrices associated with particular even permutations lead to a character χ with

$$\chi\big((1, 2)(3, 4)\big) = \operatorname{tr}(\tau) = \operatorname{tr}\begin{bmatrix} 1 & 0 & 0 \\ 0 & -1 & -1 \\ 0 & 0 & -1 \end{bmatrix} = -1,$$

$$\chi\big((1, 2, 3)\big) = \operatorname{tr}(\rho) = \operatorname{tr}\begin{bmatrix} 0 & 0 & -1 \\ -1 & 0 & 0 \\ 0 & 1 & 0 \end{bmatrix} = 0,$$

$$\chi\big((1, 2, 3, 4, 5)\big) = \operatorname{tr}(\sigma) = \operatorname{tr}\begin{bmatrix} 1 & 0 & 0 \\ 0 & \cos(2\pi/5) & -\sin(2\pi/5) \\ 0 & \sin(2\pi/5) & \cos(2\pi/5) \end{bmatrix}$$

$$= 1 + 2\cos(2\pi/5) = \frac{1 + \sqrt{5}}{2}.$$

This character agrees with ψ_2, and so the representation of A_5 as **Doc**$^+$ gives the underlying matrices.

Since A_5 is a subgroup of Σ_5, it also acts on the set $[5] = \{1, 2, 3, 4, 5\}$. The homomorphism $\rho\colon A_5 \hookrightarrow \Sigma([5]) \xrightarrow{\Xi} \mathrm{Gl}(\mathbb{F}[5])$ is a permutation representation. Its character χ_ρ satisfies $\chi_\rho(g) = \#\mathrm{Fix}(g)$. The number of fixed points of a permutation is determined by its expression as a product of disjoint cycles, namely, any missing elements in the product are fixed points. The character values are given by

	$[()]$	$[(1, 2)(3, 4)]$	$[(1, 2, 3)]$	$[(1, 2, 3, 4, 5)]$	$[(2, 1, 3, 4, 5)]$
χ_ρ	5	1	2	0	0

You can recognize this as $\mathrm{Ind}_{A_4}^{A_5}\theta_1 = \psi_1 + \psi_4$, and permutation matrices give the underlying matrices.

More generally, if $\rho\colon G \to \mathrm{Gl}(\mathbb{F}X)$ is a permutation representation, then by Burnside's Orbit Counting Theorem the character counting fixed sets satisfies

$$\langle \chi_\rho, \chi_1 \rangle = \frac{1}{\#G} \sum_{g \in G} \#\mathrm{Fix}(g) = \# \text{ number of orbits of } G \text{ in } X,$$

where χ_1 is the trivial representation. Recall that an action is transitive if for every pair $x, y \in X$, there is $g \in G$ such that $\rho(g)(x) = y$. This implies that everything in X is in a single orbit, $\langle \chi_\rho, \chi_1 \rangle = 1$, and $\chi_\rho - \chi_1$ is another character.

A group G acts doubly transitively on X if, given (x_1, y_1) and (x_2, y_2) pairs in $X \times X$, there is an element $g \in G$ such that $\rho(g)(x_1) = y_1$ and $\rho(g)(x_2) = y_2$.

Proposition 4.45 *Suppose G acts on a set X, $\rho\colon G \to \Sigma(X)$, and $\bar{\rho} = \Xi \circ \rho$ is the induced permutation representation, where $\Xi\colon \Sigma(X) \to \mathrm{Gl}(\mathbb{F}X)$. When G acts doubly transitively on a set X, the permutation character associated with the action $\chi_{\bar{\rho}}$ satisfies $\chi_{\bar{\rho}} = \chi_1 + \chi$ with $\chi \neq \chi_1$ and χ irreducible.*

Proof We have seen that $\langle \chi_{\bar{\rho}}, \chi_1 \rangle = 1$, and so $\chi_{\bar{\rho}} = \chi_1 + \chi$ for some character χ. To compute $\langle \chi, \chi \rangle$, we first expand the action of G on X to an action of G on $X \times X$ by

$$\rho_2(g)(x, y) = \big(\rho(g)(x), \rho(g)(y)\big).$$

This is easily seen to be an action of G on $X \times X$, and so the associated permutation representation has $\chi_{\rho_2}(g) = \#\mathrm{Fix}_2(g)$, where $\mathrm{Fix}_2(g) = \{(x, y) \mid \rho_2(g)(x, y) = (x, y)\}$. By the definition of ρ_2 it follows that $\mathrm{Fix}_2(g) =$

Fix$(g) \times$ Fix(g). With this observation, we have

$$\langle \chi_{\bar\rho}, \chi_{\bar\rho} \rangle = \frac{1}{\#G} \sum_{g \in G} \big(\# \mathrm{Fix}(g)\big)\big(\# \mathrm{Fix}(g)\big) = \frac{1}{\#G} \sum_{g \in G} \# \mathrm{Fix}_2(g).$$

The latter sum is the number of orbits of the expanded action of G on $X \times X$. If G acts doubly transitively on X, then the action of G on $X \times X$ has two orbits, $\Delta = \{(x, x) \mid x \in X\}$ and $X \times X - \Delta$. Thus $\langle \chi_{\bar\rho}, \chi_{\bar\rho} \rangle = 2$. Because a doubly transitive action is transitive, $\langle \chi_{\bar\rho}, \chi_1 \rangle = 1$. Write $\langle \chi, \chi \rangle = \langle \chi_{\bar\rho} - \chi_1, \chi_{\bar\rho} - \chi_1 \rangle$ and compute using linearity to see that χ is irreducible. $\qquad\square$

The action of A_5 on [5] is doubly transitive. Proposition 4.45 immediately gives us that the character of the permutation representation minus the trivial representation is an irreducible representation, ψ_4 in our table.

Exercises

4.1 What are the conjugacy classes of the group $\mathrm{Gl}_2(\mathbb{F}_p)$, where p is a prime number?

4.2 Prove that equivalence of representations is an equivalence relation.

4.3 Prove that if $\rho \colon G \to \mathrm{Gl}(V)$ is faithful and irreducible, then $Z(G) = \{g \in G \mid |\chi_\rho(g)| = \chi(e)\}$.

4.4 For the integers from 2 to 20, determine the list of degrees that are possible for the irreducible representations of groups G where $2 \le \#G \le 20$.

4.5 If $x \in G$ and $x \ne e$, then there is an irreducible character $\chi_\rho \in \mathrm{Irr}(G)$ such that $\chi_\rho(e) \ne \chi_\rho(x)$.

4.6 Suppose G is a group of order 12. Show that $\#\mathrm{Cl}(G) \ne 9$ and that possible values are 4, 6, or 12. Are there groups with these cardinalities of sets of conjugacy classes?

4.7 Suppose χ is an irreducible character of G and $\chi = \overline\chi$. Prove that $\langle \chi, \chi_1 \rangle = \frac{1}{\#G}(\chi(e) + 2\alpha)$, where α is an algebraic integer. Show that this implies that $\chi = \chi_1$, the trivial representation.

4.8 Suppose G is a group of order $2p$, where p is a prime. Show that the degrees of the irreducible representations of G are either all 1, or they are $1, 1, 2, \ldots, 2$, where there are $(p-1)/2$-many degree two representations.

4.9 Suppose G is a group of order 10 for which the number of conjugacy classes is four. Suppose you have these lines of a character table:

	[e]	[g₂]	[g₃]	[g₄)]
χ_0	1	1	1	1
χ_1	2	$\dfrac{-1+\sqrt{5}}{2}$	$\dfrac{-1-\sqrt{5}}{2}$	0

Complete the character table.

4.10 Express each normal subgroup of D_8 as an intersection of kernels of irreducible characters.

4.11 Prove that a subgroup N of a finite group G is normal if and only if $N = \bigcap_{i=1}^{k} \ker \chi_i$ for a set of distinct irreducible characters $\{\chi_1, \ldots, \chi_k\}$.

4.12 Suppose $S = \{\alpha_1, \alpha_2, \ldots, \alpha_k\}$ is a finite set of algebraic integers. If $m_i(x)$ is a monic integer polynomial with α_i as a root, then $m_i(\alpha_i) = 0$ implies $\alpha_i^d = -a_{d-1}\alpha_i^{d-1} - \cdots - a_1\alpha_i - a_0$, where $m_i(x) = x^d + a_{d-1}x^{d-1} + \cdots + a_1 x + a_0$. Let $T = \{\alpha_1^{e_1}\alpha_2^{e_2} \cdots \alpha_k^{e_k} \mid 0 \le e_i < \deg m_i(x)\}$. Suppose $\langle \mathbb{Z}, S \rangle$ denotes the smallest subring of $\mathbb{C}$ that contains S and $\mathbb{Z}$. Show that every element in $\langle \mathbb{Z}, S \rangle$ is an integral linear combination of elements of T, and hence $\langle \mathbb{Z}, S \rangle$ is finitely generated.

4.13 Suppose that $A \subset \mathbb{C}$ is a subring that contains $\mathbb{Z}$. Furthermore, suppose A consists of the integral linear combinations of the elements of a finite set U. Show that U consists of algebraic integers. (Hint: If $r \in U$, then the set $\{1, r, r^2, \ldots\}$. is linearly dependent, and so some $r^d = -(a_{r,d-1}r^{d-1} + \cdots + a_{r,1}r + a_{r,0})$ with $a_{r,i} \in \mathbb{Z}$.)

4.14 Let $[g_i]$ denote the conjugacy class in G of g_i. In Section 4.2, we showed that the linear transformation $\rho([g_i]) \colon V \to V$ is G-equivariant, and so, if ρ is irreducible, then $\rho([g_i])(v) = \lambda_i v$ for all i. The relations between conjugacy classes discussed in Section 4.2 lead to relations of the form $\lambda_i \lambda_j = \sum_k a_{ijk} \lambda_k$. Show that the set $S = \{\lambda_1, \ldots, \lambda_t\}$ and the subring $\langle S, \mathbb{Z} \rangle \subset \mathbb{C}$ is finitely generated, and hence λ_i is an algebraic integer for $i = 1, \ldots, t$.

4.15 Work out the character table for D_{10}.

4.16 Let F_{20} denote the Frobenius group of order 20 of affine transformations of the $\mathbb{F}_5$-plane. Analyze F_{20} from the viewpoints of the Sylow Theorems, conjugacy classes, and the various restrictions of representation theory to work out the character table for F_{20}.

4.17 The symmetric groups Σ_4 and Σ_5 contain A_4 and A_5, whose character tables appear in this chapter. Work out the character tables for Σ_4 and Σ_5.

4.18 Prove the converse of Proposition 4.45: If a permutation representation has a character that satisfies $\chi_{\bar{\rho}} = \chi_1 + \chi$, where χ_1 is the trivial character, and χ is irreducible, then the action of G that gave the permutation representation is doubly transitive.

5

Warmup: Fields and Polynomials

One sees moreover that the properties and the difficulties of an equation can be quite different according to the quantities which are adjoined to it. For example, the adjunction of a quantity can render an irreducible equation reducible.

ÉVARISTE GALOIS, *The First Memoir*

5.1 Constructing Fields

Although fields have appeared earlier, we give a full definition in order to apply the various properties with precision.

Definition 5.1 A *field* is a set $\mathbb{F}$ together with two operations, addition $+$ and multiplication $\cdot$, satisfying the following properties:

(1) $(\mathbb{F}, +)$ constitutes an abelian group.
(2) $(\mathbb{F}, \cdot)$ is an abelian monoid, that is, for all $a, b, c \in \mathbb{F}, a \cdot b \in \mathbb{F}, a \cdot b = b \cdot a$, $a \cdot (b \cdot c) = (a \cdot b) \cdot c$, and there is an element $1 \in \mathbb{F}$ such that $1 \cdot a = a$ for all $a \in \mathbb{F}$.
(3) for all $a, b, c \in \mathbb{F}, a \cdot (b + c) = (a \cdot b) + (a \cdot c)$.
(4) $0 \neq 1$, and for all $a \neq 0$, there is an element $a^{-1} \in \mathbb{F}$ such that $a \cdot a^{-1} = 1$.

A subring k of a field $\mathbb{F}$ is a *subfield* if $1 \in k$, and if $u \in k$ and $u \neq 0$, then $u^{-1} \in k$.

We can summarize the definition by saying that a field is a commutative ring with unit for which every nonzero element has a multiplicative inverse. When there is no chance of confusion, we write $a \cdot b = ab$.

128

We can construct fields from particular rings in two ways – by constructing multiplicative inverses and by quotients by ideals. Let us consider each method in turn.

(1) Suppose R is a commutative ring with unit for which there are no zero divisors, that is, if $ab = 0$ in R, then either $a = 0$ or $b = 0$. Such a ring is called an *integral domain*. Consider the set $X = R \times (R - \{0\})$ and define the relation $\sim$ on X by $(a, b) \sim (c, d)$ if $ad = bc$. Let us establish that $\sim$ is an equivalence relation. First, $(a, b) \sim (a, b)$ because $ab = ba$. If $(a, b) \sim (c, d)$, then $ad = bc$. This implies $cb = da$, and so $(c, d) \sim (a, b)$. Finally, if $(a, b) \sim (c, d)$ and $(c, d) \sim (e, f)$, then $ad = bc$ and $cf = de$. This implies that $adf = bcf = bde$, and so $d(af - be) = 0$. Because R is an integral domain and $d \neq 0$, we have $af = be$ and $(a, b) \sim (e, f)$.

Let Quot(R) denote the set of equivalence classes called the *field of fractions* or the *quotient field* of R. We can define two operations on Quot(R), an addition and a multiplication,

$$[(a, b)] + [(c, d)] = [(ad + bc, bd)], \quad [(a, b)] \cdot [(c, d)] = [(ac, bd)].$$

Notice that $[(0, 1)]$ is an additive identity:

$$[(0, 1)] + [(a, b)] = [(0 \cdot b + a \cdot 1, 1 \cdot b)] = [(a, b)].$$

For all $a \neq 0$, we have $[(a, a)] = [(1, 1)]$. Also, $[(1, 1)] \cdot [(a, b)] = [(a, b)]$, so $[(1, 1)]$ is a multiplicative identity. Furthermore, if $a \neq 0$ and $b \neq 0$, then

$$[(a, b)] \cdot [(b, a)] = [(ab, ba)] = [(1, 1)].$$

Thus every nonzero equivalence class has a multiplicative inverse. Although it is tedious to verify, we encourage the reader to check that these operations are well-defined, commutative, and associative and that the distributive law holds. It follows that Quot(R) is a field.

When this construction is applied to the integers $\mathbb{Z}$, we obtain a field isomorphic to the rational numbers $\mathbb{Q}$. To establish that Quot($\mathbb{Z}$) $\cong \mathbb{Q}$, identify $[(k, l)]$ with the rational number k/l. We can also apply the construction to the field of polynomials $\mathbb{F}[x]$ to obtain the *field of rational functions*,

$$\text{Quot}(\mathbb{F}[x]) = \mathbb{F}(x) = \left\{ \frac{p(x)}{q(x)} \text{ such that } p(x), q(x) \in \mathbb{F}[x], q(x) \neq 0 \right\}.$$

(2) For a ring R, an ideal $\mathfrak{m}$ is a *maximal ideal* if whenever $J \subset R$ is any other ideal for which $\mathfrak{m} \subset J \subset R$, then either $J = \mathfrak{m}$ or $J = R$. In an integral domain, an element $a \in R$ is *irreducible* if whenever $a = bc$ in R, then either b or c is a unit in R, that is, either b^{-1} or c^{-1} exists in R.

Theorem 5.2 *For an integral domain R, an ideal* $\mathfrak{m}$ *is a maximal ideal if and only if* $R/\mathfrak{m}$ *is a field. If R is a principal ideal domain and* $a \in R$ *is irreducible, then* $R/(a)$ *is a field.*

Proof Suppose that $a + \mathfrak{m}$ is in $R/\mathfrak{m}$ and $a \notin \mathfrak{m}$. Then $a + \mathfrak{m} \neq 0 + \mathfrak{m}$ in $R/\mathfrak{m}$. Let $(a, \mathfrak{m}) = \{ra + x \mid r \in R, x \in \mathfrak{m}\}$ denote the ideal generated by a and $\mathfrak{m}$. (Can you prove that $(a, \mathfrak{m})$ is an ideal of R?) By construction, $\mathfrak{m} \subset (a, \mathfrak{m}) \subset R$. Since $a \notin \mathfrak{m}$, $\mathfrak{m} \neq (a, \mathfrak{m})$, and because $\mathfrak{m}$ is a maximal ideal, $(a, \mathfrak{m}) = R$. It follows that $1 \in (a, \mathfrak{m})$, that is, there is an $s \in R$ and an $x \in \mathfrak{m}$ such that $sa + x = 1$. Then $1 + \mathfrak{m} = sa + x + \mathfrak{m} = sa + \mathfrak{m} = (s + \mathfrak{m})(a + \mathfrak{m})$, and $a + \mathfrak{m}$ has a multiplicative inverse in $R/\mathfrak{m}$.

Conversely, suppose $\mathfrak{m} \subset R$ is an ideal and $R/\mathfrak{m}$ is a field. If J is another ideal with $\mathfrak{m} \subset J \subset R$ and $\mathfrak{m} \neq J$, then there is an element $a \in J$ such that $a \notin \mathfrak{m}$. Since $a + \mathfrak{m} \neq \mathfrak{m}$, there is an element $b + \mathfrak{m}$ such that $(a + \mathfrak{m})(b + \mathfrak{m}) = 1 + \mathfrak{m}$, that is, $ab - 1 \in \mathfrak{m} \subset J$. However, $ab \in J$, and so $1 \in J$ and $J = R$. Thus $\mathfrak{m}$ is a maximal ideal.

The second statement follows from the fact that the principal ideal $(a) = \{ra \mid r \in R\}$ is a maximal ideal when a is irreducible. To see this, suppose $J \subset R$ is an ideal and $(a) \subset J \subset R$. Because R is a principal ideal domain, $J = (b)$ for some $b \in R$. Thus $a \in (b)$ and $a = bc$. Since a is irreducible, either b is a unit, or c is a unit. If c is a unit, then let c^{-1} denote its multiplicative inverse. We have $c^{-1}a = b$ and $b \in (a)$; thus $(a) = (b)$. If b is a unit, then $b^{-1}b \in (b)$ and $(1) = R \subset (b)$. Thus $(b) = R$, and (a) is a maximal ideal.   □

The prime integers p are irreducible in $\mathbb{Z}$, a principal ideal domain. So $\mathbb{Z}/(p) = \mathbb{Z}/p\mathbb{Z}$ is a field. If $\mathbb{F}$ is a field and $f(x) \in \mathbb{F}[x]$ is an irreducible polynomial, then $\mathbb{F}[x]/(f(x))$ is a field.

Definition 5.3 A *field homomorphism* is a ring homomorphism $\phi \colon \mathbb{F} \to k$ taking the multiplicative unit $1 \in \mathbb{F}$ to $1 \in k$.

A field in German is *ein Körper* (in French, *un corps*), hence the use of k and K. A key feature of the theory of fields that distinguishes it from the theory of rings is the following property.

Proposition 5.4 *A field homomorphism is always injective.*

Proof Suppose $\phi \colon \mathbb{F} \to k$ is a field homomorphism. Then $\ker \phi$ is an ideal in $\mathbb{F}$. Suppose $u \in \ker \phi$ is nonzero. Then there is a multiplicative inverse $u^{-1} \in \mathbb{F}$, and since $u^{-1}u \in \ker \phi$, we have $1 \in \ker \phi$. Then $\mathbb{F} \cdot 1 \subset \ker \phi$, and ϕ is the zero mapping. However, a field homomorphism takes 1 to 1, so ϕ cannot be the zero mapping. Thus $\ker \phi = \{0\}$, and ϕ is injective.   □

Definition 5.5 If $\mathbb{F}$ is a subfield of k, then we say that k is an *extension* of $\mathbb{F}$. When $\phi\colon \mathbb{F} \to k$ is a field homomorphism, we can identify $\mathbb{F}$ with its image in k and view the homomorphism as an inclusion and k as an extension of $\mathbb{F}$.

The classical fields present a tower of extensions $\mathbb{Q} \subset \mathbb{R} \subset \mathbb{C}$.

Every field contains a multiplicative unit 1. The unit in $\mathbb{Z}$ is also an additive generator. The interplay between the multiplication and addition in a field leads to a numerical invariant, the *characteristic* of the field. Consider the ring homomorphism $\varphi\colon \mathbb{Z} \to \mathbb{F}$ defined by $\varphi(1) = 1$. Then $\varphi(n) = \varphi(1+\cdots+1) = \varphi(1) + \cdots + \varphi(1) = n \cdot 1$. Since $mn = n + \cdots + n$ (m times), φ is a ring homomorphism. The kernel of φ is an ideal in $\mathbb{Z}$, and hence $\ker \varphi = (m)$ for some $m \in \mathbb{Z}$. If $m = 0$, then φ is injective, and we say that $\mathbb{F}$ has *characteristic zero*. If $m \neq 0$, then m cannot be composite: if $m = ab$ with $1 < a, b < m$, then $(a+m\mathbb{Z})(b+m\mathbb{Z}) = 0+m\mathbb{Z}$, and $\mathbb{Z}/m\mathbb{Z}$ contains zero divisors. However, $\mathbb{Z}/m\mathbb{Z} \cong \varphi(\mathbb{Z}) \subset \mathbb{F}$, and $\mathbb{F}$ does not contain zero divisors. Hence m must be a prime p, and we say that $\mathbb{F}$ has *characteristic p*.

When $\mathbb{F}$ has characteristic zero, $\mathbb{Z} \cong \varphi(\mathbb{Z}) \subset \mathbb{F}$, and so $\mathbb{Q} \cong \mathrm{Quot}(\mathbb{Z}) \cong \mathrm{Quot}(\varphi(\mathbb{Z})) \subset \mathbb{F}$. Thus $\mathbb{F}$ is an extension of $\mathbb{Q}$. When $\mathbb{F}$ has characteristic p, $\mathbb{F}$ is an extension of $\mathbb{Z}/p\mathbb{Z} \cong \varphi(\mathbb{Z}) \subset \mathbb{F}$. To emphasize the nature of $\mathbb{Z}/p\mathbb{Z}$ as a field, we denote the field of p elements by $\mathbb{F}_p$. The characteristic of a field determines its *prime subfield*, which is the intersection of all subfields of $\mathbb{F}$. Because every subfield k contains 1, $\varphi(\mathbb{Z})$ is a subset of k, and $\mathrm{Quot}(\varphi(\mathbb{Z})) \subset k$.

Extensions of fields come with some extra structure.

Theorem 5.6 *If k is an extension of $\mathbb{F}$, then k is a vector space over $\mathbb{F}$.*

Proof The field k is an abelian group with $+$. Multiplication by scalars in $\mathbb{F}$ is given by restricting the multiplication $\cdot\colon k \times k \to k$ to mbs: $\mathbb{F} \times k \to k$. The other vector space properties follow from the field axioms: for $a, b \in \mathbb{F}$ and $u, v \in k$, $a(u + v) = (au) + (av)$, $1 \cdot u = u$, $(a+b)u = (au) + (bu)$, and $(ab)u = a(bu)$. $\qquad\square$

Definition 5.7 An extension k of $\mathbb{F}$ is a *finite extension* if the dimension of k over $\mathbb{F}$ is finite. Otherwise, k is an *infinite extension* of $\mathbb{F}$. Let $[k : \mathbb{F}] = \dim_{\mathbb{F}} k$, the dimension of k over $\mathbb{F}$.

For example, $\mathbb{R}$ is an infinite extension of $\mathbb{Q}$, whereas $\mathbb{C}$ is a finite extension of $\mathbb{R}$ with $[\mathbb{C} : \mathbb{R}] = 2$. In the case of fields satisfying $\mathbb{F} \subset k \subset K$, we say k is an *intermediate field* of the extension $\mathbb{F} \subset K$.

Theorem 5.8 *If k is an intermediate field of $\mathbb{F}$ and K, that is, $\mathbb{F} \subset k \subset K$, then K is an extension of $\mathbb{F}$, and $[K : \mathbb{F}] = [K : k][k : \mathbb{F}]$.*

Proof Suppose $[k : \mathbb{F}] = n$ and $[K : k] = m$. Let $\{b_1, \ldots, b_n\}$ be a basis for k over $\mathbb{F}$, and let $\{c_1, \ldots, c_m\}$ be a basis for K over k.

If $u \in K$, then $u = \alpha_1 c_1 + \cdots + \alpha_m c_m$ with $\alpha_i \in k$. For each α_i, we can write $\alpha_i = \beta_{1i} b_1 + \cdots + \beta_{ni} b_n$ with $\beta_{ji} \in \mathbb{F}$. Substitution into the expression for u gives

$$u = (\beta_{11} b_1 + \cdots + \beta_{n1} b_n) c_1 + \cdots + (\beta_{1m} b_1 + \cdots + \beta_{nm} b_n) c_m$$

$$= \sum_{j=1}^{m} \sum_{i=1}^{n} \beta_{ij} b_i c_j,$$

and so the set $\{b_i c_j\}$ is a spanning set for K as a vector space over $\mathbb{F}$.

Suppose $0 = (\beta_{11} b_1 + \cdots + \beta_{n1} b_n) c_1 + \cdots + (\beta_{1m} b_1 + \cdots + \beta_{nm} b_n) c_m$. The linear independence of the set $\{c_j\}$ implies $\beta_{1i} b_1 + \cdots + \beta_{ni} b_n = 0$ for each i. Linear independence of the set $\{b_i\}$ implies $\beta_{ij} = 0$ for all i, j. Therefore $\{b_i c_j\}$ is linearly independent and a basis for K over $\mathbb{F}$. In this case,

$$[K : \mathbb{F}] = mn = [K : k][k : \mathbb{F}].$$

If $[k : \mathbb{F}]$ is infinite, then no finite set of elements in k spans k over $\mathbb{F}$. It follows that even if $[K : k] < \infty$, any finite linearly independent set of elements of k over $\mathbb{F}$ gives a linearly independent set of products of elements in K over $\mathbb{F}$. Since the set cannot span, $[K : \mathbb{F}] = \infty$. A similar argument holds when $[K : k] = \infty$.   □

Definition 5.9 If k is an extension of $\mathbb{F}$ and $S \subset k$ is any subset, let $\mathbb{F}(S)$ denote the intersection of all subfields of k that contain S. Then $\mathbb{F}(S)$ is an extension of $\mathbb{F}$ obtained by *adjoining S to $\mathbb{F}$*.

By definition, $\mathbb{F}(S) \subset k$. If S and T are subsets of k, then

$$\mathbb{F}(S \cup T) = \mathbb{F}(S)(T) = \mathbb{F}(T)(S).$$

When $S = \{a_1, \ldots, a_n\}$, we write $\mathbb{F}(S) = \mathbb{F}(a_1, \ldots, a_n)$.

For example, between $\mathbb{Q}$ and $\mathbb{R}$, let $S = \{\sqrt{2}\}$. We claim that the field $\mathbb{Q}(\sqrt{2}) = \{r + s\sqrt{2} \mid r, s \in \mathbb{Q}\}$. Clearly, $\mathbb{Q} \subset \mathbb{Q}(\sqrt{2})$. The addition and multiplication on $\mathbb{R}$ restricted to $\mathbb{Q}(\sqrt{2})$ are closed: $(r + s\sqrt{2})(t + u\sqrt{2}) = (rt + 2su) + (ru + st)\sqrt{2}$. If $r + s\sqrt{2} \neq 0$, then

$$\frac{1}{r + s\sqrt{2}} = \frac{r - s\sqrt{2}}{(r + s\sqrt{2})(r - s\sqrt{2})} = \frac{r - s\sqrt{2}}{r^2 - 2s^2} = \frac{r}{r^2 - 2s^2} - \frac{s}{r^2 - 2s^2}\sqrt{2},$$

which is in $\mathbb{Q}(\sqrt{2})$. The set $\{1, \sqrt{2}\}$ is a basis for $\mathbb{Q}(\sqrt{2})$ over $\mathbb{Q}$, and so $[\mathbb{Q}(\sqrt{2}) : \mathbb{Q}] = 2$.

Between $\mathbb{R}$ and $\mathbb{C}$, if we let $S = \{i\}$, then we have the field $\mathbb{R}(i) = \{a + bi \mid a, b \in \mathbb{R}\}$. This presentation follows because $i^2 = -1$. In this case, $\mathbb{C} = \mathbb{R}(i)$.

An extension k of $\mathbb{F}$ is called *simple* if $k = \mathbb{F}(a)$ for some $a \in k$. Both $\mathbb{R}(i)$ and $\mathbb{Q}(\sqrt{2})$ are simple extensions. We can explore simple extensions of $\mathbb{F} \subset \mathbb{F}(\alpha) \subset k$ using the polynomial ring $\mathbb{F}[x]$. Let us recall some of the important properties of this familiar ring.

Proposition 5.10 *The ring $\mathbb{F}[x]$ is a Euclidean domain.*

Proof A *Euclidean domain* R is an integral domain together with a function $v: R \to \mathbb{Z}$, the norm, satisfying (1) If $a = bc$ in R, then $v(b) \le v(a)$; (2) (*Division Algorithm*) for any $a, b \in R$ with $b \ne 0$, there are elements $q, r \in R$ for which $a = qb + r$ and $v(r) < v(b)$. For the integral domain $\mathbb{F}[x]$, the norm is the degree, $v(p(x)) = \deg p(x)$. The Division Algorithm for $\mathbb{F}[x]$ follows from long division of polynomials. $\qquad\square$

Some immediate consequences are the following:

(1) $\mathbb{F}[x]$ is a principal ideal domain and hence a unique factorization domain.

If $J \subset \mathbb{F}[x]$ is an ideal, let $m(x)$ denote a polynomial in J for which $\deg m(x) \le \deg p(x)$ for all $p(x) \in J$. Since the degree takes its values in the nonnegative integers, the Well-Ordering Principle [71] guarantees that such a polynomial exists. For any other polynomial $q(x)$ in J, the Division Algorithm implies $q(x) = Q(x)m(x) + R(x)$ with $\deg R(x) < \deg m(x)$. Since J is an ideal, $R(x) = q(x) - Q(x)m(x)$ is in J, but $R(x)$ can have degree less than the degree of $m(x)$ only if $R(x) = 0$, that is, $q(x)$ is a multiple of $m(x)$ and $J = (m(x))$.

(2) Suppose k is an extension of $\mathbb{F}$ and $p(x) \in \mathbb{F}[x]$. If $\alpha \in k$ is a root of $p(x)$, then $p(x)$ is divisible by $x - \alpha$ in $k[x]$. Furthermore, $p(x)$ can have at most $n = \deg p(x)$ many roots in k.

The Division Algorithm in $k[x]$ allows us to write $p(x) = (x-\alpha)q(x)+r(x)$ with $\deg r(x) < \deg(x - \alpha) = 1$. Hence $r(x)$ is a constant polynomial. If we evaluate both sides at $x = \alpha$, then we see that $r(x) = 0$, and so $x - \alpha$ divides $p(x)$. The second statement follows by induction on the degree of $p(x)$.

(3) If $p(x) \in \mathbb{F}[x]$ is an irreducible polynomial, then $\mathbb{F}[x]/(p(x))$ is a field and an extension of $\mathbb{F}$. In fact, the polynomial $p(x)$ has a root in the field $k = \mathbb{F}[x]/(p(x))$. If $p(x) = a_0 + a_1 x + \cdots + a_n x^n$, then

$$\begin{aligned}
p(x + (p(x))) &= a_0 + a_1(x + (p(x))) + \cdots + a_n(x + (p(x)))^n \\
&= (a_0 + a_1 x + \cdots + a_n x^n) + (p(x)) \\
&= p(x) + (p(x)) = 0 + (p(x)).
\end{aligned}$$

Thus every irreducible polynomial $f(x) \in \mathbb{F}[x]$ has a root in some extension of $\mathbb{F}$.

Get to know finite extensions

Suppose k is an extension of $\mathbb{F}$ and $\alpha \in k$. Define the *evaluation mapping* $\text{ev}_\alpha : \mathbb{F}[x] \to k$ by $\text{ev}_\alpha(p(x)) = p(\alpha)$.

1. Prove that ev_α is a ring homomorphism.
2. The kernel of the evaluation mapping is an ideal in $\mathbb{F}[x]$, and so either $\ker \text{ev}_\alpha = \{0\}$, or $\ker \text{ev}_\alpha = (m_\alpha(x))$, where $m_\alpha(x)$ may be chosen to be a monic polynomial. Show that $m_\alpha(x)$ is uniquely determined and irreducible in $\mathbb{F}[x]$. We call $m_\alpha(x)$ the *minimal polynomial* of α over $\mathbb{F}$.
3. Suppose k is an extension of $\mathbb{F}$ and $[k : \mathbb{F}] = p$, a prime integer. Show that if $\alpha \in k$ and $\alpha \notin \mathbb{F}$, then $k = \mathbb{F}(\alpha)$.
4. Explain why $\mathbb{R}$ is an infinite extension of $\mathbb{Q}$.
5. In $\mathbb{R}$, show that $\mathbb{Q}(\sqrt{2} + \sqrt{3}) = \mathbb{Q}(\sqrt{2}, \sqrt{3})$. What is the degree $[\mathbb{Q}(\sqrt{2} + \sqrt{3}) : \mathbb{Q}]$? This shows that a multiple extension like $\mathbb{Q}(\sqrt{2}, \sqrt{3})$ may be expressed as a simple extension.

Let us look at $\mathbb{Q}(\sqrt{2})$ through the eyes of the polynomial ring. The kernel of the evaluation homomorphism $\text{ev}_{\sqrt{2}} : \mathbb{Q}[x] \to \mathbb{R}$ is the ideal generated by $x^2 - 2$. The First Isomorphism Theorem for rings gives us an isomorphism $\mathbb{Q}[x]/(x^2 - 2) \cong \text{ev}_{\sqrt{2}}(\mathbb{Q}[x])$. We write $\text{ev}_{\sqrt{2}}(\mathbb{Q}[x])$ as $\mathbb{Q}[\sqrt{2}]$. Notice that

$$\text{ev}_{\sqrt{2}}(p(x)) = p(\sqrt{2}) = a_0 + a_1\sqrt{2} + \cdots + a_n(\sqrt{2})^n = r + s\sqrt{2}$$

with $r, s \in \mathbb{Q}$. This is because $(\sqrt{2})^{2k} = 2^k \in \mathbb{Q}$ and $(\sqrt{2})^{2k+1} = 2^k\sqrt{2}$. Thus

$$\mathbb{Q}[x]/(x^2 - 2) \cong \mathbb{Q}[\sqrt{2}] = \{r + s\sqrt{2} \mid r, s \in \mathbb{Q}\}.$$

Since $\mathbb{Q}(\sqrt{2})$ is the smallest subfield of $\mathbb{R}$ that contains $\mathbb{Q}$ and $\sqrt{2}$, and $\mathbb{Q}[\sqrt{2}]$ is a field, we have $\mathbb{Q}[\sqrt{2}] = \mathbb{Q}(\sqrt{2})$. In this form, we see that $\mathbb{Q}(\sqrt{2})$ is an extension of $\mathbb{Q}$ of degree two. This is an example of a more general fact.

Proposition 5.11 *If $p(x) \in \mathbb{F}[x]$ is an irreducible polynomial of degree n, then $\mathbb{F}[x]/(p(x))$ is an extension of $\mathbb{F}$ in which $p(x)$ has a root. Also, $\mathbb{F}[x]/(p(x))$ is a finite extension of $\mathbb{F}$ of degree n, that is, $[\mathbb{F}[x]/(p(x)) : \mathbb{F}] = n$.*

Proof Let $B = \{1 + (p(x)), x + (p(x)), \dots, x^{n-1} + (p(x))\} \subset \mathbb{F}[x]/(p(x))$. We claim that B is a basis for $\mathbb{F}[x]/(p(x))$ over $\mathbb{F}$. Let $q(x) + (p(x)) \in \mathbb{F}[x]/(p(x))$. By the Division Algorithm for $\mathbb{F}[x]$, we can write $q(x) =$

$Q(x)p(x) + r(x)$ with $\deg r(x) < \deg p(x) = n$. In the quotient field, $q(x) + (p(x)) = r(x) + (p(x))$ and $r(x) + (p(x)) \in \mathrm{Span}(B)$. So B spans $\mathbb{F}[x]/(p(x))$.

Suppose $b_0 + b_1 x + \cdots + b_{n-1} x^{n-1} + (p(x)) = 0 + (p(x))$. Then $b_0 + b_1 x + \cdots + b_{n-1} x^{n-1} \in (p(x))$, and so $b_0 + b_1 x + \cdots + b_{n-1} x^{n-1} = Q(x)p(x)$ for some polynomial $Q(x)$. However, the degree of $b_0 + b_1 x + \cdots + b_{n-1} x^{n-1}$ is $n - 1 < \deg p(x) + \deg Q(x)$, so this equation only holds if $Q(x) = 0$. Then $b_0 + b_1 x + \cdots + b_{n-1} x^{n-1} = 0$, which implies that $b_0 = b_1 = \cdots = b_{n-1} = 0$, and B is linearly independent. Thus $[\mathbb{F}[x]/(p(x)) : \mathbb{F}] = n$. $\qquad\square$

It is not always the case that the kernel of $\mathrm{ev}_\alpha : \mathbb{F}[x] \to k$ is nontrivial. For example, for $\alpha = e$ or π in $\mathbb{R}$, ev_e and $\mathrm{ev}_\pi : \mathbb{Q}[x] \to \mathbb{R}$ have trivial kernels because neither e nor π is a root of a polynomial with coefficients in $\mathbb{Q}$ [51]. If $\alpha \in k$ is not the root of any polynomial with coefficients in $\mathbb{F}$, we say that α is a *transcendental number* over $\mathbb{F}$ or that α is *transcendental* over $\mathbb{F}$. When $\ker \mathrm{ev}_\alpha = (p(x)) \neq \{0\}$, we say that α is an *algebraic number* over $\mathbb{F}$ or that α is *algebraic* over $\mathbb{F}$. We derived some properties of algebraic numbers over $\mathbb{Q}$ in Chapter 4. An extension k of $\mathbb{F}$ is said to be an *algebraic extension* if every element in k is algebraic over $\mathbb{F}$.

Theorem 5.12 *If k is a finite extension of $\mathbb{F}$, then every element $\alpha \in k$ is algebraic over $\mathbb{F}$.*

Proof Suppose $[k : \mathbb{F}] = n < \infty$. It follows that any subset of k with more than n elements is linearly dependent. If $\alpha \in k$, consider $1, \alpha, \alpha^2, \ldots, \alpha^n$ in k. This set is linearly dependent, so there are values $a_0, a_1, \ldots, a_n$ in $\mathbb{F}$, not all zero, such that $a_0 + a_1 \alpha + \cdots + a_n \alpha^n = 0$. Then α is a root of the polynomial $p(x) = a_0 + a_1 x + \cdots + a_n x^n$ in $\mathbb{F}[x]$. $\qquad\square$

When we solved $x^2 - 2$ in $\mathbb{Q}[x]/(x^2 - 2)$, we obtained a root $x + (x^2 - 2)$. However, a quadratic polynomial has two roots. It is not hard to guess that $-x + (x^2 - 2)$ is the other root, but let us find it from first principles. Since $p(X) = X^2 - 2$ has the root x in $k = \mathbb{Q}[x]/(x^2 - 2)$, $p(X)$ is divisible by $X - x$. Let us divide:

$$
\begin{array}{r}
X \;+\; x \\
X - x \;)\; \overline{X^2 -2} \\
\underline{X^2 \;-\; xX } \\
xX \quad -2 \\
\underline{xX \quad -x^2}
\end{array}
$$

where the difference of the last lines vanishes because $x^2 - 2 = 0$ in k. Therefore $X^2 - 2 = (X - x)(X + x)$ in $k[X]$.

Let us try a more complicated example. Let $f(x) = x^3 - 2$. Because $\sqrt[3]{2} \notin \mathbb{Q}$, $f(x)$ is irreducible over $\mathbb{Q}$. By construction, $K = \mathbb{Q}[x]/(x^3 - 2)$ contains one root of $f(x)$, $x + (x^3 - 2)$. To find the others, we can carry out polynomial division over $K[X]$:

$$
\begin{array}{r}
X^2 \;+\; xX \;+\; x^2 \\[2pt]
\hline
X - x \,)\; X^3 -2 \\
X^3 \;-\; xX^2 \\
\hline
xX^2 \\
xX^2 \;-\; x^2 X \\
\hline
x^2 X \quad -2 \\
x^2 X \quad -x^3 \\
\hline
\end{array}
$$

The difference of the last lines vanishes because $x^3 - 2 = 0$ in K.

The quadratic formula holds for fields in which $2 \neq 0$, and so we can solve for X satisfying the equation $X^2 + xX + x^2 = 0$:

$$
X = \frac{-x \pm \sqrt{x^2 - 4x^2}}{2} = \frac{x(-1 \pm \sqrt{-3})}{2} \notin \mathbb{Q}[x]/(x^3 - 2).
$$

To obtain all the roots – there are three – we need to adjoin $\omega = \dfrac{-1 + \sqrt{-3}}{2}$ to K.

Let $L = K(\omega)$. In the ring $L[X]$, we have $X^3 - 2 = (X - x)(X - x\omega)(X - x\omega^2)$ because

$$
\omega^2 = \left(\frac{-1 + \sqrt{-3}}{2}\right)^2 = \frac{1 - 2\sqrt{-3} - 3}{4}
$$

$$
= \frac{-2 - 2\sqrt{-3}}{4} = \frac{-1 - \sqrt{-3}}{2}.
$$

We also know that $\sqrt[3]{2}$ is a real number, so we can identify K with the image of $\mathrm{ev}_{\sqrt[3]{2}} \colon \mathbb{Q}[x] \to \mathbb{R}$ as $\mathbb{Q}[\sqrt[3]{2}] = \mathbb{Q}(\sqrt[3]{2})$. The field L can be constructed in $\mathbb{C}$ by adjoining ω to $\mathbb{Q}(\sqrt[3]{2})$, giving $L \cong \mathbb{Q}(\sqrt[3]{2})(\omega) = \mathbb{Q}(\sqrt[3]{2})\left(\dfrac{-1 + \sqrt{-3}}{2}\right) = \mathbb{Q}\left(\sqrt[3]{2}, \dfrac{-1 + \sqrt{-3}}{2}\right)$. In $L[x]$, $x^3 - 2 = (x - \sqrt[3]{2})(x - \omega\sqrt[3]{2})(x - \omega^2 \sqrt[3]{2})$. We call L a *splitting field* for $x^3 - 2$ because the polynomial factors into linear factors in $L[x]$. Notice that $[L : \mathbb{Q}] = 6$.

Definition 5.13 Let $f(x) \in \mathbb{F}[x]$ be a nonconstant polynomial. A finite extension K of $\mathbb{F}$ is a *splitting field* for $f(x)$ over $\mathbb{F}$ if $f(x) = a(x - \alpha_1) \cdots (x - \alpha_n)$ in $K[x]$ with $a \in \mathbb{F}$ and $\alpha_i \in K$, and if $f(x)$ does not split into linear factors in any intermediate field between $\mathbb{F}$ and K.

For the examples of $x^2 - 2$ and $x^3 - 2$, splitting fields over $\mathbb{Q}$ were shown to be $\mathbb{Q}(\sqrt{2})$ and $\mathbb{Q}\left(\sqrt[3]{2}, \dfrac{1 + \sqrt{-3}}{2} \right)$, respectively.

Proposition 5.11 tells us that an irreducible polynomial has a root in an extension of the field of the coefficients. By factoring a given polynomial into irreducible factors the proposition can be applied iteratively to construct a splitting field. This construction was proved by LEOPOLD KRONECKER (1823–1891) [55]:

Kronecker's Theorem *If $f(x)$ is a nonconstant polynomial with coefficients in a field $\mathbb{F}$, there is an extension K of $\mathbb{F}$ for which $f(x) = a(x - \alpha_1) \cdots (x - \alpha_n)$ in $K[x]$ with $a \in \mathbb{F}$ and $\alpha_i \in K$. Furthermore, K can be chosen to be the smallest such extension giving a splitting field of $f(x)$.*

Proof We proceed by induction on the degree of $f(x)$. In the case $f(x) = ax + b$ with $a \neq 0$, we have $f(x) = a(x - b/a)$, and the theorem holds. Suppose the theorem holds for all polynomials of degree less than n over any field. Because $\mathbb{F}[x]$ is a unique factorization domain, we can factor $f(x) = u_1(x)u_2(x) \cdots u_s(x)$ with each $u_i(x)$ irreducible in $\mathbb{F}[x]$. If every $u_i(x)$ is linear, then $K = \mathbb{F}$, and we are done. If not, let $u_j(x)$ have degree greater than one. Over the field $k_1 = \mathbb{F}[x]/(u_j(x))$, $u_j(x)$ has a root $r = x + (u_j(x))$, and so also does $f(x)$. We can write $f(x) = (x - r)Q(x)$ with $\deg Q(x) < n$. Note that $Q(x) \in k_1[x]$. We apply the induction hypothesis and $Q(x) = a(x - \alpha_2) \cdots (x - \alpha_n)$ in an extension K of k_1 with $a \in k_1$ and $\alpha_i \in K$. Then $f(x) = a(x - r)(x - \alpha_2) \cdots (x - \alpha_n)$ in $K[x]$. The coefficient of the leading term of $f(x)$ is a, and so $a \in \mathbb{F}$.

Inside K, consider the subfield $L = \mathbb{F}(\alpha_1, \ldots, \alpha_n)$. Then $f(x)$ splits completely over L, and any subfield of L that is strictly smaller must leave out some root of $f(x)$, and so $f(x)$ fails to split completely in that extension. Hence L is a splitting field of $f(x)$. $\qquad\qquad\square$

When L is a splitting field of a polynomial $f(x) \in \mathbb{F}[x]$, we adopt the notation E_f for L.

A computationally difficult step in the argument of Kronecker's Theorem is the factorization of $f(x)$ into irreducible factors. We take up this difficulty next.

5.2 Irreducibility Criteria

When is a polynomial irreducible? Of course, this depends on the domain in which the coefficients are found. For example, $x^2 + 1$ has no roots in $\mathbb{R}$

because the square of a real number is positive. The degree of $x^2 + 1$ is two. If it were reducible, it would be a product of linear polynomials $x^2 + 1 = (ax + b)(cx + d)$. Since this would imply that $-b/a$ and $-d/c$ were roots of $x^2 + 1$ in $\mathbb{R}$, no such decomposition exists. Over the complex numbers, $x^2 + 1 = (x + i)(x - i)$, and thus $x^2 + 1$ is reducible; over the field $\mathbb{F}_5$, $x^2 + 1 = (x - 2)(x + 2)$ is reducible, whereas $x^2 + 1$ is irreducible over $\mathbb{F}_7$.

To be able to use the tools of divisibility, let us consider polynomials with coefficients in a unique factorization domain R whose field of quotients $\mathrm{Quot}(R)$ will play a role. In such a ring, the *greatest common divisor* of a pair of elements $a, b \in R$ is an element $d \in R$ with $a = rd$ and $b = sd$ in R, that is, d divides both a and b, and if d' divides both a and b, then d' divides d, that is, $d = cd'$. The greatest common divisor is uniquely defined up to multiplication by a unit.

By induction we can define the greatest common divisor of more than two elements:

$$\gcd(a_0, a_1, \ldots, a_n) = \gcd\big(\gcd(a_0, a_1, \ldots, a_{n-1}), a_n\big).$$

When R is also a principal ideal domain, we have the ideal $J = \{a_0x_0 + \cdots + a_nx_n \mid x_i \in R\}$. Then $J = (d)$ for some $d \in R$ and $d = \gcd(a_0, \ldots, a_n)$. (Can you prove this?) For a unique factorization domain that is not a principal ideal domain, we can factor a and b as

$$a = u_1^{e_1} u_2^{e_2} \cdots u_n^{e_n}, \quad b = u_1^{f_1} u_2^{f_2} \cdots u_n^{f_n},$$

where all u_i are distinct irreducible elements of R, and e_i, f_j are nonnegative. Then $\gcd(a, b) = u_1^{m_1} \cdots u_n^{m_n}$ where $m_i = \min\{e_i, f_i\}$.

Definition 5.14 Let $f(x) = a_0 + a_1x + \cdots + a_nx^n$ with $a_i \in R$ and $a_n \neq 0$. The *content* of $f(x)$, denoted $c(p(x))$, is defined as $c(p(x)) = \gcd(a_0, a_1, \ldots, a_n)$. A polynomial $P(x)$ whose content is one is called a *primitive polynomial*.

Notice that if $u \in R$ divides $c(f(x))$, then u divides every coefficient of $f(x)$, and we can write $f(x) = ug(x)$ with $g(x) \in R[x]$.

A key property of content was proved by CARL-FRIEDRICH GAUSS (1777–1855) in §42 of [33].

Gauss's Lemma *Let $p(x)$ and $q(x)$ be polynomials in $R[x]$ with R a unique factorization domain. Then*

$$c\big(p(x)q(x)\big) = c\big(p(x)\big)c\big(q(x)\big).$$

Proof By the properties of the greatest common divisor, if $d = \gcd(a_0, a_1, \ldots, a_n)$, then $\gcd(a_0/d, a_1/d, \ldots, a_n/d) = 1$. If $p(x) = a_0 + a_1x + \cdots a_nx^n$, then

$$p(x) = d\big((a_0/d) + (a_1/d)x + \cdots + (a_n/d)x^n\big)$$
$$= c\big(p(x)\big)p_1(x) \text{ with } c\big(p_1(x)\big) = 1.$$

Similarly, $q(x) = c(q(x))q_1(x)$ with $c(q_1(x)) = 1$. Suppose $p(x) = dp_1(x) = DP_1(x)$ with $c(p_1(x)) = c(P_1(x)) = 1$. If l is an irreducible in R that divides D, then either l divides d, or it divides all the coefficients of $p_1(x)$. Since the latter condition fails, l divides d. Checking this for all the irreducible factors of D to whatever multiplicity, we deduce that D divides d. Then $(d/D)p_1(x) = P_1(x)$. The same argument with any divisor of d/D having to divide all the coefficients of $P_1(x)$ shows us that d/D is a unit in R, and so $p_1(x)$ and $P_1(x)$ differ at most by a multiple of a unit.

We next observe that the content of $p_1(x)q_1(x)$ is also 1. Write $p_1(x) = b_0 + b_1x + \cdots + b_nx^n$ and $q_1(x) = c_0 + c_1x + \cdots + c_mx^m$. The product $p_1(x)q_1(x)$ has the coefficients given by $d_i = \sum_{l=0}^{i} b_l c_{i-l}$ for $0 \le i \le m + n$. Suppose u is an irreducible in R. Because the content of $p_1(x)$ and of $q_1(x)$ is one, there are indices j and k for which u divides $b_0, \ldots, b_{j-1}$ but u does not divide b_j and u divides $c_0, \ldots, c_{k-1}$ but not c_k. If we consider $d_{j+k} = \sum_{l=0}^{j+k} b_l c_{j+k-l}$, the summand $b_j c_k$ is not divisible by u, whereas u divides $b_l c_{j+k-l}$ for $l < j$ and for $l > j$. Hence u does not divide d_{j+k}, and $c(p_1(x)q_1(x)) = 1$. With this fact in hand, we can write $p(x)q(x) = c(p(x))p_1(x)c(q(x))q_1(x) = c(p(x))c(q(x))p_1(x)q_1(x)$.

If $p(x)q(x) = c(p(x)q(x))r(x)$ with $c(r(x)) = 1$, then we have that $r(x) = \alpha p_1(x)q_1(x)$ for α a unit in R. Thus $c(p(x)q(x)) = c(p(x))c(q(x))$. $\square$

Since the polynomial ring $R[x]$ is a subring of $\mathrm{Quot}(R)[x]$, which is a principal ideal domain, we ask if factorization in $R[x]$ differs significantly from factorization in $\mathrm{Quot}(R)[x]$.

Theorem 5.15 *Let R denote a unique factorization domain. A polynomial $f(x)$ in $R[x]$ of degree greater than one is reducible in $\mathrm{Quot}(R)[x]$ if and only if $f(x)$ factors in $R[x]$ as a product of polynomials of degree greater than or equal to one.*

Proof Suppose the polynomial $f(x)$ factors as $p(x)q(x)$ in $\mathrm{Quot}(R)[x]$. By clearing the denominators we can write $p(x) = (a/b)p_1(x)$ and $q(x) = (A/B)q_1(x)$ with $p_1(x)$, $q_1(x)$ in $R[x]$ and $c(p_1(x)) = c(q_1(x)) = 1$, $\gcd(a, b) = 1$, and $\gcd(A, B) = 1$. This leads to the relation $bBf(x) = aAp_1(x)q_1(x)$ in $R[x]$. By Gauss's Lemma, $bBc(f(x)) = uaA$ with u, a unit in R. This gives us the factorization $f(x) = u^{-1}\dfrac{uaA}{bB}p_1(x)q_1(x) = u^{-1}c(f(x))p_1(x)q_1(x)$ in $R[x]$. The converse is clear. $\square$

Corollary 5.16 *Let R denote a unique factorization domain. Then $R[x]$ is a unique factorization domain. Furthermore, if $\mathbb{F}$ is a field, then $R[x_1, \ldots, x_n]$ and $\mathbb{F}[x_1, \ldots, x_n]$ are unique factorization domains.*

Proof It follows from Theorem 5.15 that an irreducible polynomial $u(x)$ in $R[x]$ is also irreducible in $\mathrm{Quot}(R)[x]$. Since $\mathrm{Quot}(R)[x]$ is a unique factorization domain, if $f(x) = u_1(x) \cdots u_n(x) = v_1(x) \cdots v_m(x)$ in $R[x]$ with irreducible $u_i(x)$ and $v_j(x)$, then these factorizations hold in $\mathrm{Quot}(R)[x]$, and so the polynomials $u_i(x)$ and $v_j(x)$ differ by at most a factor of a unit and indexing. Hence $R[x]$ is a unique factorization domain.

If $R[x]$ is a unique factorization domain, then $R[x, y] = R[x][y]$ is also a unique factorization domain. By induction $R[x_1, \ldots, x_n]$ is a unique factorization domain for a positive integer n. For a field $\mathbb{F}$, we start with $\mathbb{F}[x_1]$ a unique factorization domain, and the corollary for $\mathbb{F}[x_1, \ldots, x_n]$ follows by induction. $\qquad\square$

In the case of $R = \mathbb{Z}$, Theorem 5.15 has useful consequences for integral polynomials of small degree. If $f(x)$ is a quadratic polynomial with integer coefficients, then

$$f(x) = a_0 + a_1 x + a_2 x^2 = (a + bx)(c + dx) = ac + (ad + bc)x + bdx^2.$$

The values of a, b, c, and d are fashioned out of divisors of a_0 and a_2. If no combination of such factors is possible, then $f(x)$ is irreducible. More generally, if $f(x) = a_0 + a_1 x + \cdots + a_n x^n$ is in $\mathbb{Z}[x]$ and $f(x)$ has a rational root a/b, then a must divide a_0, and b must divide a_n.

An effective criterion was proved by a student of Gauss, GOTTHOLD EISENSTEIN (1823–1852) in [26].

Eisenstein's criterion *Suppose $P(x) = a_0 + a_1 x + \cdots + a_n x^n$ is an integral polynomial and there is a prime integer p for which p divides $a_0, a_1, \ldots, a_{n-1}$, but p does not divide a_n. Furthermore, suppose p^2 does not divide a_0. Then $P(x)$ is irreducible in $\mathbb{Q}[x]$.*

Proof Suppose $P(x) = F(x)G(x)$ in $\mathbb{Z}[x]$, where $F(x) = b_0 + b_1 x + \cdots + b_k x^k$ and $G(x) = c_0 + c_1 x + \cdots + c_l x^l$. Consider the homomorphism of rings $\pi : \mathbb{Z} \to \mathbb{F}_p$ given by $\pi(n) = n + (p) = [n]$, the mod p congruence class of n. We can extend π to a ring homomorphism $\hat{\pi} : \mathbb{Z}[x] \to \mathbb{F}_p[x]$ where $\hat{\pi}(a_0 + a_1 x + \cdots + a_n x^n) = [a_0] + [a_1]x + \cdots + [a_n]x^n$. Because π is a homomorphism of rings, $\hat{\pi}$ takes products to products. The conditions of Eisenstein's criterion imply that $\hat{\pi}(P(x)) = [a_n]x^n$. Thus $\hat{\pi}(F(x))\hat{\pi}(G(x)) = [a_n]x^n$. Furthermore, $\mathbb{F}_p[x]$ is a unique factorization domain. The factors

$\hat{\pi}(F(x))$ and $\hat{\pi}(G(x))$ must be a unit in $\mathbb{F}_p$ times a power of x. Checking degrees, it can only be that $[a_n]x^n = [b_k]x^k[c_l]x^l$ in $\mathbb{F}_p[x]$ as this summand of $F(x)G(x)$ is the only one to contribute to $[a_n]x^n$. This implies that p divides $b_0, \ldots, b_{k-1}$ and $c_0, \ldots, c_{l-1}$. Since both b_0 and c_0 are divisible by p, p^2 divides $a_0 = b_0 c_0$, which contradicts our assumptions, and so $P(x)$ is irreducible over $\mathbb{Z}$. By Theorem 5.15, $P(x)$ is irreducible in $\mathbb{Q}[x]$. $\quad\square$

For example, $P(x) = 14 - 8x + 6x^2 - 4x^3 + x^5$ satisfies Eisenstein's criterion for the prime 2, and so $P(x)$ is irreducible.

Get to know irreducibility criteria

1. Suppose m is an integer that is not a square. Show that $x^2 - m$ is irreducible in $\mathbb{Q}[x]$.

2. For an integer c, consider the function $T_c \colon \mathbb{Z}[x] \to \mathbb{Z}[x]$ given by $T_c(p(x)) = p(x + c)$. Prove that T_c is a ring homomorphism. Show that T_c^{-1} is given by $T_c^{-1} = T_{-c}$. Finally, show that if $p(x + c)$ is irreducible in $\mathbb{Z}[x]$, then so is $p(x)$.

3. Use the previous exercise to prove a theorem of Gauss: Let p be a prime number and define

$$\Phi_p(x) = \frac{x^p - 1}{x - 1} = x^{p-1} + x^{p-2} + \cdots + x + 1.$$

Show that $\Phi_p(x)$ is irreducible by showing that $\Phi_p(x + 1)$ is irreducible. This polynomial is a *cyclotomic polynomial*. More generally, a cyclotomic polynomial is an irreducible factor of $x^n - 1$ for a positive integer n. (There is more on cyclotomic polynomials in the next chapter.)

4. Prove that a version of Eisenstein's criterion works where you reverse the roles of the leading coefficient and the constant term of a polynomial. (Hint: for a polynomial $p(x)$ of degree n, consider the polynomial $x^n p(x^{-1})$.) How would you state Eisenstein's criterion for a unique factorization domain R?

5. Using Eisenstein' criterion, come up with a half-dozen examples of irreducible polynomials in $\mathbb{Z}[x]$. In the proof of Eisenstein's criterion, we considered the ring homomorphism $\hat{\pi} \colon \mathbb{Z}[x] \to \mathbb{F}_p[x]$. Suppose we restrict our attention to monic polynomials in $\mathbb{Z}[x]$. Prove a criterion for irreducibility of monic polynomials in $\mathbb{Z}[x]$ that uses mod p reduction $\hat{\pi} \colon \mathbb{Z}[x] \to \mathbb{F}_p[x]$.

5.3 Separability

When an arbitrary polynomial $f(x) \in \mathbb{F}[x]$ splits completely over an extension k of $\mathbb{F}$, it can happen that some linear factors repeat themselves. We say that an associated root has *multiplicity m* if $f(x) = (x - \alpha)^m q(x) \in k[x]$ with $q(\alpha) \neq 0$. We say that a polynomial is *separable* if its roots are distinct, that is, it has no multiple roots. Otherwise, $f(x)$ is *inseparable*. To determine whether a polynomial is separable or not, we turn to an algebraic operation on polynomials taken from the calculus.

Definition 5.17 The *derivative* of a polynomial $p(x) = a_0 + a_1 x + \cdots + a_n x^n \in \mathbb{F}[x]$ is given by $p'(x) = a_1 + 2a_2 x + \cdots + n a_n x^{n-1}$.

This definition makes sense over every field and does not depend upon limits, though it may lead to some unfamiliar results. For example, in $\mathbb{F}_2[x]$, $(x^2 + x + 1)' = 2x + 1 = 1$.

Proposition 5.18 *The derivative operation satisfies* (a) $(p(x) + q(x))' = p'(x) + q'(x)$, (b) $(ap(x))' = ap'(x)$, *and* (c) $(p(x)q(x))' = p'(x)q(x) + p(x)q'(x)$.

Proof Properties (a) and (b) are straightforward and left to the reader. As for (c), suppose $p(x) = a_0 + a_1 x + \cdots + a_n x^n$ and $q(x) = b_0 + b_1 x + \cdots + b_m x^m$. Then $p(x)q(x) = \sum_{i=0}^{m+n} c_i x^i$, where $c_i = \sum_{j=0}^{i} a_j b_{i-j}$, and $(p(x)q(x))' = \sum_{i=1}^{m+n} i c_i x^{i-1}$. We can write $i c_i x^{i-1} = \sum_{j=0}^{i} i a_j b_{i-j} x^{i-1}$ and, using $i = j + (i - j)$,

$$i a_j b_{i-j} x^{i-1} = \left(j a_j x^{j-1} \right) \left(b_{i-j} x^{i-j} \right) + \left(a_j x^j \right) \left((i - j) b_{i-j} x^{(i-j)-1} \right).$$

We put this into the derivative of a product:

$$\left(p(x)q(x) \right)' = \sum_{i=1}^{m+n} i c_i x^{i-1}$$

$$= \sum_{i=1}^{m+n} \sum_{j=0}^{i} \left(j a_j x^{j-1} \right) \left(b_{i-j} x^{i-j} \right)$$

$$+ \sum_{i=1}^{m+n} \sum_{j=0}^{i} \left(a_j x^j \right) \left((i - j) b_{i-j} x^{(i-j)-1} \right)$$

$$= p'(x)q(x) + p(x)q'(x). \qquad \square$$

Although factoring $p(x)$ may differ in $\mathbb{F}[x]$ and in $k[x]$ when k is an extension of $\mathbb{F}$, the values of $p(x)$ and $p'(x)$ are the same. Over a field, we can compute the greatest common divisor of a pair of polynomials $p(x)$ and $q(x)$ in the same manner as we would in the integers – apply the Division Algorithm repeatedly:

$$p(x) = q(x)Q_1(x) + r_1(x) \qquad \text{where } \deg r_1(x) < \deg q(x),$$
$$q(x) = r_1(x)Q_2(x) + r_2(x) \qquad \text{where } \deg r_2(x) < \deg r_1(x),$$

$$\vdots \quad \vdots$$

$$r_{n-2}(x) = r_{n-1}(x)Q_{n-1}(x) + r_n(x) \quad \text{where } \deg r_n(x) < \deg r_{n-1}(x),$$
$$r_{n-1}(x) = r_n(x)Q_n(x).$$

Because $\gcd(p(x), q(x)) = \gcd(q(x), r_1(x)) = \cdots = \gcd(r_{i-1}(x), r_i(x))$, we have $r_n(x) = \gcd(p(x), q(x))$. The Division Algorithm is carried out by polynomial long division, and so, beginning with polynomials in $\mathbb{F}[x]$, we arrive at a gcd in $\mathbb{F}[x]$. When we perform the Division Algorithm in $k[x]$ with two polynomials in $\mathbb{F}[x]$, because polynomial long division only depends on the coefficients, the same values from $\mathbb{F}$ appear, and the gcd will be the same. With this observation, we prove the following fact.

Proposition 5.19 *A polynomial $p(x) \in \mathbb{F}[x]$ has a multiple root in k, an extension of $\mathbb{F}$, if and only if $\gcd(p(x), p'(x)) \neq 1$, that is, if $p(x)$ and $p'(x)$ are relatively prime in $\mathbb{F}[x]$.*

Proof Suppose $p(x) = (x - \alpha)^l q(x)$ in $k[x]$. Then

$$p'(x) = l(x-\alpha)^{l-1}q(x) + (x-\alpha)^l q'(x) = (x-\alpha)^{l-1}\big(lq(x) + (x-\alpha)q'(x)\big),$$

and so $(x - \alpha)^{l-1}$ divides $\gcd(p(x), p'(x))$. For $l > 1$, $\gcd(p(x), p'(x)) = f(x) \neq 1$ in $\mathbb{F}[x]$, and $(x - \alpha)^{l-1}$ divides $f(x)$ in $k[x]$.

Suppose $\gcd(p(x), p'(x)) = f(x)$ in $\mathbb{F}[x]$. Then $p(x) = f(x)q(x)$ and $p'(x) = f(x)Q(x)$. Suppose $\alpha \in k$ is a root of $f(x)$. Then $(x - \alpha)$ divides $f(x)$, $p(x)$, and $p'(x)$. Let $f(x) = (x - \alpha)g(x)$ in $k[x]$. We can write $p(x) = (x - \alpha)g(x)q(x)$ and $p'(x) = (x - \alpha)g(x)Q(x)$. It follows that

$$p'(x) = g(x)q(x) + (x - \alpha)\big(g(x)q(x)\big)' = (x - \alpha)g(x)Q(x).$$

This implies that $g(x)q(x) = (x - \alpha)(g(x)Q(x) - (g(x)q(x))')$, and so α is a root of $g(x)q(x)$. Let $g(x)q(x) = (x - \alpha)F(x)$. Then $p(x) = (x - \alpha)^2 F(x)$, and $p(x)$ has a multiple root in $k[x]$. $\qquad \square$

For fields of characteristic zero, a polynomial of nonzero degree has a nonzero derivative. If the polynomial is known to be irreducible, then $\gcd(f(x), f'(x)) = 1$ because the only divisors of an irreducible polynomial are itself and units. The only possible case of $\gcd(f(x), f'(x)) = f(x)$ is when

$f'(x) = 0$. This is impossible for polynomials of nonzero degree over a field of characteristic zero. However, if the characteristic of $\mathbb{F}$ is a prime, then it is possible that a polynomial of nonzero degree can have a zero derivative. We discuss this case in the next section.

5.4 Finite Fields

When p is a prime integer and $f(x)$ is an irreducible polynomial in $\mathbb{F}_p[x]$, we obtain a field $\mathbb{F}_p[x]/(f(x))$. The degree of this extension of $\mathbb{F}_p$ is given by the degree of $f(x)$. Finite extensions of $\mathbb{F}_p$ are finite fields, and so we obtain two finite abelian groups – one with addition $(\mathbb{F}_p[x]/(f(x)), +)$ and one with multiplication $(\mathbb{F}_p[x]/(f(x)) - \{0\}, \cdot)$. For example, over $\mathbb{F}_3$ the polynomial $x^2 + x + 2$ is irreducible because it does not have a root in $\mathbb{F}_3$. Here is the table of the multiplicative group of units in $\mathbb{F}_p[x]/(x^2 + x + 2)$:

·	1	2	x	$x+1$	$x+2$	$2x$	$2x+1$	$2x+2$
1	1	2	x	$x+1$	$x+2$	$2x$	$2x+1$	$2x+2$
2	2	1	$2x$	$2x+2$	$2x+1$	x	$x+2$	$x+1$
x	x	$2x$	$2x+1$	1	$x+1$	$x+2$	$2x+2$	2
$x+1$	$x+1$	$2x+2$	1	$x+2$	$2x$	2	x	$2x+1$
$x+2$	$x+2$	$2x+1$	$x+1$	$2x$	2	$2x+2$	1	x
$2x$	$2x$	x	$x+2$	2	$2x+2$	$2x+1$	$x+1$	1
$2x+1$	$2x+1$	$x+2$	$2x+2$	x	1	$x+1$	2	$2x$
$2x+2$	$2x+2$	$x+1$	2	$2x+1$	x	1	$2x$	$x+2$

Suppose k is any field of characteristic p. Then k is an extension of $\mathbb{F}_p = \mathbb{Z}/p\mathbb{Z}$, the prime subfield of k. If k is a finite extension of $\mathbb{F}_p$, then k is a finite-dimensional vector space over $\mathbb{F}_p$, and it is isomorphic to $\mathbb{F}_p^{\times n}$ with $n = [k : \mathbb{F}_p]$. The cardinality of k is p^n. *It is customary to denote p^n by q.*

An important number-theoretic result due to Lagrange is that for each prime p, there is a *primitive root*, a residue $[a] \in \mathbb{Z}/p\mathbb{Z}$ for which $\{[a], [a^2], \ldots, [a^{p-1}]\} = \mathbb{Z}/p\mathbb{Z} - \{0\}$. That is, the multiplicative group of units $(\mathbb{Z}/p\mathbb{Z})^\times$ is a cyclic group with generator $[a]$. For example, $\mathbb{Z}/23\mathbb{Z}^\times = \langle[5]\rangle$. This phenomenon extends to finite fields. A dance through the group table for $(\mathbb{F}_3[x]/(x^2 + x + 2))^\times$ shows it is generated by the element $x + (x^2 + x + 2)$.

Theorem 5.20 *If k is a finite extension of $\mathbb{F}_p$, then $k^\times = k - \{0\}$ is a cyclic group.*

Proof We know that $k^\times$ is an abelian group of order $q - 1$. Suppose θ is an element of maximal order in $k^\times$ and that θ has order M. If $M < q - 1$, then there must be another element α in $k^\times$ that is not a power of θ. The powers

θ^i are distinct for $0 < i \leq M$, and they all satisfy the equation $x^M - 1 = 0$. Since k is a field, $x^M - 1$ can have at most M many roots. They are, in this case, $\{\theta, \theta^2, \ldots, \theta^M\}$. If $\alpha^M = 1$, then α is among the roots of $x^M - 1$; but $\alpha \notin \{\theta, \theta^2, \ldots, \theta^M\}$, so $\alpha^M \neq 1$. Let N denote the order of α. Since $\alpha^M \neq 1$, N does not divide M, and there is a prime ℓ for which ℓ^s divides N but does not divide M. Let ℓ^t be the largest power ($t \geq 0$) of ℓ that divides M. Notice that $s > t$. Define $\theta' = \theta^{\ell^t}$ and $\alpha' = \alpha^{N/\ell^s}$. Because ℓ does not divide M/ℓ^t, $\gcd(M/\ell^t, \ell^s) = 1$. The order of $\theta'\alpha'$ is the least common multiple of M/ℓ^t and ℓ^s, which is $M\ell^{s-t} > M$, a contradiction to the choice of θ of maximal order. With this contradiction, we conclude that $M = q - 1$ and $k^\times = \langle \theta \rangle$. □

Corollary 5.21 *If k is a finite field of characteristic p and order $q = p^l$, then $k = \mathbb{F}_p(\theta)$ where the powers of $\theta \in k$ give all of the roots of $x^{q-1} - 1$ in k.*

In fields of characteristic p, $p = 1 + 1 + \cdots + 1$ (p times) $= 0$, from which it follows that $\binom{p}{i} = 0$ for $0 < i < p$. The binomial $(a + b)^p = a^p + \sum_{i=1}^{p-1} \binom{p}{i} a^i b^{p-i} + b^p = a^p + b^p$ in k, sometimes called *naive exponentiation*. Of course, $(ab)^p = a^p b^p$ and $1^p = 1$, so the function Fr: $k \to k$ given by $\mathrm{Fr}(a) = a^p$ is a field homomorphism from k to k. Field homomorphisms are injective, so Fr is injective. Because k is a finite field, Fr is also surjective and thus an isomorphism. An isomorphism from a field to itself is called an *automorphism*, and Fr is known as the *Frobenius automorphism*. Notice that this implies that every element of a finite field is a pth power. When we iterate the Frobenius automorphism, we raise elements to higher powers of p: $\mathrm{Fr}^{\circ m}(a) = a^{p^m}$.

If we add zero to the roots of $x^{q-1} - 1$, then we obtain the polynomial $x^q - x$, which has q roots in k. Therefore k is a splitting field of $x^q - x$. Furthermore, the polynomial is separable, $x^q - x = \prod_{\alpha \in k} (x - \alpha)$.

We could also work from the polynomial $x^q - x$ in $\mathbb{F}_p[x]$. We know that a splitting field for $x^q - x$ exists, say K. Then K is a finite extension of $\mathbb{F}_p$. The set of roots of $x^q - x$ has cardinality q in K, and each solution α satisfies $\alpha^q = \alpha$. Notice that sums and products of roots satisfy, for $q = p^n$,

$$(\alpha + \beta)^q = \mathrm{Fr}^{\circ n}(\alpha + \beta) = \mathrm{Fr}^{\circ n}(\alpha) + \mathrm{Fr}^{\circ n}(\beta) = \alpha + \beta,$$
$$\mathrm{Fr}^{\circ n}(\alpha\beta) = \mathrm{Fr}^{\circ n}(\alpha)\,\mathrm{Fr}^{\circ n}(\beta) = \alpha\beta.$$

The set of roots is closed under addition and multiplication. Also, $(\beta^{-1})^q = (\beta^q)^{-1} = \beta^{-1}$, and the roots of $x^q - x$ form a subfield k of K of q elements. We conclude:

Proposition 5.22 *For every $n > 0$, there is a finite field k of characteristic p with p^n elements.*

Alternatively, we could apply the evaluation mapping $ev_\theta : \mathbb{F}_p[x] \to K$ with $ev_\theta(f(x)) = f(\theta)$. The kernel of ev_θ is a principal ideal $(m(x))$ with $m(x) \in \mathbb{F}_p[x]$ and $m(x)$ irreducible in this ring. Hence $\mathbb{F}_p[x]/(m(x)) \cong \mathbb{F}_p(\theta)$. It is a combinatorial exercise to prove that there are irreducible polynomials of any degree in $\mathbb{F}_p[x]$, and the proposition would follow from this observation. (In fact, there are $\dfrac{1}{n} \sum_{d|n} \mu(d) p^{n/d}$ many irreducible polynomials of degree n in $\mathbb{F}_p[x]$, where $\mu(d)$ is the *Möbius function*. A nice account of this identity is found in [58].)

If k is a finite extension of $\mathbb{F}_p$, then k is a finite field of p^n elements with $n = [k : \mathbb{F}_p]$. Suppose L is an intermediate field, $\mathbb{F}_p \subset L \subset k$. Theorem 5.8 tells us that $[k : \mathbb{F}_p] = [k : L][L : \mathbb{F}_p]$. Since L is also a finite field of characteristic p, L contains p^m elements with $m = [L : \mathbb{F}_p]$. Theorem 5.8 implies that m must divide n. In this case, we can write $n = ml$ and

$$p^n - 1 = p^{ml} - 1 = \left(p^m\right)^l - 1 = \left(p^m - 1\right)\left(\left(p^m\right)^{l-1} + \cdots + p^m + 1\right) = \left(p^m - 1\right)t,$$

and so $x^{p^n-1} - 1 = (x^{p^m-1})^l - 1 = (x^{p^m-1} - 1)((x^{p^m-1})^{l-1} + \cdots + (x^{p^m-1}) + 1)$. It follows that the roots of $x^{p^m-1} - 1$ lie among the roots of $x^{p^n-1} - 1$ and the subfield L is a splitting field for $x^{p^m-1} - 1$. Thus for every divisor of n, there is an intermediate field L between $\mathbb{F}_p$ and k. This may remind the reader that for every divisor of n, there is a subgroup of $\mathbb{Z}/n\mathbb{Z}$ with that cardinality. More on this suggestion appears in the next chapter.

We can use the Frobenius automorphism to explore separability of polynomials over $\mathbb{F}_p$. Recall that a polynomial $P(x)$ in $\mathbb{F}[x]$ is separable if and only if $\gcd(P(x), P'(x)) = 1$. As we showed, if $\mathbb{F}$ has characteristic zero and $P(x)$ is irreducible and has degree greater than zero, then $P(x)$ is separable. For fields k of characteristic p, it is possible for a nonconstant polynomial to have zero derivative. In this case, we must have

$$P(x) = a_0 + a_1 x^p + a_2 x^{2p} + \cdots + a_n x^{np}.$$

We can rewrite $P(x)$:

$$
\begin{aligned}
P(x) &= a_0 + a_1 x^p + a_2 (x^2)^p + \cdots + a_n (x^n)^p \\
&= (b_0)^p + (b_1)^p (x^2)^p + \cdots + (b_n)^p (x^n)^p \\
&= \mathrm{Fr}\left(b_0 + b_1 x + b_2 x^2 + \cdots + b_n x^n\right).
\end{aligned}
$$

This follows because each coefficient $a_i = \mathrm{Fr}(b_i)$. Thus $P(x) = (Q(x))^p$ with $Q(x)$ in $k[x]$, and $P(x)$ is reducible. The contrapositive of this observation implies the following result.

Proposition 5.23 *If k is a finite field of characteristic p, an irreducible polynomial $P(x)$ is separable.*

Let k be a finite field of characteristic p. This choice leads to the finite groups:

$$\mathrm{Gl}_m(k) = \{m \times m \text{ invertible matrices with entries in } k\},$$
$$\mathrm{Sl}_m(k) = \{m \times m \text{ matrices with determinant one}\},$$
$$\mathrm{PSl}_m(k) = \mathrm{Sl}_m(k)/Z\big(\mathrm{Sl}_m(k)\big).$$

The orders of these groups are computable exactly as in the case of $\mathbb{F}_p$: Suppose $[k : \mathbb{F}_p] = n$ and $q = p^n$. Then

$$\# \mathrm{Gl}_m(k) = \big(q^m - 1\big)\big(q^m - q\big)\big(q^m - q^2\big) \cdots \big(q^m - q^{m-1}\big),$$
$$\# \mathrm{Sl}_m(k) = \# \mathrm{Gl}_m(k)/(q-1),$$
$$\# \mathrm{PSl}_m(k) = \# \mathrm{Gl}_m(k)/(q-1)\gcd(q-1, m).$$

The arguments in Chapter 3 and in the exercises at the end of that chapter carry over to prove that $\mathrm{PSl}_n(k)$ is a finite simple group for fields $k = \mathbb{F}_q$. In the case of $\mathbb{F}_{2^2}$, $\mathrm{PSl}_2(\mathbb{F}_{2^2})$ has 60 elements. The projective plane $\mathbb{F}_{2^2}P^2$ consists of five lines that are permuted doubly transitively by $\mathrm{PSl}_2(\mathbb{F}_{2^2})$. The reader can use this fact to show that $\mathrm{PSl}_2(\mathbb{F}_{2^2})$ is isomorphic to A_5. Among the groups of order 168, we find $\mathrm{PSl}_2(\mathbb{F}_7)$ and $\mathrm{PSl}_3(\mathbb{F}_2)$. In fact, these groups are isomorphic. A proof may be found in [46]. In the case of $\mathrm{PSl}_2(\mathbb{F}_{3^2})$, we get a group of order $360 = 6!/2 = \#A_6$. In fact, $\mathrm{PSl}_2(\mathbb{F}_{3^2}) \cong A_6$, a fact that has many proofs [46].

5.5 Algebraic Closures

The existence of a splitting field for each nonconstant polynomial in $\mathbb{F}[x]$ suggests that there might be an algebraic extension L of $\mathbb{F}$ over which *every* polynomial in $\mathbb{F}[x]$ splits completely over L. Such a field would be maximal with respect to algebraic extensions of $\mathbb{F}$ in the following sense: If M is any algebraic extension of L, and $M = L(\alpha_1, \ldots, \alpha_n)$, then each α_i is a root of a polynomial with coefficients in L. By the following lemma the α_i also satisfy polynomials with coefficients in $\mathbb{F}$ and hence are algebraic over $\mathbb{F}$. Then $\alpha_i \in L$ for all i, and it follows that $M = L$. Thus there are no further algebraic extensions of L.

Lemma 5.24 *Suppose $\mathbb{F} \subset L \subset K$ is a tower of extensions with L algebraic over $\mathbb{F}$ and K algebraic over L. Then K is algebraic over $\mathbb{F}$.*

Proof Let $\alpha \in K$. Then α is a root of a polynomial $f(x) \in L[x]$, $f(x) = a_0 + a_1 x + \cdots + a_n x^n$. Consider the intermediate field $M = \mathbb{F}(a_0, \ldots, a_n)$ that is algebraic over $\mathbb{F}$ and $[M : \mathbb{F}] < \infty$. Since $f(x) \in M[x]$, α is algebraic over M, and the minimal polynomial of α, $m_\alpha(x) \in M[x]$, is a divisor of $f(x)$. Then $[M(\alpha) : M] = \deg m_\alpha(x) < \infty$, and $[M(\alpha) : \mathbb{F}] = [M(\alpha) : M][M : \mathbb{F}]$ is finite. Therefore α is algebraic over $\mathbb{F}$, and K is algebraic over $\mathbb{F}$. $\square$

Definition 5.25 A field k is *algebraically closed* if any polynomial in $k[x]$ splits completely over k. An extension K of $\mathbb{F}$ is an *algebraic closure* of $\mathbb{F}$ if K is algebraically closed and every element of K is algebraic over $\mathbb{F}$.

Since $\mathbb{F}[x] \subset K[x]$, every polynomial in $\mathbb{F}[x]$ splits completely over the algebraic closure K of $\mathbb{F}$. Because K is algebraically closed, K is maximal among algebraic extensions of $\mathbb{F}$. But do algebraic closures exist?

Theorem 5.26 *For a field $\mathbb{F}$, there is an extension L of $\mathbb{F}$ that is an algebraic closure of $\mathbb{F}$.*

Proof (This theorem was first proved by ERNST STEINITZ (1871–1928) in [81]. The proof here is attributed to Emil Artin, following [70].) We first construct an extension of $\mathbb{F}$ in which every nonconstant polynomial $p(x)$ splits completely. Associate with $p(x)$ a variable Y_p. Let $R = \mathbb{F}[\{Y_p\}]$ denote the ring of all polynomials in the variables Y_p with coefficients in $\mathbb{F}$. Let J denote the ideal in R generated by the set $p(Y_p)$ for all nonconstant $p(x) \in \mathbb{F}[x]$. We claim that $J \neq R$. If $J = R$, then there is a finite collection of polynomials $q_1(Y_{q_1}), \ldots, q_m(Y_{q_m})$ in J with polynomials $f_1, \ldots, f_m$ in R for which $1 = f_1 q_1(Y_{q_1}) + \cdots + f_m q_m(Y_{q_m})$. By Kronecker's Theorem we can find a finite extension k of $\mathbb{F}$ in which $q_i(x)$ has a root for $i = 1, \ldots, m$. Let $r_i \in k$ be a root of $q_i(x)$. Define the function ϕ from $\{Y_p\}$ to k by sending Y_{q_i} to r_i and the other Y_p mapped to any arbitrary element of k. A set-theoretic function $\phi: \{Y_p\} \to k$ extends to a homomorphism $\hat{\phi}: R = \mathbb{F}[\{Y_p\}] \to k$. Then $\phi(1) = 1 = \phi(f_1)q_1(r_1) + \cdots \phi(f_m)q_m(r_m) = 0$, a contradiction. Hence $1 \notin J$ and $J \neq R$.

A standard result in ring theory [24] tells us that there is a maximal ideal $\mathfrak{m}$ that contains J. Let $K_1 = \mathbb{F}[\{Y_p\}]/\mathfrak{m}$, a field. Notice that every nonconstant polynomial $p(x)$ has a root in K_1, namely, $p(Y_p + \mathfrak{m}) = p(Y_p) + \mathfrak{m} = \mathfrak{m}$. Hence every variable Y_p is algebraic over $\mathbb{F}$, and polynomials in the variables $\{Y_p\}$ are algebraic over $\mathbb{F}$. Thus K_1 is an algebraic extension of $\mathbb{F}$.

We can repeat this construction for K_1 to obtain a field K_2 in which every nonconstant polynomial with coefficients in K_1 has a root in K_2. Continuing in this manner, we construct a tower of field extensions

$$\mathbb{F} \subset K_1 \subset K_2 \subset \cdots .$$

The union of the fields K_i is a field $L = \bigcup_i K_i$, which is an extension of $\mathbb{F}$. If $P(x)$ is a polynomial in $L[x]$, then it lies in some $K_j[x]$, and hence there is a root of $P(x)$ in $K_{j+1} \subset L$. Thus L is algebraically closed. Furthermore, any element α of L is algebraic over $\mathbb{F}$ because α lies in some K_i, and α is algebraic over K_{i-1}. By Lemma 5.24, α is algebraic over $\mathbb{F}$. Therefore L is an algebraic closure of $\mathbb{F}$.   □

The *Fundamental Theorem of Algebra* states that $\mathbb{C}$ is an algebraic closure of $\mathbb{R}$. Proofs of this fact abound, beginning with Gauss' doctoral thesis [32]. Most proofs involve some analytic or topological facts. A proof using algebraic methods (and a little analysis) will be given in the next chapter.

Exercises

5.1   For a polynomial $a_0 + a_1 x + \cdots + a_n x^n \in \mathbb{Z}[x]$ having a root $m/n \in \mathbb{Q}$, show that m is a divisor of a_0 and n is a divisor of a_n.

5.2   Let p denote a prime integer, and let $r_p \colon \mathbb{Z} \to \mathbb{F}_p$ be reduction modulo p. Extending r_p to the coefficients of a polynomial gives a mapping $r_p \colon \mathbb{Z}[x] \to \mathbb{F}_p[x]$ defined by $r_p(a_0 + a_1 x + \cdots + a_n x^n) = [a_0] + [a_1]x + \cdots + [a_n]x^n$. Suppose $\deg f(x) = \deg r_p(f(x))$ and $r_p(f(x))$ is irreducible in $\mathbb{F}_p[x]$. Show that $f(x)$ is irreducible in $\mathbb{Z}[x]$.

On compass and straightedge constructions

Among the classical Greek problems of antiquity, there are *doubling the cube*: given a cube on side s, construct using compass and straightedge the side of a cube with twice the volume; and *trisecting an angle*: given an angle $\angle ABC$, construct using compass and straightedge an angle which is one-third of the given angle. As it turns out, these problems can be reframed in terms of extensions of fields. In the following exercises, we translate the notion of compass-and-straightedge constructions into field theory and then explore the classical problems.

5.3   Let us begin with the *rational plane*, that is, the set of points in $\mathbb{R}^2$ of the form (a, b) with both a and b rational numbers. The tools, straightedge and compass, allow us to join any two points that have been constructed so far (straightedge) and form a circle with center any point constructed so far and radius a length among the pairs of points already constructed (compass). To construct new points, we can intersect constructed lines and lines, lines and circles, and circles with circles. For example, Euclid's Proposition I.1 shows how we can construct an equilateral triangle on any base. If the base has rational length, then show that the opposite

point has coordinates one of which is irrational. If we add that irrational
to the plane, then show that we have constructed the extension of $\mathbb{Q}$ given
by $\mathbb{Q}(\sqrt{3})$.

5.4 A real number r is called *constructible* if there is a sequence of straightedge and compass constructions that arrive at a point P with r as one of
its coordinates. This gives us a tower of extensions

$$\mathbb{Q} \subset K_1 \subset K_2 \subset \cdots \subset K_n = K$$

with $K_i = K_{i-1}(a_{i1}, b_{i1}, a_{i2}, b_{i2}, \ldots, a_{in_i}, b_{in_i})$ with (a_{ij}, b_{ij}) the coordinates of the new points resulting from the ith construction based on the
previous constructions from $\mathbb{Q}$. Prove that $[K_i : K_{i-1}]$ takes its values
among $\{1, 2, 4\}$. Prove further that the extension K has degree a power
of 2 over $\mathbb{Q}$.

5.5 To solve the problem of doubling the cube, we have to construct $\sqrt[3]{2}$. If
we could construct a finite sequence of straightedge and compass con-
structions $\mathbb{Q} \subset K_1 \subset \cdots \subset K_n = K$ with K containing $\sqrt[3]{2}$, then we
could consider the sequence of field extensions

$$\mathbb{Q} \subset K_1 \cap \mathbb{Q}(\sqrt[3]{2}) \subset K_2 \cap \mathbb{Q}(\sqrt[3]{2}) \subset \cdots \subset K_n \cap \mathbb{Q}(\sqrt[3]{2}) = \mathbb{Q}(\sqrt[3]{2}).$$

Prove that $(K_{i-1} \cap \mathbb{Q}(\sqrt[3]{2}))(a_{i1}, b_{i1}, \ldots) = K_i \cap \mathbb{Q}(\sqrt[3]{2})$ and that this is
an extension of degree 1, 2, or 4. Deduce that $[\mathbb{Q}(\sqrt[3]{2}) : \mathbb{Q}]$ is a power of
2 if a construction of $\sqrt[3]{2}$ is possible. However, we know that $[\mathbb{Q}(\sqrt[3]{2}) :
\mathbb{Q}] = 3$. Thus it is impossible to double the cube using straightedge and
compass.

5.6 If we construct an angle $\angle ABC$ with B at the origin, then the rays meet
the unit circle in a pair of points. If the ray $\vec{BA}$ lies along the x-axis,
then $\vec{BC}$ meets the unit circle (constructible) in a point $(\cos(\angle ABC),
\sin(\angle ABC))$. Prove that $\cos(3\theta)$ satisfies the equation $\cos(3\theta) =
4\cos^3(\theta) - 3\cos(\theta)$. For $\theta = \pi/9$ (that is, 20°), $\cos(3\theta) = \cos(\pi/3) =
1/2$. Thus the trisected angle of 60° has a cosine that is a root of $8x^3 -
6x - 1$. Show that this polynomial is irreducible over $\mathbb{Q}$ and the relevant
root lives in an extension of $\mathbb{Q}$ of degree three. Deduce that it cannot be
constructed with straightedge and compass.

5.7 Suppose k is an extension of $\mathbb{F}$ and $[k : \mathbb{F}] = 1$. Show that $k = \mathbb{F}$.

5.8 Let k be an extension of $\mathbb{F}$ with $[k : \mathbb{F}] = n$. Suppose $f(x) \in \mathbb{F}[x]$ is a
polynomial of degree l and $\gcd(l, n) = 1$. Show that $f(x)$ has no roots
in k.

5.9 Suppose $f(x)$ is a polynomial in $\mathbb{F}_p[x]$ of degree greater than zero. In
the field $\mathbb{F}_q$ for $q = p^n$ and $n \geq 1$, prove that the number of roots of $f(x)$
in $\mathbb{F}_q$ is equal to the degree of $\gcd(f(x), x^q - x)$. (Hint: if you compute

the gcd in $\mathbb{F}_p[x]$ or in $\mathbb{F}_q[x]$, then you get the same answer. Notice that $x^q - x$ splits completely over $\mathbb{F}_q$. How would you write $f(x)$ in terms of the roots in $\mathbb{F}_q$?)

5.10 Show that the field of $9 = 3^2$ elements is isomorphic to $\mathbb{F}_3[i]$ where $i^2 = -1$. Notice that $x^2 + 1$ is irreducible over $\mathbb{F}_3$ and is a divisor of $x^9 - x$.

5.11 Prove that $\mathrm{PSl}_2(\mathbb{F}_{2^2})$ is isomorphic to A_5.

6

Galois Theory

The proposed goal is to determine the characteristics for the solubility
of equations by radicals. We can affirm that in pure analysis there does
not exist any material that is more obscure and perhaps more isolated
from all the rest. The novelty of this material has required the use of new
terminology, of new symbols.

ÉVARISTE GALOIS, *Dossier 9: Preliminary discussion*

In the previous chapter, we developed the theory of fields. We next explore
the relevant mappings between fields. A field homomorphism $\phi: k \to K$ is
a ring homomorphism satisfying $\phi(1) = 1$. The key property of field homomorphisms is that they are always injective (Proposition 5.4). We denote by
$\mathrm{Hom}(k, K)$ the set of field homomorphisms from k to K. If we identify k with
its image $\phi(k) \subset K$, then we can regard k as a subfield of K. It follows that
k and K share a prime subfield $\mathbb{F} \subset k \subset K$, making k and K vector spaces
over $\mathbb{F}$ (Theorem 5.6). The field $\mathbb{F}$ is the field of quotients of the subring of
k generated by 1. Since $\phi: k \to K$ takes 1 to 1, if $u \in \mathbb{F}$, then there are
integers n and m such that $u = n \cdot m^{-1}$ for $n = 1 + 1 + \cdots + 1$ (n times),
and similarly for m. Then $\phi(u) = \phi(n)\phi(m)^{-1} = n \cdot m^{-1} = u$. For $v \in k$,
we have $\phi(uv) = \phi(u)\phi(v) = u\phi(v)$, and so every field homomorphism
$\phi: k \to K$ is a linear transformation of vector spaces over $\mathbb{F}$. When we want
to emphasize this vector space structure, we write $\mathrm{Hom}_{\mathbb{F}}(k, K)$ for the set of
field homomorphisms as linear transformations over $\mathbb{F}$.

6.1 Automorphisms

Definition 6.1 Suppose k is a field. Let $\mathrm{Aut}(k)$ denote the *group of automorphisms* $\phi: k \to k$, field isomorphisms between k and itself, with the binary
operation given by composition.

152

For example, let $\mathbb{Q}[i]$ denote the subfield of $\mathbb{C}$ generated by $\mathbb{Q}$ and i. We have shown that $\mathbb{Q}[i] \cong \mathbb{Q}[x]/(x^2+1)$. Complex conjugation $c\colon \mathbb{Q}[i] \to \mathbb{Q}[i]$, $c(z) = \bar{z}$, is an automorphism of $\mathbb{Q}[i]$: Observe that $c(u + v) = \overline{u + v} = \bar{u} + \bar{v} = c(u) + c(v)$ and $c(uv) = \overline{uv} = \bar{u} \cdot \bar{v} = c(u)c(v)$. The prime field of $\mathbb{Q}[i]$ is $\mathbb{Q}$, and complex conjugation is the identity on $\mathbb{Q}$. As a linear transformation over $\mathbb{Q}$, conjugation can be represented as the matrix $\left[\begin{smallmatrix} 1 & 0 \\ 0 & -1 \end{smallmatrix}\right]$, where we write an element $a + bi$ in $\mathbb{Q}[i]$ by $\left[\begin{smallmatrix} a \\ b \end{smallmatrix}\right]$ with $a, b \in \mathbb{Q}$.

To generalize the automorphism group of a field to an extension K of a field k, we introduce an important subgroup of $\text{Aut}(K)$.

Definition 6.2 Let K be an extension of k. The *Galois group* of the extension is given by

$$\mathcal{G}al(K/k) = \{\phi\colon K \to K \mid \phi \in \text{Aut}(K) \text{ and } \phi|_k = \text{Id}_k\},$$

that is, for $\phi \in \mathcal{G}al(K/k)$ and $a \in k$, we have $\phi(a) = a$.

The Galois group $\mathcal{G}al(K/k)$ is clearly a subgroup of $\text{Aut}(K)$. Another way to think of $\mathcal{G}al(K/k)$ is as $\text{Hom}_k(K, K)$, the automorphisms of K that are k-linear transformations.

An automorphism of a simple extension $\phi\colon k(\alpha) \to k(\alpha)$ is determined by knowing $\phi(\alpha)$: Suppose $\sigma \in \mathcal{G}al(k(\alpha)/k)$. The evaluation mapping $\text{ev}_\alpha\colon k[x] \to k(\alpha)$ is surjective when α is algebraic. Because $k(\alpha) = k[\alpha]$, every element of $k(\alpha)$ can be written as $f(\alpha)$ for a polynomial $f(x) \in k[x]$. The automorphism ϕ is the identity on k and when evaluated on $f(\alpha)$ satisfies

$$\phi\big(f(\alpha)\big) = \phi\big(a_0 + a_1\alpha + \cdots + a_s\alpha^s\big)$$
$$= a_0 + a_1\phi(\alpha) + \cdots + a_s\big(\phi(\alpha)\big)^s = f\big(\phi(\alpha)\big).$$

Thus ϕ is determined by its value on α. This computation has an important consequence.

Proposition 6.3 *Suppose K is an extension of k and $\sigma \in \mathcal{G}al(K/k)$. If $\alpha \in K$ is a root of $f(x) \in k[x]$, then $\sigma(\alpha)$ is also a root of $f(x)$.*

Proof Because $\sigma(f(\alpha)) = f(\sigma(\alpha))$ and $f(\alpha) = 0$, we have $f(\sigma(\alpha)) = 0$, and $\sigma(\alpha)$ is also a root of $f(x)$. $\qquad\square$

More generally, we have the following:

Lemma 6.4 *Suppose K is a finite extension of k with $K = k(\alpha_1, \ldots, \alpha_n)$. An automorphism $\phi\colon K \to K$ in $\mathcal{G}al(K/k)$ is uniquely determined by the values $\{\phi(\alpha_1), \ldots, \phi(\alpha_n)\}$.*

Proof Let $k[x_1, x_2, \ldots, x_n]$ denote the ring of polynomials in the variables $x_1, \ldots, x_n$, and coefficients in k. Consider the evaluation homomorphism ev: $k[x_1, x_2, \ldots, x_n] \rightarrow k(\alpha_1, \ldots, \alpha_n)$ given by $\text{ev}(f) = f(\alpha_1, \ldots, \alpha_n)$. If evaluation is surjective, then every element of K can be written as $\text{ev}(F)$ for some $F(x_1, \ldots, x_n) \in k[x_1, \ldots, x_n]$. An automorphism $\phi: K \rightarrow K$ in $\mathcal{G}al(K/k)$ satisfies

$$\phi\big(F(\alpha_1, \ldots, \alpha_n)\big) = F\big(\phi(\alpha_1), \ldots, \phi(\alpha_n)\big),$$

and so ϕ is uniquely determined by its values $\{\phi(\alpha_i)\}$.

It remains to show that ev: $k[x_1, \ldots, x_n] \rightarrow k(\alpha_1, \ldots, \alpha_n)$ is surjective. We have shown that the statement holds for $n = 1$. Let us assume that it holds for extensions with fewer generators than n. Since $k(\alpha_1, \ldots, \alpha_n) = k(\alpha_1, \ldots, \alpha_{n-1})(\alpha_n)$, and α_n is algebraic over $k(\alpha_1, \ldots, \alpha_{n-1})$, we can write a typical element in $k(\alpha_1, \ldots, \alpha_{n-1})(\alpha_n)$ as $f_n(\alpha_n)$ where $f_n(x) \in k(\alpha_1, \ldots, \alpha_{n-1})[x]$. By the induction hypothesis we can write

$$f_n(x) = q_0(\alpha_1, \ldots, \alpha_{n-1}) + q_1(\alpha_1, \ldots, \alpha_{n-1})x + \cdots + q_t(\alpha_1, \ldots, \alpha_{n-1})x^t$$

with $q_i(x_1, \ldots, x_{n-1}) \in k[x_1, \ldots, x_{n-1}]$. Let

$$P(x_1, \ldots, x_n) = q_0(x_1, \ldots, x_{n-1}) + q_1(x_1, \ldots, x_{n-1})x_n$$
$$+ \cdots + q_t(x_1, \ldots, x_{n-1})x_n^t,$$

a polynomial in $k[x_1, \ldots, x_n]$. Furthermore, $f_n(x) = P(\alpha_1, \ldots, \alpha_{n-1}, x)$, so $f_n(\alpha_n) = \text{ev}(P(x_1, \ldots, x_n))$, and the evaluation homomorphism is surjective. $\square$

A particular example of a Galois group displays many of the features that motivate the general case. The example is the field $\mathbb{F}_q$, a finite field of characteristic p with $q = p^n$, as an extension of $\mathbb{F}_p$. In Chapter 5, we showed that $\mathbb{F}_q$ is a splitting field of $x^q - x$ over $\mathbb{F}_p$. We also introduced the Frobenius automorphism Fr: $\mathbb{F}_q \rightarrow \mathbb{F}_q$ given by $\text{Fr}(u) = u^p$. Because $x^q - x = 0$ for all $x \in \mathbb{F}_q$, the n-fold composition of the Frobenius automorphism with itself is the identity: $\text{Fr}^{on}(u) = u^{p^n} = u^q = u$ for all $u \in \mathbb{F}_q$. The cyclic subgroup $\langle \text{Fr} \rangle \subset \text{Aut}(\mathbb{F}_q) = \mathcal{G}al(\mathbb{F}_q/\mathbb{F}_p)$ has order n: if $\text{Fr}^{om} = \text{Id}$ for some $m < n$, then $u^{p^m} - u = 0$ for all $u \in \mathbb{F}_q$. However, $x^{p^m} - x$ can have only $p^m < p^n$ roots, so $\text{Fr}^{om} \neq \text{Id}$ for $m < n$.

We also showed in Chapter 5 that the group of units in $\mathbb{F}_q$ is a cyclic group, $\mathbb{F}_q^\times = \mathbb{F}_q - \{0\} = \langle \theta \rangle$ with generator θ. The evaluation map $\text{ev}_\theta: \mathbb{F}_p[x] \rightarrow \mathbb{F}_q$ has a kernel generated by $m_\theta(x)$, the minimal polynomial of θ. Suppose $m_\theta(x) = a_0 + a_1 x + \cdots + a_n x^n$. Applying Fr to $m_\theta(\theta)$, we get

$$0 = \text{Fr}\big(m_\theta(\theta)\big) = \text{Fr}(a_0) + \text{Fr}(a_1)\,\text{Fr}(\theta) + \cdots + \text{Fr}(a_n)\,\text{Fr}(\theta)^n$$
$$= a_0 + a_1\theta^p + \cdots + a_n\big(\theta^p\big)^n = m_\theta(\theta^p).$$

Thus θ^p is also a root of $m_\theta(x)$. Iterating Fr obtains that $\theta, \theta^p, \ldots, \theta^{p^{n-1}}$ are roots of $m_\theta(x)$. Since $\mathbb{F}_q \cong \mathbb{F}_p[x]/(m_\theta(x))$, and $\deg m_\theta(x) = n$, we have found all the roots of $m_\theta(x)$.

Suppose $\sigma \in \mathcal{G}al(\mathbb{F}_q/\mathbb{F}_p)$. Then $\sigma(\theta)$ is also a root of $m_\theta(x)$. We have established that the roots of $m_\theta(x)$ are of the form θ^{p^j}, and so $\sigma(\theta) = \theta^{p^j}$ for some $0 \le j < n$. Since θ is a generator for $\mathbb{F}_q^\times$, the value of σ on θ determines σ, and so $\sigma = \mathrm{Fr}^{\circ j}$. Therefore the Galois group $\mathcal{G}al(\mathbb{F}_q/\mathbb{F}_p) = \langle \mathrm{Fr} \rangle \cong C_n$ is isomorphic to a cyclic group of order n.

As discussed in Chapter 5, notice that the subfields of $\mathbb{F}_q$ are isomorphic to $\mathbb{F}_{p^m}$, where m divides n. Recall that the subgroups of the cyclic group C_n are isomorphic to C_m for m dividing n. Thus there is a correspondence between subgroups of $\mathcal{G}al(\mathbb{F}_q/\mathbb{F}_p)$ and subfields of $\mathbb{F}_q$. As we will see further along, this correspondence is not exactly the one that you might first guess. However, it is a general feature of certain field extensions. In this diagram for $n = 12$, the inclusions go upward in the diagram of groups, downward for fields.

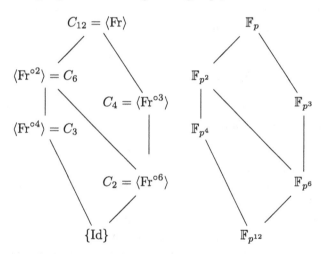

Figure 6.1

To investigate the Galois group $\mathcal{G}al(K/k)$ for a finite extension K of a field k, we can use the fact that the elements of K are algebraic over k (Theorem 5.12). Let $\alpha \in K$. The evaluation mapping $\mathrm{ev}_\alpha \colon k[x] \to K$ has a kernel $(m_\alpha(x))$ with α a root of $m_\alpha(x)$, its minimal polynomial, a monic and irreducible polynomial in $k[x]$. There is an isomorphism induced by ev_α, $\overline{\mathrm{ev}_\alpha} \colon k[x]/(m_\alpha(x)) \to k(\alpha)$, the smallest subfield of K containing k and α.

If $\beta \in K$ is another root of $m_\alpha(x)$, then the minimal polynomial of β is equal to $m_\alpha(x)$ because $m_\beta(x)$ would divide $m_\alpha(x)$, an irreducible polynomial. Then there is an isomorphism $\overline{\text{ev}_\beta}: k[x]/(m_\alpha(x)) \to k(\beta)$. Putting these isomorphisms together, we get an isomorphism of fields $\overline{\text{ev}_\beta} \circ (\overline{\text{ev}_\alpha})^{-1}: k(\alpha) \to k(\beta)$. This isomorphism is the identity mapping on $k \subset k(\alpha)$ to $k \subset k(\beta)$.

Suppose $\phi \in \mathcal{G}al(K/k)$, that is, we have an automorphism $\phi: K \to K$ that is the identity on the subfield k. Then $\phi(\alpha)$ is also a root of $m_\alpha(x)$ by Proposition 6.3. By the previous paragraph, $\phi|_{k(\alpha)}: k(\alpha) \to k(\phi(\alpha))$ is an isomorphism. If every pair of roots induced an automorphism of K, then this process would be transparent: automorphisms of a finite extension K of k are controlled by certain polynomials and their roots in the extension K. With appropriate assumptions on the extension, we will find this to be true.

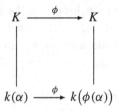

Lemma 6.5 *Suppose K is a finite extension of k and $\alpha \in K$ has minimal polynomial $m_\alpha(x) \in k[x]$. Then the set $\text{Hom}_k(k(\alpha), K)$ of field homomorphisms $\psi: k(\alpha) \to K$ that restrict to the identity on k is in one-to-one correspondence with the roots of $m_\alpha(x)$ in K.*

Proof If $\psi: k(\alpha) \to K$ is a field homomorphism that fixes k, then $0 = \psi(m_\alpha(\alpha)) = m_\alpha(\psi(\alpha))$, and we can associate a root of $m_\alpha(x)$ with ψ by $\psi \mapsto \psi(\alpha)$. Define the function $R: \text{Hom}_k(k(\alpha), K) \to \{\text{roots of } m_\alpha(x)\}$ given by $R(\psi) = \psi(\alpha)$. Because $\overline{\text{ev}_\beta} \circ (\overline{\text{ev}_\alpha})^{-1}: k(\alpha) \to k(\beta) \subset K$, as constructed above, maps α to β, the mapping from homomorphisms to roots is surjective. By Lemma 6.4 a field homomorphism $\phi: k(\alpha) \to K$ that restricts to the identity on k is determined by the value of $\phi(\alpha)$. Therefore the mapping $\phi \mapsto \phi(\alpha)$ is injective. $\qquad\square$

For example, consider the extension $\mathbb{Q}[\sqrt[3]{2}]$ of $\mathbb{Q}$. The minimal polynomial for $\sqrt[3]{2}$ is $x^3 - 2$, and the only root in $\mathbb{Q}[\sqrt[3]{2}]$ is $\sqrt[3]{2}$. It follows that $\mathcal{G}al(\mathbb{Q}[\sqrt[3]{2}]/\mathbb{Q})$ is the trivial group $\{\text{Id}\}$. A splitting field for $x^3 - 2$ over $\mathbb{Q}$ is $\mathbb{Q}(\sqrt[3]{2}, \omega)$ with $\omega = \dfrac{-1 + \sqrt{3}}{2}$ (see the example after Theorem 5.12). The tower of extensions $\mathbb{Q} \subset \mathbb{Q}[\sqrt[3]{2}] \subset \mathbb{Q}(\sqrt[3]{2}, \omega)$ allows us to apply Lemma 6.5: the number of field homomorphisms $\mathbb{Q}[\sqrt[3]{2}] \to \mathbb{Q}(\sqrt[3]{2}, \omega)$ is 3, namely $\sqrt[3]{2}$ maps to one of $\sqrt[3]{2}$, $\omega\sqrt[3]{2}$, or $\omega^2\sqrt[3]{2}$. The group elements may be displayed in a table where the each row gives the image of the first row:

$\sqrt[3]{2}$	$\omega\sqrt[3]{2}$	$\omega^2\sqrt[3]{2}$	Id	(1)(2)(3)
$\omega\sqrt[3]{2}$	$\omega^2\sqrt[3]{2}$	$\sqrt[3]{2}$	σ	(123)
$\omega^2\sqrt[3]{2}$	$\sqrt[3]{2}$	$\omega\sqrt[3]{2}$	σ^2	(132)
$\sqrt[3]{2}$	$\omega^2\sqrt[3]{2}$	$\omega\sqrt[3]{2}$	τ	(13)
$\omega\sqrt[3]{2}$	$\sqrt[3]{2}$	$\omega^2\sqrt[3]{2}$	$\sigma\tau$	(12)
$\omega^2\sqrt[3]{2}$	$\omega\sqrt[3]{2}$	$\sqrt[3]{2}$	$\sigma^2\tau$	(13)

The same logic can be applied to show that there are two field homomorphisms (isomorphisms in this case) $\mathbb{Q}[\sqrt[3]{2}](\omega) \to \mathbb{Q}(\sqrt[3]{2}, \omega)$ that extend a choice of field homomorphism $\mathbb{Q}[\sqrt[3]{2}] \to \mathbb{Q}(\sqrt[3]{2}, \omega)$, namely ω maps to one of ω or ω^2. Thus $\#\mathcal{G}al(\mathbb{Q}(\sqrt[3]{2}, \omega)/\mathbb{Q}) = 3 \cdot 2 = 6$. Since each automorphism of $\mathbb{Q}(\sqrt[3]{2}, \omega)$ takes the roots of $x^3 - 2$ to each other, this group is a subgroup of $\Sigma(\{\sqrt[3]{2}, \omega\sqrt[3]{2}, \omega^2\sqrt[3]{2}\}) \cong \Sigma_3$, and being of order 6, it is the entire group.

We employ Lemma 6.5 in the case that $K = E_f$, a splitting field of a polynomial. To gain the most flexible formulation, we follow Artin [3] and introduce possibly different fields and isomorphisms between them. The toehold in this theorem and the next proposition is induction, making these results appear a little mysterious. However, their consequences are rich and useful.

Theorem 6.6 *Suppose $\phi_0: k_1 \to k_2$ is an isomorphism of fields. Given a polynomial $p(x) \in k_1[x]$, let $f(x) = \phi_0(p(x)) \in k_2[x]$. Suppose E_p is a splitting field of $p(x)$ over k_1 and E_f is a splitting field of $f(x)$ over k_2. Then there is a field isomorphism to $\phi: E_p \to E_f$ that is an extension of ϕ_0, that is, $\phi|_{k_1} = \phi_0$.*

Proof We proceed by induction on $[E_p : k_1] = n$. If $n = 1$, then $E_p = k_1$ and $p(x) = a(x - a_1) \cdots (x - a_m)$ with a and $a_i \in k_1$ for all i. It follows that $f(x) = \phi_0(a)(x - \phi_0(a_1)) \cdots (x - \phi_0(a_m))$, so $f(x)$ also splits completely in $k_2[x]$, and $E_f = k_2$. The desired isomorphism is ϕ_0.

Suppose the theorem is true for all cases in which the number of roots of a polynomial in $k_1[x]$ in its splitting field and outside of k_1 is less than $n > 1$. Suppose that $p(x)$ has n roots in E_p not contained in k_1. Factor $p(x) = u_1(x)u_2(x) \cdots u_s(x)$ into a product of irreducible polynomials in $k_1[x]$. Not all of the $u_i(x)$ are linear because we would again be in the case $k_1 = E_p$. Suppose $u_1(x)$ has the degree greater than one. Because E_p is a splitting field for $p(x)$, we can factor $p(x) = a(x - \alpha_1) \cdots (x - \alpha_t)$ with $a \in k_1$ and $\alpha_i \in E_p$. Since $p(x) = u_1(x)q(x)$, the roots of $u_1(x)$ are among the roots of $p(x)$. Let α_1 be a root of $u_1(x)$.

The corresponding polynomial is $f(x) = \phi_0(p(x)) = v_1(x)v_2(x)\cdots v_s(x)$ with $v_i(x) = \phi_0(u_i(x))$. Because ϕ_0 is an isomorphism, it induces an isomorphism of rings $\hat{\phi}_0: k_1[x] \rightarrow k_2[x]$, and the $v_i(x)$ are irreducible because the $u_i(x)$ are. The evaluation mapping $\mathrm{ev}_{\alpha_1}: k_1[x] \rightarrow E_p$ has kernel $(u_1(x))$, and so $k_1(\alpha_1) \cong k_1[x]/(u_1(v))$. The isomorphism ϕ_0 takes $k_1[x]/(u_1(x))$ to $k_2[x]/(\phi_0(u_1(x))) = k_2[x]/(v_1(x)) \cong k_2(\beta_1)$, where β_1 is a root of both $v_1(x)$ and $f(x)$. Using the isomorphisms afforded by evaluations, we have

$$k_1(\alpha_1) \cong k_1[x]/(u_1(x)) \cong k_2[x]/(v_1(x)) \cong k_2(\beta_1).$$

The polynomial $p(x)$ is also in $k_1(\alpha_1)[x]$, and E_p is a splitting field of $p(x)$ over $k_1(\alpha_1)$. Likewise, $f(x) \in k_2(\beta_1)[x]$, and E_f is a splitting field of $f(x)$ over $k_2(\beta_1)$. There are fewer than n roots of $p(x)$ outside $k_1(\alpha_1)$ in E_p, and so the induction hypothesis gives us an isomorphism $\phi: E_p \rightarrow E_f$ that extends the isomorphism $k_1(\alpha_1) \rightarrow k_2(\beta_1)$, and hence it extends $\phi_0: k_1 \rightarrow k_2$. $\square$

Corollary 6.7 *Splitting fields of a given polynomial are isomorphic.*

Proof Suppose K and L are splitting fields of a polynomial $f(x) \in k[x]$. In the theorem, let $\phi_0: k \rightarrow k$ be the identity mapping. Then the identity extends to an isomorphism $\phi: K \rightarrow L$. $\square$

With this corollary, we can speak of *the* splitting field of $p(x)$, *the* meaning up to isomorphism. A finite field of order $q = p^n$ for a prime p is the splitting field of the polynomial $f(x) = x^q - x$. It follows that all fields of order q are isomorphic. This justifies the notation $\mathbb{F}_q$ for such a field.

Proposition 6.8 *If $f(x) \subset k[x]$ is a separable polynomial and E_f is its splitting field over k, then $\#\mathcal{G}al(E_f/k) = [E_f : k]$.*

Proof We proceed by induction on the degree of the extension. When $[E_f : k] = 1$, we have $E_f = k$, so $\mathcal{G}al(E_f/k) = \{\mathrm{Id}\}$, and the proposition holds.

Suppose $[E_f : k] > 1$ and $\alpha \in E_f$ is a root of $f(x)$ with $\alpha \notin k$. Then the minimal polynomial $m_\alpha(x)$ divides $f(x)$, and we can express $f(x) = m_\alpha(x)q(x)$ in $k[x]$. Suppose $\deg m_\alpha(x) = d$. Since $m_\alpha(x)$ splits completely over E_f and $f(x)$ is separable, there are d distinct field isomorphisms $k(\beta) \rightarrow E_f$ for each root β of $m_\alpha(x)$. By Theorem 6.6 there are d automorphisms $\phi: E_f \rightarrow E_f$ extending each $k(\beta) \rightarrow E_f$. Each such automorphism fixes the subfield k, and so $\phi \in \mathcal{G}al(E_f/k)$.

We know that $[E_f : k] = [E_f : k(\alpha)][k(\alpha) : k] = d[E_f : k(\alpha)]$. Since E_f is the splitting field of $q(x)$ over $k(\alpha)$ and over $k(\beta)$, each of the d isomorphisms has $[E_f : k(\alpha)] = [E_f : k]/d$ many extensions to E_f. Hence $\#\mathcal{G}al(E_f/k) = d\#\mathcal{G}al(E_f/k(\alpha)) = d[E_f : k(\alpha)] = [E_f : k]$. $\square$

In the case that $k_1 = k_2 = k$ and $E_p = E_f$, we obtain an action of the Galois group on the roots of $p(x)$. The group action

$$\rho: \mathcal{G}al(E_p/k) \to \Sigma(\{\alpha_1, \ldots, \alpha_n\}) \text{ is defined by } \rho(\sigma)(\alpha_i) = \sigma(\alpha_i).$$

Numbering the roots gives an isomorphism $\Sigma(\{\alpha_1, \ldots, \alpha_n\}) \cong \Sigma_n$. Composing with this isomorphism determines a group action $\hat{\rho}: \mathcal{G}al(E_p/k) \to \Sigma_n$, where $\hat{\rho}(\sigma)(i) = j$ if $\rho(\sigma)(\alpha_i) = \sigma(\alpha_i) = \alpha_j$.

Proposition 6.9 *Let $p(x) \in k[x]$ be a separable polynomial, and let E_p be the splitting field of $p(x)$ over k. The action of the Galois group $\mathcal{G}al(E_p/k)$ on the roots of $p(x)$ is faithful. The set of roots may be written as a union of orbits under this action with $n = m_1 + m_2 + \cdots + m_s$, where m_i is the cardinality of the ith orbit. Then $p(x) = u_1(x) \cdots u_s(x)$ in $k[x]$ with $\deg u_i = m_i$. Hence the action is transitive if and only if $p(x)$ is irreducible.*

Proof Suppose $p(x)$ is irreducible. Recall that an action is faithful if $\ker \rho = \{e\}$ and transitive if for any pair of roots α and β of $p(x)$, there is an element σ in the Galois group with $\sigma(\alpha) = \beta$. Suppose $\rho(\sigma) = \rho(\tau)$ for $\sigma, \tau \in \mathcal{G}al(E_p/k)$. Then $\rho(\sigma^{-1}\tau) = \text{Id}$, and $\sigma^{-1}\tau(\alpha_i) = \alpha_i$ for all i. Therefore $\sigma(\alpha_i) = \tau(\alpha_i)$ for all i. Since $E_p = k(\alpha_1, \ldots, \alpha_n)$, Lemma 6.4 implies $\sigma = \tau$. Hence ρ is injective, and the action is faithful.

Suppose α and β are roots of $p(x)$. Because $p(x)$ is irreducible, $k(\alpha) \cong k[x]/(p(x)) \cong k(\beta)$, giving an isomorphism $\phi_0: k(\alpha) \to k(\beta)$. Since E_p is the splitting field of $p(x)$ over k, it is also the splitting field of $p(x)$ over $k(\alpha)$ and over $k(\beta)$. Theorem 6.6 determines an automorphism $\phi: E_p \to E_p$ extending ϕ_0. Furthermore, $\phi|_k = \text{Id}$, so $\phi \in \mathcal{G}al(E_p/k)$ with $\phi(\alpha) = \beta$. Hence the action of the Galois group on the roots of $p(x)$ is transitive.

To prove the converse, suppose $\alpha \in E_p$ is a root of $p(x)$ and $m_\alpha(x) \in k[x]$ is the minimal polynomial of α. If $\deg p(x) = n$, then we can factor $p(x) = a(x - \alpha)(x - \alpha_2) \cdots (x - \alpha_n)$ over E_p. Because we have a transitive action, there are automorphisms $\phi_i \in \mathcal{G}al(E_f/k)$ with $\phi_i(\alpha) = \alpha_i$. Because $p(\alpha) = 0$, $m_\alpha(x)$ divides $p(x)$. By Proposition 6.3, $\phi_i(\alpha)$ is also a root of $m_\alpha(x)$. Hence $a^{-1}p(x) = (x - \alpha)(x - \alpha_2) \cdots (x - \alpha_n)$ divides $m_\alpha(x)$. Therefore $p(x)$ and $m_\alpha(x)$ differ by at most a unit. Since $m_\alpha(x)$ is irreducible, so is $p(x)$.

When $p(x)$ is reducible, we can write $p(x) = u_1(x) \cdots u_s(x)$ with each $u_i(x)$ irreducible. Let the set of roots of $p(x)$ be $\{\alpha_1, \ldots, \alpha_n\} \subset E_p$. Then α_1 is a root of some $u_i(x)$, say $u_1(x)$. The value of $\phi(\alpha_1)$ for $\phi \in \mathcal{G}al(E_p/k)$ is also a root of $u_1(x)$, and so the orbit of α_1 is the set of roots of $u_1(x)$. If β is in the orbit of α, then β cannot be a root of $u_i(x)$ for $i \neq 1$, since if it were, then β would be a double root of $p(x)$, contradicting separability. Thus the set of roots of $p(x)$ is a disjoint union of orbits under the action of $\mathcal{G}al(E_p/k)$, and each orbit is acted on transitively by $\mathcal{G}al(E_p/k)$ because the automorphisms that permute roots of $u_i(x)$ extend to automorphisms of E_p by Theorem 6.6.   □

Proposition 6.10 *Let E_p be an extension of a field k that is the splitting field of a separable polynomial $p(x) \in k[x]$. Then $\mathcal{G}al(E_p/k)$ is a finite group.*

Proof   Suppose $p(x) = u_1(x) \cdots u_s(x)$ is a factorization of $p(x)$ in $k[x]$ into a product of irreducible polynomials. We can construct a tower of splitting fields, indexed on the products $u_1(x) \cdots u_i(x)$ for $1 \le i \le s$,

$$k \subset E_{u_1} \subset (E_{u_1})_{u_2} \subset \cdots \subset (E_{u_1 \cdots u_{s-2}})_{u_{s-1}} \subset E_p.$$

By Lemma 6.5, for each root α of $u_1(x)$, there are field inclusions $k(\alpha) \to E_{u_1}$, as many as $\deg u_1(x)$. By Theorem 6.6 each of these extends to an automorphism $E_{u_1} \to E_{u_1}$. Apply Lemma 6.5 again to each such automorphism to obtain finitely many automorphisms $(E_{u_1})_{u_2} \to (E_{u_1})_{u_2}$. Continuing in this manner, we obtain finitely many automorphisms from E_p to E_p fixing k.   □

In fact, the argument for Proposition 6.10 shows that the order of $\mathcal{G}al(E_p/k)$ is bounded by $n!$ for $n = \deg p(x)$. By Theorem 5.8 we have

$$[E_p : k] =$$
$$\big[E_p : (E_{u_1 \cdots u_{s-2}})_{u_{s-1}} \big]\big[(E_{u_1 \cdots u_{s-2}})_{u_{s-1}} : (E_{u_1 \cdots u_{s-3}})_{u_{s-2}} \big] \cdots [E_{u_1} : k].$$

By Proposition 6.9

$$\big[(E_{u_1 \cdots u_{t-1}})_{u_t} : (E_{u_1 \cdots u_{t-2}})_{u_{t-1}} \big] = \#\mathcal{G}al\big((E_{u_1 \cdots u_{t-1}})_{u_t}/(E_{u_1 \cdots u_{t-2}})_{u_{t-1}} \big),$$

and this Galois group acts transitively on the roots of $u_t(x)$. If $\deg u_t(x) = d_t$, then $d_t \le \#\mathcal{G}al((E_{u_1 \cdots u_{t-1}})_{u_t}/(E_{u_1 \cdots u_{t-2}})_{u_{t-1}}) \le d_t!$, where the lower bound is due to transitivity, and the upper bound to the faithful representation in Σ_{d_t}. Because $d_1! d_2! \cdots d_s! \le (d_1 + \cdots + d_s)! = n!$, we have an upper bound for the order of $\mathcal{G}al(E_p/k)$.

Proposition 6.9 does not extend to reducible polynomials. For roots α and β from different irreducible factors of a polynomial, $k(\alpha) \cong k[x]/(u_i(x))$ and $k(\beta) \cong k[x]/(u_j(x))$, which are not isomorphic fields unless $(u_i(x)) = (u_j(x))$. Because $u_i(x)$ and $u_j(x)$ are irreducible, this means that if $(u_i(x)) = (u_j(x))$, then they can differ at most by a factor of a unit. Taking both to be monic implies $u_i(x) = u_j(x)$. If we consider the orbit $\mathcal{O}_\alpha$ of α under the action of $\mathcal{G}al(E_p/k)$, Proposition 5.11 shows that $\mathcal{O}_\alpha$ consists of the roots of $m_\alpha(x) = u_i(x)$. The set of roots of $p(x)$ is a disjoint union of orbits of $\mathcal{G}al(E_p/k)$, and each automorphism $\phi \colon E_p \to E_p$ acting on the roots preserves the orbit structure. When we number the roots $\{\alpha_1, \alpha_2, \ldots, \alpha_n\}$, the group of permutations $\Sigma(\{\alpha_1, \ldots, \alpha_n\})$ is isomorphic to Σ_n. The representation of ϕ in Σ_n has a cycle structure determined by the orbit structure, and so it is possible to get some idea of which subgroup is determined by $\mathcal{G}al(E_p/k)$ in Σ_n. More on this idea later.

For an extension K of k, the subgroup structure of $\mathcal{G}al(K/k)$ can be related to the collection of subfields of K that contain k.

Definition 6.11 For a subgroup H of $\mathcal{G}al(K/k)$, define the *fixed field* of H by $K^H = \{\alpha \in K \mid \phi(\alpha) = \alpha \text{ for all } \phi \in H\}$.

Lemma 6.12 K^H *is a subfield of K.*

Proof If α and β are in K^H and $\phi \in H$, then $\phi(\alpha + \beta) = \phi(\alpha) + \phi(\beta) = \alpha + \beta$ and $\alpha + \beta \in K^H$. Also, $\phi(\alpha\beta) = \phi(\alpha)\phi(\beta) = \alpha\beta$ and $\alpha\beta \in K^H$. Finally, $1 = \phi(1) = \phi(\alpha\alpha^{-1}) = \phi(\alpha)\phi(\alpha^{-1}) = \alpha\phi(\alpha^{-1})$, from which it follows that $\alpha^{-1} \in K^H$. $\square$

For example, let us take $H = \langle \text{Fr}^{\circ m} \rangle$ in $\mathcal{G}al(\mathbb{F}_q/\mathbb{F}_p)$ with $q = p^n$. The fixed field of the subgroup H is $\mathbb{F}_q^H = \{u \in \mathbb{F}_q \mid u^{p^m} = u\}$. If m divides n, then $\langle \text{Fr}^{\circ m} \rangle$ is a group of order n/m, and $x^{p^m} - x$ divides $x^{p^n} - x$. The field $\mathbb{F}_q^H$ consists of the p^m roots of $x^{p^m} - x$ in $\mathbb{F}_q$, where $x^{p^m} - x$ splits completely. Because $\mathbb{F}_q^H$ is a splitting field over $\mathbb{F}_p$, $\mathbb{F}_q^H$ is isomorphic to $\mathbb{F}_{p^m}$. The subgroup $C_{n/m} \cong \langle \text{Fr}^{\circ m} \rangle$ corresponds to the subfield $\mathbb{F}_{p^m}$ of $\mathbb{F}_q$. When $n = 12$, we see in Fig. 6.1 how the Hasse diagram for C_{12} corresponds to the graph of inclusions of subfields of $\mathbb{F}_{p^{12}}$.

For an intermediate field of an extension K of k, we can associate a subgroup of $\mathcal{G}al(K/k)$ in the following manner: Let $k \subset L \subset K$ be a subfield of K containing k. The Galois group $\mathcal{G}al(K/L) = \{\phi \in \text{Aut}(K) \mid \phi(\alpha) = \alpha \text{ for all } \alpha \in L\}$ is a subgroup of $\mathcal{G}al(K/k)$ because $\phi|_L = \text{Id}$ implies $\phi|_k = \text{Id}$.

This leads to a pair of functions: Let

$$\mathcal{I}(K, k) = \{L, \text{ a subfield of } K \mid k \subset L \subset K\},$$

$$\mathcal{T}_G = \{H \mid H, \text{ a subgroup of } G\}.$$

Both these sets are ordered by inclusion. For $G = \mathcal{G}al(K/k)$, define

$$\Gamma \colon \mathcal{I}(K, k) \to \mathcal{T}_G, \quad \Gamma(L) = \mathcal{G}al(K/L),$$

$$\Phi \colon \mathcal{T}_G \to \mathcal{I}(K, k), \quad \Phi(H) = K^H.$$

Proposition 6.13 *The functions Γ and Φ are order-reversing, and the compositions satisfy $\Gamma \circ \Phi(H) \supset H$ and $\Phi \circ \Gamma(L) \supset L$.*

Proof Suppose $H \subset H' \subset \mathcal{G}al(K/k)$. Then $K^H \supset K^{H'}$ because elements of K fixed by H' are automatically fixed by H. If $k \subset L \subset L' \subset K$, then

$\mathcal{G}al(K/L) \supset \mathcal{G}al(K/L')$ because an automorphism that fixes L' automatically fixes L. We have

$$\Phi \circ \Gamma \colon L \mapsto H = \mathcal{G}al(K/L) \mapsto K^H = K^{\mathcal{G}al(K/L)} \supset L,$$

because $u \in L$ is fixed by all $\phi \in H = \mathcal{G}al(K/L)$, and so $u \in K^H$. Moreover,

$$\Gamma \circ \Phi \colon H \mapsto K^H \mapsto \mathcal{G}al\big(K/K^H\big) \supset H.$$

The inclusion $\Gamma \circ \Phi(H) \supset H$ follows because, for all $\sigma \in H$ and $\beta \in K^H$, $\sigma(\beta) = \beta$, so $\sigma \in \mathcal{G}al(K/K^H)$. □

If it were always the case that $\Gamma \circ \Phi = \mathrm{Id}$ and $\Phi \circ \Gamma = \mathrm{Id}$, there would be a perfect correspondence between intermediate fields and subgroups of the Galois group. This correspondence would relate the structure of intermediate fields to the subgroup lattice of the Galois group, and vice versa, providing an invaluable tool, especially to study roots of polynomials. Then we ask

When does $L = K^{\mathcal{G}al(K/L)}$? When does $H = \mathcal{G}al(K/K^H)$?

For the extension $\mathbb{F}_q$ of $\mathbb{F}_p$, the correspondence is perfect. In this case the extension is always the splitting field of a separable polynomial. For the extension $\mathbb{Q}[\sqrt[3]{2}]$ of $\mathbb{Q}$, the minimal polynomial $m_{\sqrt[3]{2}}(x) = x^3 - 2$ has only one root in $\mathbb{Q}[\sqrt[3]{2}]$, and so there is only one element in $\mathcal{G}al(\mathbb{Q}[\sqrt[3]{2}]/\mathbb{Q})$, whereas $[\mathbb{Q}[\sqrt[3]{2}] : \mathbb{Q}] = 3$. Not much information can be obtained from the trivial group.

In his work on number theory, RICHARD DEDEKIND (1831–1916) [57] introduced many important reformulations of ideas in the theory of fields and of Galois theory, including the focus on the automorphisms of a field. In the case of an abelian group G, he developed the notion of characters as homomorphisms $G \to \mathbb{C}$; these characters were the precursor to the general notion of characters in representation theory. Any homomorphism of fields $k \to K$ is a multiplicative homomorphism $k^\times \to K$ of the units in the field $k^\times = k - \{0\}$, an abelian group, to an extension field K. This generalizes Dedekind's abelian characters to fields. Anticipating the extension of characters to more general settings (see Chapter 4), Dedekind proved the following key result.

Dedekind's Independence Theorem Let $f_1, f_2, \ldots, f_n$ denote distinct homomorphisms of fields $k \to K$. Then the set $\{f_1, \ldots, f_n\}$ is linearly independent in the sense that if there are $a_i \in K$ such that $a_1 f_1(u) + \cdots + a_n f_n(u) = 0$ for all $u \in k$, then $a_i = 0$ for all i.

The similar independence result for the set of irreducible representations of a finite group G follows from the First Orthogonality Theorem, proved by Frobenius, who was inspired by Dedekind [14, 39].

Proof We proceed by induction on n, the number of distinct homomorphisms. If $n = 1$ and $a_1 f_1(u) = 0$ for all $u \in k$, then $a_1 f_1(1) = a_1 = 0$, and the theorem holds.

Suppose the theorem holds for sets of distinct homomorphisms of cardinality less than n. Suppose there exist $a_1, \ldots, a_n$ in K, not all zero, such that $a_1 f_1(u) + \cdots + a_n f_n(u) = 0$ for all $u \in k$. Notice that all $a_i \neq 0$, because if some $a_j = 0$, then the relation holds for fewer than n homomorphisms, and the induction hypothesis applies. If we multiply the relation by a_n^{-1}, then we get the relation $b_1 f_1(u) + \cdots + b_{n-1} f_{n-1}(u) + f_n(u) = 0$ for all $u \in k$. Since $f_1 \neq f_n$, there is $x \in k$ such that $f_1(x) \neq f_n(x)$. In this case, $x \neq 0$ and $x \neq 1$. For $y \in k$, we have $b_1 f_1(xy) + \cdots + b_{n-1} f_{n-1}(xy) + f_n(xy) = 0$. Because every f_i is a field homomorphism,

$$b_1 f_1(x) f_1(y) + \cdots + b_{n-1} f_{n-1}(x) f_{n-1}(y) + f_n(x) f_n(y) = 0.$$

We can divide by $f_n(x) \neq 0$:

$$b_1 \frac{f_1(x)}{f_n(x)} f_1(y) + \cdots + b_{n-1} \frac{f_{n-1}(x)}{f_n(x)} f_{n-1}(y) + f_n(y) = 0.$$

Subtract this relation from $b_1 f_1(y) + \cdots + b_{n-1} f_{n-1}(y) + f_n(y) = 0$ to get

$$b_1 \left(1 - \frac{f_1(x)}{f_n(x)}\right) f_1(y) + \cdots + b_{n-1} \left(1 - \frac{f_{n-1}(x)}{f_n(x)}\right) f_{n-1}(y) = 0.$$

We know that $1 - \dfrac{f_1(x)}{f_n(x)} \neq 0$ and $b_1 \neq 0$, so this is a linear dependence among $f_1, \ldots, f_{n-1}$, a contradiction to the induction hypothesis. It follows that $a_i = 0$ for all i and the subset $\{f_1, \ldots, f_n\} \subset \text{Hom}(k, K)$ is linearly independent. $\qquad\square$

Corollary 6.14 *If $\phi_1, \phi_2, \ldots, \phi_n$ are distinct field isomorphisms in* $\text{Aut}(k)$, *then $\{\phi_1, \ldots, \phi_n\}$ is linearly independent.*

We can apply these independence results to subgroups of the Galois group.

Theorem 6.15 *Suppose K is a finite extension of k and H is a subgroup of* $\mathcal{G}al(K/k)$. *Then $[K : K^H] = \#H$.*

Proof (Artin [3]) Let $H = \{\text{Id} = \sigma_1, \sigma_2, \ldots, \sigma_n\}$. Because the elements of H are distinct automorphisms of K, as a set of automorphisms, H is linearly independent. Suppose $[K : K^H] = r < n$. Let $\{\alpha_1, \ldots, \alpha_r\}$ be a basis for K over K^H. Consider the system of linear equations in K:

$$\alpha_1 x_1 + \sigma_2(\alpha_1) x_2 + \cdots + \sigma_n(\alpha_1) x_n = 0,$$
$$\alpha_2 x_1 + \sigma_2(\alpha_2) x_2 + \cdots + \sigma_n(\alpha_2) x_n = 0,$$

$$\vdots \quad \vdots \qquad \vdots \quad \vdots$$

$$\alpha_r x_1 + \sigma_2(\alpha_r) x_2 + \cdots + \sigma_n(\alpha_r) x_n = 0.$$

This is a system of r linear equations in n unknowns, and by Proposition 3.12, with $r < n$, there is a nonzero solution, say $u_1, \ldots, u_n$. For any $y \in K$, we can write $y = c_1 \alpha_1 + c_2 \alpha_2 + \cdots + c_r \alpha_r$ with $c_i \in K^H$. Then $\sigma_j(c_i) = c_i$ for all i and j. Since $\sum_{i=1}^{n} u_i \sigma_i(\alpha_j) = 0$ for all j, we have

$$\sum_{i=1}^{n} u_i \sigma_i(y) = \sum_{j=1}^{r} \sum_{i=1}^{n} u_i c_j \sigma_i(\alpha_j) = \sum_{j=1}^{r} \left(c_j \sum_{i=1}^{n} u_i \sigma_i(\alpha_j) \right) = 0.$$

Thus $\sum_{i=1}^{n} u_i \sigma_i(y) = 0$ for all $y \in K$. This contradicts Dedekind's Independence Theorem, and so $[K : K^H] \geq n$.

Suppose $[K : K^H] > n$ and so there is a set $\beta_1, \ldots, \beta_{n+1}$ of elements of K that are linearly independent over K^H. Consider the system of linear equations

$$\beta_1 y_1 + \beta_2 y_2 + \cdots + \beta_{n+1} y_{n+1} = 0,$$
$$\sigma_2(\beta_1) y_1 + \sigma_2(\beta_2) y_2 + \cdots + \sigma_2(\beta_{n+1}) y_{n+1} = 0,$$

$$\vdots \quad \vdots \qquad \vdots \quad \vdots$$

$$\sigma_n(\beta_1) y_1 + \sigma_n(\beta_2) y_2 + \cdots + \sigma_n(\beta_{n+1}) y_{n+1} = 0.$$

This is a system of n linear equations in $n + 1$ unknowns, and since the system of equations is homogeneous, there are solutions of the form $v_1, \ldots, v_{n+1}$, not all zero. Among the solutions, choose one with the least number of nonzero entries and rearrange it if necessary to be $(v_1, \ldots, v_s, 0, \ldots, 0)$. We can divide through by v_1 to make it equal to 1. So we have, for all j,

$$\sigma_j(\beta_1) + \sigma_j(\beta_2) v_2 + \cdots + \sigma_j(\beta_s) v_s = 0.$$

Apply σ_k to the equation to get

$$\sigma_k \sigma_j(\beta_1) + \sigma_k \sigma_j(\beta_2) \sigma_k(v_2) + \cdots + \sigma_k \sigma_j(\beta_s) \sigma_k(v_s) = 0.$$

Because H is a group, there is some $\sigma_i = \sigma_k \sigma_j$, and so $\sigma_i(\beta_1) + \sigma_i(\beta_2) \sigma_k(v_2) + \cdots + \sigma_i(\beta_s) \sigma_k(v_s) = 0$. When we subtract this relation from the equation $\sigma_i(\beta_1) + \sigma_i(\beta_2) v_2 + \cdots + \sigma_i(\beta_{n+1}) v_{n+1} = 0$, we get

$$\sigma_i(\beta_2) \big[v_2 - \sigma_k(v_2) \big] + \cdots + \sigma_s(\beta_s) \big[v_s - \sigma_k(v_s) \big] = 0.$$

If for all i and k, $v_i = \sigma_k(v_i)$, then $v_i \in K^H$ for all i, and the first equation ($\sigma_1 = \text{Id}$) becomes a linear dependence among the β_i, contradicting their linear independence. Thus $v_r \neq \sigma_k(v_r)$ for some r and k, and we obtain a solution with a smaller number of nonzero entries than the least number, a contradiction. Thus $[K : K^H] = \#H$. □

This relation suggests how the failure of a perfect correspondence between the set of intermediate fields $\mathcal{I}(k, K)$ and the set of subgroups $\mathcal{T}_G$ of $G = \mathcal{G}al(K/k)$ can happen. If $L \neq K^{\mathcal{G}al(K/L)}$, then this is because there are too few automorphisms in $\mathcal{G}al(K/L)$ to precisely fix L. Likewise, if H is a subgroup of $\mathcal{G}al(K/k)$ for which $H \neq \mathcal{G}al(K/K^H)$, then H fixes a larger subfield than is fixed by $\mathcal{G}al(K/K^H)$,

$$\left[K : K^{\mathcal{G}al(K/K^H)}\right] = \#\mathcal{G}al(K/K^H) > \#H = \lfloor K : K^H \rfloor.$$

To obtain a condition that allows us to prove $\Gamma \circ \Phi = \mathrm{Id}$ and $\Phi \circ \Gamma = \mathrm{Id}$, we introduce a property of an extension that is a generalization of a splitting field.

Definition 6.16 An extension K of a field k is a *normal extension* if, whenever an irreducible polynomial $u(x) \in k[x]$ has a root in K, then $u(x)$ splits completely over K. A finite extension K of k is *separable* if for all $\alpha \in K$, the minimal polynomial $m_\alpha(x)$ is separable.

If K is a finite extension, then every element of K is algebraic over k. Suppose K is a normal extension of k and $\alpha \in K$. The minimal polynomial $m_\alpha(x) \in k[x]$ then splits over K, and so K contains all the roots of $m_\alpha(x)$. It follows that the splitting field E_{m_α} of $m_\alpha(x)$ is contained in K. Thus a finite and normal extension contains the splitting fields of every element in K.

If k is a field of characteristic zero, then finite extensions of k are separable. When $k = \mathbb{F}_p$ for a prime integer p, finite extensions of k are also separable.

The following important result summarizes how the normal condition interacts with the Galois group.

Theorem 6.17 *For a finite extension K of k, let $G = \mathcal{G}al(K/k)$. The following conditions are equivalent:*

(1) The fixed field of G satisfies $K^G = k$.

(2) K is a normal and separable extension of k.

(3) K is the splitting field over k of a separable polynomial $f(x) \in k[x]$.

Proof We first show that (3) implies (1). Suppose $k \subset K^G \subset K$ with K a splitting field of $f(x) \in k[x]$, a separable polynomial. Then, since $k \subset K^G$, K is a splitting field of $f(x) \in K^G[x]$. Because K^G is an intermediate field, $\mathcal{G}al(K/K^G)$ is a subgroup of $\mathcal{G}al(K/k)$. Suppose $\sigma \in G = \mathcal{G}al(K/k)$. Then $\sigma(u) = u$ for all $u \in K^G$, and so $\sigma \in \mathcal{G}al(K/K^G)$. Therefore $\mathcal{G}al(K/K^G) = \mathcal{G}al(K/k)$. However, by Proposition 6.8 and Theorem 5.8, $\#\mathcal{G}al(K/k) = [K :$

$k] = [K : K^G][K^G : k]$. By Theorem 6.15, $[K : K^G] = \#\mathcal{G}al(K/k)$, so we
have $[K^G : k] = 1$ and $k = K^G$.

To prove that (1) implies (2), suppose $\alpha \in K$ and $m_\alpha(x)$ is the minimal poly-
nomial of α in $k[x]$. Since G acts on K, let $\mathcal{O}_\alpha = \{\sigma(\alpha) \mid \sigma \in \mathcal{G}al(K/k)\} =$
$\{\alpha = \alpha_1, \alpha_2, \ldots, \alpha_r\}$ denote the orbit of α. Let $g(x) = (x - \alpha_1) \cdots (x - \alpha_r)$.
If $\sigma \in G$, then σ permutes the elements in $\mathcal{O}_\alpha$, and so $\sigma(g(x)) = g(x)$. This
means that the coefficients of $g(x)$ are fixed by G, and so $g(x) \in K^G[x] =$
$k[x]$. Since α is a root of $g(x)$, $m_\alpha(x)$ divides $g(x)$. However, for any $\sigma \in G$,
$\sigma(m_\alpha(\alpha)) = m_\alpha(\sigma(\alpha)) = 0$, and so the elements of $\mathcal{O}_\alpha$ are roots of $m_\alpha(x)$,
and $g(x)$ divides $m_\alpha(x)$. Thus $g(x) = m_\alpha(x)$ is irreducible and separable and
splits over K. Therefore K is a normal extension of k.

To show that (2) implies (3), suppose K is normal and separable. Because
K is a finite extension of k, $K = k(\alpha_1, \ldots, \alpha_n)$. Let $m_{\alpha_i}(x) \in k[x]$ denote the
minimal polynomial for α_i. From the set $\{m_{\alpha_i}(x)\}$ take the distinct polynomials and form their product $F(x) = m_{\alpha_1}(x) \cdots m_{\alpha_r}(x)$. Then $F(x)$ is separable
because there are no shared roots among minimal polynomials. Since K is
normal over k, $F(x)$ splits completely in $K[x]$. The splitting field of $F[x]$
contains all of the α_i, and so $E_F \subset K \subset E_F$. Therefore K is the splitting field
of a separable polynomial in $k[x]$.   □

Definition 6.18 A finite extension K of k is called a *Galois extension* if the
fixed field of $\mathcal{G}al(K/k)$ satisfies $K^{\mathcal{G}al(K/k)} = k$.

By Theorem 6.17 a Galois extension K is also a finite, normal, and separable
extension of k, and K is a splitting field of a separable polynomial $f(x) \in k[x]$.

Corollary 6.19 *If K is a Galois extension of k and L is an intermediate field
$k \subset L \subset K$, then K is a Galois extension of L. Also, $[K : L] = \#\mathcal{G}al(K/L)$,
and $L = K^{\mathcal{G}al(K/L)}$.*

Proof   Since K is the splitting field of a separable polynomial $f(x) \in k[x] \subset$
$L[x]$, by Proposition 6.8 and Theorem 6.15 the corollary follows.   □

This statement proves that $\Phi \circ \Gamma = \mathrm{Id}$ for the mappings in the correspondence between subgroups of $\mathcal{G}al(K/k)$ and intermediate fields of the extension
K of k. To finish the proof of the correspondence for Galois extensions, we
need to show that $H = \mathcal{G}al(K/K^H)$ for any subgroup H of $\mathcal{G}al(K/k)$. We
know that $[K : K^H] = \#H$ and that $\mathcal{G}al(K/K^H) \supset H$. We also know from
the corollaries that K is a Galois extension of K^H, and so $K^{\mathcal{G}al(K/K^H)} = K^H$.
Then $\#H = [K : K^H] = [K : K^{\mathcal{G}al(K/K^H)}] = \#\mathcal{G}al(K/K^H)$, and so
$H = \mathcal{G}al(K/K^H)$.

The Fundamental Theorem of Galois Theory *For a Galois extension K of a field k, let G denote $\mathcal{G}al(K/k)$, $\mathcal{I}(k, K)$ the set of intermediate fields between k and K, and $\mathcal{T}_G$ the set of subgroups of G. The mappings $\Gamma: \mathcal{I}(k, K) \to \mathcal{T}_G$ and $\Phi: \mathcal{T}_G \to \mathcal{I}(k, K)$ given by $\Gamma(L) = \mathcal{G}al(K/L)$ and $\Phi(H) = K^H$ are inverses of each other, and they establish an order-reversing bijection. Furthermore, $[\mathcal{G}al(K/k) : \mathcal{G}al(K/L)] = [L : k]$ for all intermediate fields L.*

Proof We need to prove the final statement of the theorem. Let $G = \mathcal{G}al(K/k)$ and $H = \mathcal{G}al(K/L)$. Then K is a Galois extension of k and of L, and $k = K^G$, $L = K^H$. It follows from the Theorem 5.8 that $[K : L][L : k] = [K : k]$, so

$$\left[K^H : K^G\right] = [L : k] = \left[K : K^G\right]/\left[K : K^H\right] = \#G/\#H = [G : H]. \quad \square$$

The Fundamental Theorem is the warp and woof of Galois theory. Questions about intermediate fields, which may concern the constructions of roots of a polynomial can be turned into questions about the subgroups of a finite group. For example, we refine the correspondence between $\mathcal{I}(k, K)$ and $\mathcal{T}_G$ by extending an analogy between the groups $\mathcal{G}al(K/L)$ for an intermediate field L and the stabilizer subgroup G_x of a group action $\rho: G \to \Sigma(X)$. Lemma 2.10 showed that $G_{\rho(g)(x)} = gG_xg^{-1}$. Suppose $\sigma \in \mathcal{G}al(K/k)$. Then, for all $u \in L$ and $h \in \mathcal{G}al(K/L)$,

$$\left(\sigma \circ h \circ \sigma^{-1}\right)\left(\sigma(u)\right) = \sigma\left(h(u)\right) = \sigma(u),$$

because $h(u) = u$ for all $u \in L$. Thus $\sigma \circ h \circ \sigma^{-1} \in \mathcal{G}al(K/\sigma(L))$. The fields $\sigma(L)$ are called *conjugate fields* of L, and the subgroups of the Galois group that fix each conjugate field are conjugate to one another.

Proposition 6.20 *Suppose K is a Galois extension of k and $k \subset L \subset K$ is an intermediate field. The subgroup $\mathcal{G}al(K/L)$ of $\mathcal{G}al(K/k)$ is a normal subgroup if and only if L is a normal extension of k.*

Proof It suffices to prove that $\sigma(L) = L$ for all σ in $\mathcal{G}al(K/k)$ if and only if L is a normal extension of k. (By the way, this does not mean that σ acts as the identity on L.) Suppose $f(x)$ is a separable polynomial with coefficients in k for which $L = E_f$. By the equivalences proved in Theorem 6.15, L is a normal extension of k. We can express $L = k(\alpha_1, \ldots, \alpha_s)$ for the roots $\alpha_1, \ldots, \alpha_s$ of $f(x)$ in L. If $\sigma \in \mathcal{G}al(K/k)$, then $\sigma(\alpha_i)$ is also a root of $f(x)$ by Proposition 6.3. Then $\sigma(\alpha_i) \in L$ for all i, and $\sigma(L) = L$ by Lemma 6.4.

Suppose $\sigma(L) = L$ for all $\sigma \in \mathcal{G}al(K/k)$. Then

$$\mathcal{G}al(K/L) = \mathcal{G}al\left(K/\sigma(L)\right) = \sigma\mathcal{G}al(K/L)\sigma^{-1},$$

and $\mathcal{G}al(K/L)$ is normal in $\mathcal{G}al(K/k)$. $\quad \square$

Corollary 6.21 *Suppose K is a Galois extension of k and $k \subset L \subset K$ is an intermediate field with L a normal extension of k. Then $\mathcal{G}al(L/k) \cong \mathcal{G}al(K/k)/\mathcal{G}al(K/L)$.*

Proof Because L is normal, if $\sigma \in \mathcal{G}al(K/k)$, then $\sigma(L) = L$. Consider the homomorphism $r \colon \mathcal{G}al(K/k) \to \mathcal{G}al(L/k)$ given by $r(\sigma) = \sigma|_L \colon L \to L$. By Theorem 6.6 this mapping is surjective. The kernel of r is $\mathcal{G}al(K/L)$, and so $\mathcal{G}al(L/k) \cong \mathcal{G}al(K/k)/\mathcal{G}al(K/L)$. □

Not every finite extension is normal. However, we can extend the field to make it so.

Definition 6.22 A finite extension K of k has a field L as its *normal closure* if L is a normal extension of k with $K \subset L$ and, for any intermediate field F such that $k \subset K \subset F \subset L$, if F is normal over k, then $F = L$.

The normal closure of the extension $k \subset K$ is the smallest extension of K that is normal over k.

Proposition 6.23 *Suppose K is a finite extension of k. Then there is a finite extension L of K that is a normal closure of $k \subset K$. Furthermore, any two normal closures of $k \subset K$ are isomorphic.*

Proof Because K is a finite extension of k, we can express K as the result of finitely many adjunctions of elements to k, that is, $K = k(\alpha_1, \alpha_2, \ldots, \alpha_r)$. Let $\{m_{\alpha_1}(x), m_{\alpha_2}(x), \ldots, m_{\alpha_t}(x)\}$ denote the set of distinct minimal polynomials for the α_i. Define $f(x) = m_{\alpha_1}(x)m_{\alpha_2}(x) \cdots m_{\alpha_t}(x)$. The splitting field E_f of $f(x)$ over k contains α_i for all i and so contains K. By Theorem 6.15, E_f is normal over k. Suppose L is another field such that $k \subset K \subset L \subset E_f$ and L normal over k. Then $m_{\alpha_i}(x)$ splits completely in L for all i. It follows that $f(x)$ splits completely in L, and so $E_f \subset L$ and $L = E_f$. Thus E_f is a normal closure of $k \subset K$.

Suppose M is another field such that $k \subset K \subset M$ and M a normal closure of $k \subset K$. Once again, it follows that $f(x)$ splits completely over M, and so $E_f \subset M$ up to isomorphism. Because E_f is normal over k, by the definition of the normal closure, $E_f \cong M$. □

Suppose $k \subset L \subset K$ for a Galois extension K of k and an intermediate field L. If L is a normal extension of k, then for all $\sigma \in \mathcal{G}al(K/k)$, $\sigma(L) = L$. Suppose this is not true. Then there is a set of intermediate fields $\{\sigma(L) \mid \sigma \in \mathcal{G}al(K/k)\} = \mathcal{O}_L$. The action of $\mathcal{G}al(K/k)$ permutes the collection $\mathcal{O}_L$. If there were a smallest intermediate field that contained every $\sigma(L)$, then it would be invariant under the action of $\mathcal{G}al(K/k)$. In pursuit of this idea, we introduce the following:

Definition 6.24 Suppose E_1 and E_2 are subfields of a field K and each is an extension of k. The *compositum* of E_1 and E_2, denoted E_1E_2, is the smallest subfield of K that contains E_1 and E_2.

As usual, by the smallest subfield we mean the intersection of all subfields of K that contain both E_1 and E_2. If the extensions E_1 and E_2 are finite, then $E_1 = k(\alpha_1, \ldots, \alpha_s)$ and $E_2 = k(\beta_1, \ldots, \beta_t)$. Then $E_1E_2 = k(\alpha_1, \ldots, \alpha_s, \beta_1, \ldots, \beta_t)$. This is clear because any field that contains E_1 and E_2 must contain all the α_i and β_j. The extension of k obtained by adjoining these elements is by definition the smallest subfield of K containing them.

Proposition 6.25 *Let $k \subset L \subset K$ be a tower of field extensions with $k \subset K$ a Galois extension. Then the compositum $\prod_{\sigma \in \mathcal{G}al(K/k)} \sigma(L)$ of the conjugates of L is the normal closure of L in K.*

Proof If M is the normal closure of L in K over k, then for all $\sigma \in \mathcal{G}al(K/k)$, $\sigma(M) = M$. Because $L \subset M$, we have $\sigma(L) \subset M$. It follows that $\prod_{\sigma \in \mathcal{G}al(K/k)} \sigma(L) \subset M$. However, for all $\tau \in \mathcal{G}al(K/k)$, $\tau \prod_{\sigma \in \mathcal{G}al(K/k)} \sigma(L) = \prod_{\sigma \in \mathcal{G}al(K/k)} \tau\sigma(L) = \prod_{\zeta \in \mathcal{G}al(K/k)} \zeta(L)$, and so the compositum is normal over k. It follows that $\prod_{\sigma \in \mathcal{G}al(K/k)} \sigma(L) = M$. $\qquad\square$

Get to know the Fundamental Theorem of Algebra

With all the basic theorems of Galois theory in hand, we can give a proof of the Fundamental Theorem of Algebra: If $f(x)$ is a nonconstant complex polynomial, then $f(x)$ has a root in $\mathbb{C}$.

1. Let $a + bi$ be any complex number. Show that there is another complex number $c + di$ for which $(c + di)^2 = a + bi$. Using this fact, show that any quadratic polynomial $f(x) = a_0 + a_1x + a_2x^2$ with $a_i \in \mathbb{C}$ has a root in $\mathbb{C}$. (You may assume that any positive real number has a square root, a fact that can be proved from the least upper bound property of the real numbers.)
2. Conclude from the previous exercise that there are no extensions L of $\mathbb{C}$ of degree two.
3. Suppose $f(x)$ is a real polynomial of odd degree. Recall that an odd degree polynomial satisfies $f(-t) = -f(t)$. Show that $f(x)$ has a root in $\mathbb{R}$. (This result is an analytic fact that follows from the Intermediate Value Theorem.)
4. Show that there are no extensions of $\mathbb{R}$ of odd degree greater than one.

5. Suppose $f(x) \in \mathbb{C}[x]$ is a nonconstant polynomial. If $f(x) = a_0 + a_1 x + \cdots + a_{n-1} x^{n-1} + a_n x^n$, then let $\bar{f}(x) = \bar{a}_0 + \bar{a}_1 x + \cdots + \bar{a}_n x^n$. Show that $F(x) = f(x)\bar{f}(x)$ is a real polynomial whose roots coincide with the roots of $f(x)$. Thus it suffices to prove that every nonconstant real polynomial has a root in $\mathbb{C}$.

6. Suppose $g(x) \in \mathbb{R}[x]$ is irreducible. Let $h(x) = (x^2 + 1)g(x)$, and let E_h be its splitting field over $\mathbb{R}$. Let $G = \mathcal{G}al(E_h/\mathbb{R})$ and suppose $\#G = 2^m n$ with odd n. By the Sylow Theorems, G has a 2-subgroup H of order 2^m. Let $L = (E_h)^H$ with $\mathbb{R} \subset L \subset E_h$. By the Fundamental Theorem of Galois Theory, $[L : \mathbb{R}] = [G : H] = n$ is odd. By Exercise 4 above, conclude that $n = 1$, and so $\mathcal{G}al(E_h/\mathbb{R})$ is a 2-group.

7. Since $i \in E_h$, E_h contains $\mathbb{C}$ as a subfield. As a subgroup of $\mathcal{G}al(E_h/\mathbb{R})$, $\mathcal{G}al(E_h/\mathbb{C})$ is a 2-group, and so it is solvable. The composition series for a 2-group always contains a subgroup of index two, say Γ. Then $[(E_h)^\Gamma : \mathbb{C}] = [\mathcal{G}al(E_h/\mathbb{C}) : \Gamma] = 2$. However, there are no extensions of $\mathbb{C}$ of degree two. It follows that $2^m = 1$, and so $\mathcal{G}al(E_h/\mathbb{C}) = \{\text{Id}\}$ and $E_h = \mathbb{C}$, that is, $h(x)$ has all of its roots in $\mathbb{C}$.

6.2 Cyclotomy

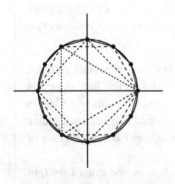

Figure 6.2

The Galois correspondence between the subgroups of the Galois group $\mathcal{G}al(K/k)$ and the intermediate fields between k and K provides a tool to analyze the roots of polynomials and other constructions that can be framed in terms of field extensions. The earliest exploration of field extensions in which the correspondence is suggested is found in Gauss' *Disquisitiones Arithmeticae* [33], Chapter 7, which treats the theory of *cyclotomy*, that is, circle division. The problem undertaken by Gauss (aged 19 at the time of discovery [34]) is the question of constructing a regular n-gon using only straightedge and compass (see the exercises of Chapter 5).

In the complex plane the vertices of a regular n-gon inscribed in the unit circle are given by the points $e^{2\pi i \ell/n} = \cos(2\pi\ell/n) + i\sin(2\pi\ell/n)$, $\ell = 0, 1, \ldots, n - 1$. Each vertex is also a root of the polynomial $x^n - 1$, and so these points are called the *nth roots of unity*. We denote this set of points by $\mu_n = \{e^{2\pi i \ell/n} \mid 0 \le \ell < n\}$. In Fig. 6.2 the 12th roots of unity are shown on the unit circle. Notice how the regular 12-gon contains the regular hexagon, square, and equilateral triangle as subsets of the 12th roots of unity.

Proposition 6.26 *The set μ_n of nth roots of unity over a field k, equipped with the field multiplication, is a finite cyclic group.*

Proof Let E_n denote the splitting field of $x^n - 1$ over k and μ_n the set of roots of $x^n - 1$ in E_n. If $\alpha, \beta \in \mu_n$, then $\alpha^n = 1 = \beta^n$. It follows that $(\alpha\beta)^n = \alpha^n \beta^n = 1$ and μ_n is closed under multiplication. Since $\alpha \cdot (\alpha^{n-1}) = 1$, each $\alpha \in \mu_n$ has an inverse. Thus μ_n is a finite abelian subgroup of $E_n^\times$. Suppose N is the maximal order of the elements of μ_n and $\omega \in \mu_n$ has order N. If $\alpha \in \mu_n$ has order m, then the product $\alpha\omega$ has order $\mathrm{lcm}(m, N) \ge N$. But N is the maximal order of elements in μ_n, so $\mathrm{lcm}(m, N) \le N$ and $\mathrm{lcm}(m, N) = N$, which implies that m divides N. Thus every element of μ_n is a root of $x^N - 1$, and so $\#\mu_n \le N$. However, the order of the subgroup generated by ω is N, and so $\mu_n = \langle\omega\rangle$, that is, μ_n is a cyclic group.

If the characteristic of k is zero or a prime p that does not divide n, then $x^n - 1$ is separable, and $\#\mu_n = n$. If the characteristic of k is a prime p that divides n, then $x^n - 1 = x^{p^e j} - 1 = (x^j - 1)^{p^e}$, where p does not divide j. The roots of $x^n - 1$ are the roots of $x^j - 1$, each with multiplicity p^e. $\square$

Let us assume that k has characteristic zero or a prime p that does not divide n, so that $x^n - 1$ is separable. A generator ω of μ_n is called a *primitive nth root of unity*, and every nth root of unity is a power of ω. Each ω^ℓ generates a subgroup of μ_n of some order d that divides n. Because μ_n is cyclic, $\langle\omega^\ell\rangle = \langle\omega^{n/d}\rangle$. The set μ_n is partitioned into disjoint subsets according to the order of each element. Let $\Omega_d = \{\omega^l \mid \langle\omega^l\rangle = \langle\omega^{n/d}\rangle\}$. For example, when $n = 12$, the partition takes the form

$$\mu_{12} = \Omega_1 \cup \Omega_2 \cup \Omega_3 \cup \Omega_4 \cup \Omega_6 \cup \Omega_{12},$$

where $\Omega_1 = \{1\}$, $\Omega_2 = \{\omega^6\}$, $\Omega_3 = \{\omega^4, \omega^8\}$, $\Omega_4 = \{\omega^3, \omega^9\}$, $\Omega_6 = \{\omega^2, \omega^{10}\}$, and $\Omega_{12} = \{\omega, \omega^5, \omega^7, \omega^{11}\}$. In Fig. 6.2 the subgroups correspond to inscribed d-gons inside the regular 12-gon for $d = 2, 3, 4,$ and 6. Define the polynomial $\Phi_d(x)$ whose roots lie in μ_n and have order d:

$$\Phi_d(x) = \prod_{\alpha \in \mu_n, \#\langle\alpha\rangle = d} x - \alpha.$$

Then $x^n - 1 = (x - 1)(x - \omega) \cdots (x - \omega^n) = \prod_{d \mid n} \Phi_d(x)$. The monic polynomials $\Phi_d(x)$ are called the *cyclotomic polynomials*; their roots are the primitive dth roots of unity. The $\Phi_d(x)$ do not depend on n because when d

divides n, $x^d - 1$ divides $x^n - 1$, and $\omega^{n/d}$ is a primitive dth root of unity. The first few cyclotomic polynomials over $\mathbb{Q}$ are $\Phi_1(x) = x - 1$, $\Phi_2(x) = x + 1$, $\Phi_3(x) = x^2 + x + 1$, and $\Phi_4(x) = x^2 + 1$. For prime p, $\Phi_p(x) = \dfrac{x^p - 1}{x - 1} = x^{p-1} + x^{p-2} + \cdots + x + 1$.

The main properties of cyclotomic polynomials are listed here.

Proposition 6.27 (1) *The degree of $\Phi_d(x)$ is $\phi(d) = \#\{1 \le \ell < d \mid \gcd(\ell, d) = 1\}$, the Euler ϕ-function. (2) $\Phi_d(x)$ is a monic integral polynomial for all $d > 0$. (3) $\Phi_d(x)$ is irreducible over $\mathbb{Q}$.*

Proof (1) Suppose $\mu_d = \langle \omega \rangle$. Then the powers ω^ℓ that are also generators of μ_d are those for which $\gcd(\ell, d) = 1$ (see [38]). The number of such powers in μ_d is $\phi(d)$. By the definition of $\Phi_d(x)$ its degree is $\phi(d)$.

(2) We prove that $\Phi_d(x)$ is an integral polynomial by induction on d. We have $\Phi_1(x) = x - 1$ and $\Phi_2(x) = x + 1$, so the induction is underway. Suppose $\Phi_d(x) \in \mathbb{Z}[x]$ for $d < n$. Since $x^n - 1 = \Phi_n(x) \prod_{d \mid n, d < n} \Phi_d(x)$, and the product $\prod_{d \mid n, d < n} \Phi_d(x)$ and $x^n - 1$ are monic integral polynomials, by Theorem 5.15, $\Phi_n(x)$ is also a monic integral polynomial.

(3) We follow the 1857 proof of Dedekind [20], also presented by van der Waerden [88]. For other proofs, see [95]. Suppose that $\Phi_d(x) = f(x)g(x)$ in $\mathbb{Z}[x]$, where ω is a root of $f(x)$, an irreducible polynomial in $\mathbb{Z}[x]$. By Theorem 5.15 a factorization of $f(x)$ in $\mathbb{Q}[x]$ implies a factorization in $\mathbb{Z}[x]$, and $f(x)$ is irreducible in $\mathbb{Q}[x]$. Because $\Phi_d(x)$ is monic, this implies that $f(x)$ is monic, and so $f(x) = m_\omega(x)$, the minimal polynomial for ω in $\mathbb{Q}[x]$.

Suppose ℓ is a positive integer with $\gcd(\ell, d) = 1$. Then ω^ℓ is a primitive dth root of unity. By the celebrated theorem of Dirichlet [2] there are infinitely many primes q with $q \equiv \ell \pmod{d}$. Every prime q that does not divide d has ω^q as a root of $\Phi_d(x)$. If ω^q is also a root of $f(x)$, then $\Phi_d(x)$ divides $f(x)$, from which it follows that $\Phi_d(x) = f(x)$, an irreducible polynomial. If there is q for which ω^q is not a root of $f(x)$ but of $g(x)$, then ω is a root of $g(x^q)$. It follows that $m_\omega(x)$ divides $g(x^q)$.

Consider the ring homomorphism $r \colon \mathbb{Z}[x] \to \mathbb{F}_q[x]$ where we take the coefficients of an integral polynomial to their residues mod q. We can write $g(x^q) = m_\omega(x)h(x) = f(x)h(x)$ in $\mathbb{Z}[x]$, and so $r(g(x^q)) = r(f(x))r(h(x))$. Suppose $u(x) \in \mathbb{F}_q[x]$ is an irreducible factor of $r(f(x))$. In $\mathbb{F}_q[x]$, then we can write $r(f(x)) = u(x)l(x)$ in $\mathbb{F}_q[x]$. By Fermat's Little Theorem [2], $a^q \equiv a \pmod{q}$ for all $a \in \mathbb{F}_q$, and so $(r(g(x)))^q \equiv r(g(x^q))$ in $\mathbb{F}_q[x]$. Thus

$$\left(r\big(g(x)\big)\right)^q \equiv r\big(g(x^q)\big) \equiv r\big(f(x)\big)r\big(h(x)\big) \equiv u(x)l(x)r\big(h(x)\big) \pmod{q}.$$

Therefore $u(x)$ divides $r(g(x))$ in $\mathbb{F}_q[x]$. Since $x^d - 1$ is separable over $\mathbb{F}_q$ and $r(\Phi_d(x))$ is a factor of $x^d - 1$, $r(\Phi_d(x))$ is separable. However,

$r(\Phi_d(x)) = r(f(x))r(g(x)) = (u(x))^2 l(x)r(h(x))$, which contradicts the separability of $r(\Phi_d(x))$. This contradiction implies $\Phi_d(x)$ is irreducible and provides a minimal polynomial for ω. □

When the field k has the appropriate characteristic, and ω is a primitive nth root of unity, then $\Phi_n(x)$ and $x^n - 1 = (x - 1)(x - \omega) \cdots (x - \omega^{n-1})$ split completely in $k(\omega)$. The extension $E_n = k(\omega)$ is a Galois extension of k of degree $\phi(n) = \deg \Phi_n(x)$. Suppose $\sigma \in \mathcal{G}al(k(\omega)/k)$. Then $\sigma(\omega)$ is also a root of $\Phi_n(x)$, and so $\sigma(\omega) = \omega^t$ with $\gcd(n, t) = 1$.

The group of units $(\mathbb{Z}/n\mathbb{Z})^\times$ in the ring $\mathbb{Z}/n\mathbb{Z}$ consists of the set of residues mod n that are relatively prime to n. If $\gcd(n, t) = 1$, then there are integers a and b for which $an + bt = 1$. This relation implies $bt \equiv 1 \pmod{n}$, and so the congruence class of t is a unit in $\mathbb{Z}/n\mathbb{Z}$. Because $\gcd(ab, n) = 1$ when $\gcd(a, n) = 1 = \gcd(b, n)$, the residues in $(\mathbb{Z}/n\mathbb{Z})^\times$ are closed under multiplication. On the other hand, a unit in $\mathbb{Z}/n\mathbb{Z}$ has an inverse and so determines a remainder mod n that is relatively prime to n. Putting these observations together, we obtain the following result.

Theorem 6.28 *Let $E_n = k(\omega)$ be the splitting field of $x^n - 1$ over a field k, and let ω be a primitive nth root of unity. The mapping $P: \mathcal{G}al(k(\omega)/k) \to (\mathbb{Z}/n\mathbb{Z})^\times$, given by $P(\sigma) = [t]$ when $\sigma(\omega) = \omega^t$, is an injective homomorphism. When $k = \mathbb{Q}$, P is an isomorphism.*

Proof Suppose ω and ζ are primitive nth roots of unity in $k(\omega)$. Then $\zeta = \omega^\ell$ for some $1 \le \ell < n$ for which $\gcd(\ell, n) = 1$. If $\sigma(\omega) = \omega^t$, then $\sigma(\zeta) = \sigma(\omega^\ell) = (\omega^t)^\ell = (\omega^\ell)^t = (\zeta)^t$; therefore the mapping P is independent of the choice of which primitive nth root of unity we choose.

To see that P is a homomorphism, suppose $\sigma(\omega) = \omega^t$ and $\tau(\omega) = \omega^s$. Then

$$(\sigma \circ \tau)(\omega) = \sigma(\omega^s) = \omega^{st}, \text{ and so } P(\sigma \circ \tau) = [st] = P(\sigma)P(\tau).$$

By Lemma 6.4 an automorphism $\sigma: k(\omega) \to k(\omega)$ is determined by its value on ω, and so $P(\sigma) = P(\tau)$ if and only if $\sigma(\omega) = \omega^t = \tau(\omega)$, and so $\sigma = \tau$, that is, P is injective.

In the case $k = \mathbb{Q}$, we have shown that $\Phi_n(x)$ is irreducible over $\mathbb{Q}$ of degree $\phi(n)$. The extension $E_n = \mathbb{Q}(\omega)$ has degree $[E_n : \mathbb{Q}] = \phi(n) = \#\mathcal{G}al(E_n/\mathbb{Q})$. Thus $P: \mathcal{G}al(E_n/\mathbb{Q}) \to (\mathbb{Z}/n\mathbb{Z})^\times$ is an injective homomorphism of groups of the same order and therefore is bijective. □

With this structure theorem, we can use the Galois correspondence to explore the set of intermediate fields of a cyclotomic extension. A classic case is Gauss' analysis of the construction of a regular 17-gon (a heptadecagon) by straightedge and compass. Let $\omega = e^{2\pi i/17}$ denote the primitive 17th root of unity. The Galois group $\mathcal{G}al(\mathbb{Q}(\omega)/\mathbb{Q})$ is isomorphic to $(\mathbb{Z}/17\mathbb{Z})^\times$, a group of order 16. Since $\mathbb{Z}/17\mathbb{Z}$ is a field, its group of units is a cyclic group, and a short computation (for example, if you have studied number theory [38], the Legendre symbol $\left(\dfrac{3}{17}\right) = -1$) shows that $(\mathbb{Z}/17\mathbb{Z})^\times = \langle 3 \rangle$. The cyclic group of order 16 has exactly three proper nontrivial subgroups of orders 8, 4, and 2. The Galois correspondence may be pictured as a ladder:

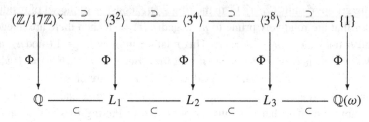

Since $x^{17} - 1 = (x - 1)(x^{16} + x^{15} + \cdots + x + 1) = \Phi_1(x)\Phi_{17}(x)$, the primitive 17th root of unity ω satisfies

$$\omega + \omega^2 + \cdots + \omega^{15} + \omega^{16} = -1.$$

Let $\sigma \in \mathcal{G}al(\mathbb{Q}(\omega)/\mathbb{Q})$ denote the generator corresponding to the congruence class of [3] in $(\mathbb{Z}/17\mathbb{Z})^\times$, which is determined by $\sigma(\omega) = \omega^3$. In §343 of [33], Gauss introduced the use of *periods*: Let

$$\alpha = \omega^{3^1} + \omega^{3^3} + \omega^{3^5} + \omega^{3^7} + \omega^{3^9} + \omega^{3^{11}} + \omega^{3^{13}} + \omega^{3^{15}} \text{ and}$$

$$\beta = \omega^{3^2} + \omega^{3^4} + \omega^{3^6} + \omega^{3^8} + \omega^{3^{10}} + \omega^{3^{12}} + \omega^{3^{14}} + \omega^{3^{16}}.$$

Because $\sigma(\omega^{3^m}) = (\omega^3)^{3^m} = \omega^{3^{m+1}}$, these elements in $\mathbb{Q}(\omega)$ satisfy $\sigma(\alpha) = \beta$ and $\sigma(\beta) = \alpha$. It follows that both are fixed by σ^2. Let L_1 denote the subfield of $\mathbb{Q}(\omega)$ fixed by the subgroup $\langle \sigma^2 \rangle$, that is, $L_1 = \mathbb{Q}(\omega)^{\langle \sigma^2 \rangle}$. Then α and β are in L_1. Because $\alpha + \beta = -1$, we have $\beta = -1 - \alpha$.

To determine α and β precisely, we apply an ancient method. Multiplying α and β, we get a $\mathcal{G}al(\mathbb{Q}(\omega)/\mathbb{Q})$-invariant element of $\mathbb{Q}(\omega)$. A straightforward computation gives $\alpha\beta = -4$. The sum $\alpha + \beta = -1$, and so α and β are roots of the quadratic polynomial

$$(x - \alpha)(x - \beta) = x^2 - (\alpha + \beta)x + \alpha\beta = x^2 + x - 4.$$

The quadratic formula gives us $\alpha, \beta = \dfrac{-1 \pm \sqrt{17}}{2}$. Since α and β are not in $\mathbb{Q}$, $x^2 + x - 4$ is irreducible over $\mathbb{Q}$. Hence $L_1 = \mathbb{Q}(\alpha) = \mathbb{Q}(\omega)^{\langle \sigma^2 \rangle}$, an extension of degree two over $\mathbb{Q}$ generated by α. Because $\langle 3^2 \rangle$ is normal in

$(\mathbb{Z}/17\mathbb{Z})^\times$, we find

$$\mathcal{G}al(L_1/\mathbb{Q}) \cong \mathcal{G}al\big(\mathbb{Q}(\omega)/\mathbb{Q}\big)/\mathcal{G}al\big(\mathbb{Q}(\omega)/L_1\big) \cong (\mathbb{Z}/17\mathbb{Z})^\times/\langle 3^2 \rangle \cong C_2.$$

The next subgroup in the ladder is of order four generated by σ^4. The analogous construction of periods suggests

$$\alpha_1 = \omega^{3^1} + \omega^{3^5} + \omega^{3^9} + \omega^{3^{13}}, \qquad \alpha_2 = \omega^{3^3} + \omega^{3^7} + \omega^{3^{11}} + \omega^{3^{15}},$$
$$\beta_1 = \omega^{3^2} + \omega^{3^6} + \omega^{3^{10}} + \omega^{3^{14}}, \qquad \beta_2 = \omega^{3^4} + \omega^{3^8} + \omega^{3^{12}} + \omega^{3^{16}}.$$

The action of σ gives the cycle $\sigma : \alpha_1 \mapsto \beta_1 \mapsto \alpha_2 \mapsto \beta_2 \mapsto \alpha_1$. It follows that the elements $\alpha_1, \alpha_2, \beta_1, \beta_2$ are fixed by σ^4. By construction $\alpha_1 + \alpha_2 = \alpha$, and a computation gives $\alpha_1\alpha_2 = \sum_{j=1}^{16} \omega^{3^j} = -1$. Thus α_1 and α_2 are the roots of the quadratic polynomial $x^2 - \alpha x - 1 \in L_1[x]$. Let $L_2 = L_1(\alpha_1) = \mathbb{Q}(\alpha, \alpha_1)$. Because $\mathcal{G}al(\mathbb{Q}(\omega)/\mathbb{Q})$ is abelian and L_2 is an intermediate field, L_2 is normal over $\mathbb{Q}$. The periods β_1, β_2 satisfy $\beta_1 + \beta_2 = \beta$ and $\beta_1\beta_2 = -1$, making them the roots of $x^2 - \beta x - 1 \in L_1[x]$. Because $\mathcal{G}al(\mathbb{Q}(\omega)/\mathbb{Q}) = \langle \sigma \rangle$, the product $(x - \alpha_1)(x - \alpha_2)(x - \beta_1)(x - \beta_2)$ is invariant under the action of $\mathcal{G}al(\mathbb{Q}(\omega)/\mathbb{Q})$ and hence is in $\mathbb{Q}[x]$:

$$(x - \alpha_1)(x - \alpha_2)(x - \beta_1)(x - \beta_2) = \big(x^2 - \alpha x - 1\big)\big(x^2 - \beta x - 1\big)$$
$$= x^4 + x^3 - 6x^2 - x + 1.$$

The Galois group of L_2 over $\mathbb{Q}$ is given by $\mathcal{G}al(L_2/\mathbb{Q}) \cong \mathcal{G}al(\mathbb{Q}(\omega)/\mathbb{Q})/\mathcal{G}al(\mathbb{Q}(\omega)/L_2) \cong \langle \sigma \rangle/\langle \sigma^4 \rangle \cong C_4$. Since the coset $\sigma\langle \sigma^4 \rangle$ generates the group $\mathcal{G}al(L_2/\mathbb{Q})$, we see that the group acts transitively and faithfully on the roots of $x^4 + x^3 - 6x^2 - x + 1$, and this polynomial is irreducible. Since it has a root α_1 in L_2, by normality all the roots appear in L_2, and so we find $L_2 = \mathbb{Q}(\alpha, \alpha_1) = \mathbb{Q}(\omega)^{\langle \sigma^4 \rangle}$.

Again the quadratic formula lets us express α_1 explicitly. Since α_1 is a root of $x^2 - \alpha x - 1$, we have $\alpha_1 = \dfrac{\alpha + \sqrt{\alpha^2 + 4}}{2}$. It follows from $\alpha^2 - \alpha - 4 = 0$ that $\alpha^2 + 4 = 8 - \alpha = \dfrac{17 - \sqrt{17}}{2}$. Substituting this into the expression for α_1, we get

$$\alpha_1 = \frac{-1 + \sqrt{17} + \sqrt{34 - 2\sqrt{17}}}{4}.$$

The last intermediate field L_3 corresponds to the subgroup $\langle \sigma^8 \rangle$. Following the same strategy, we define

$$\gamma_1 = \omega^{3^1} + \omega^{3^9}, \quad \gamma_2 = \omega^{3^3} + \omega^{3^{11}}, \quad \gamma_3 = \omega^{3^5} + \omega^{3^{13}}, \quad \gamma_4 = \omega^{3^7} + \omega^{3^{15}},$$
$$\eta_1 = \omega^{3^2} + \omega^{3^{10}}, \quad \eta_2 = \omega^{3^4} + \omega^{3^{12}}, \quad \eta_3 = \omega^{3^6} + \omega^{3^{14}}, \quad \eta_4 = \omega^{3^8} + \omega^{3^{16}}.$$

Then $\gamma_1 + \gamma_3 = \alpha_1$ and $\gamma_1\gamma_3 = (\omega^{3^1} + \omega^{3^9})(\omega^{3^5} + \omega^{3^{13}}) = (\omega^3 + \omega^{14})(\omega^5 + \omega^{12}) = \omega^8 + \omega^{15} + \omega^2 + \omega^9 = \beta_1$. Thus γ_1 and γ_3 are roots of the quadratic polynomial $x^2 - \alpha_1 x + \beta_1 \in L_2[x]$. To see that we have identified elements that are left fixed by σ^8, consider the pattern $\sigma : \gamma_1 \mapsto \eta_1 \mapsto \gamma_2 \mapsto \eta_2 \mapsto \gamma_3 \mapsto \eta_3 \mapsto \gamma_4 \mapsto \eta_4 \mapsto \gamma_1$. Applying σ to the quadratic polynomial, we see that pairs η_1, η_3 are roots of $x^2 - \beta_1 x + \alpha_2$; γ_2, γ_4 are roots of $x^2 - \alpha_2 x + \beta_2$; and η_2, η_4 are roots of $x^2 - \beta_2 x + \alpha_1$. The product of these four polynomials is invariant under σ and hence in $\mathbb{Q}[x]$. The product can be computed to be $x^8 + x^7 - 7x^6 - 6x^5 + 15x^4 + 10x^3 - 10x^2 - 4x + 1$. The same argument by normality finds all γ_i and η_j in $L_3 = L_2(\gamma_1)$.

The elements γ_i and η_j all take the form $\omega^{3^k} + \omega^{3^{k+8}}$, and $3^k + 3^{k+8} = 3^k(1 + 3^8) \equiv 0 \,(\mathrm{mod}\, 17)$. Thus $\omega^{3^k} + \omega^{3^{k+8}} = \omega^{3^k} + \omega^{-3^k}$, and so ω^{3^k} and ω^{-3^k} are the roots of the quadratic polynomial $x^2 - (\omega^{3^k} + \omega^{-3^k})x + 1 \in L_3[x]$. When $\omega = e^{2\pi i/17}$, a generator for μ_{17}, we have $\omega + \omega^{-1} = \omega^{3^{16}} + \omega^{3^8} = \eta_4$.

The value for ω is given by the quadratic equation: $\omega = \dfrac{\eta_4 + \sqrt{\eta_4^2 - 4}}{2}$. As observed by Gauss [33], §365, the real part of ω, $\cos(2\pi/17)$, can be expressed as $\dfrac{\eta_4}{2} = \dfrac{\omega + \omega^{-1}}{2}$. Passing through all the quadratic equations discussed above, we get

$$\cos(2\pi/17) = -\frac{1}{16} + \frac{1}{16}\sqrt{17} + \frac{1}{16}\sqrt{34 - 2\sqrt{17}}$$
$$+ \frac{1}{8}\sqrt{17 + 3\sqrt{17} - \sqrt{34 - 2\sqrt{17}} - 2\sqrt{34 + 2\sqrt{17}}}.$$

The fact that the angle $2\pi/17$ is constructible by straightedge and compass emerges from the structure of $\mathcal{G}al(\mathbb{Q}(\omega)/\mathbb{Q})$, a 2-group, with a composition series all of whose subquotients are isomorphic to $\mathbb{Z}/2\mathbb{Z}$. As we saw in the exercises at the end of Chapter 5, the tower of intermediate fields corresponding to the composition series determines constructibility using straightedge and compass. For a prime p, the isomorphism $\mathcal{G}al(\mathbb{Q}(\theta)/\mathbb{Q}) \cong (\mathbb{Z}/p\mathbb{Z})^\times$ for θ a primitive pth root of unity restricts a straightedge and compass construction of a regular p-gon to primes of the form $p = 2^{2^j} + 1$, the *Fermat primes*. There are only five such primes known, 3, 5, 17, 257, and 65537. The bisection of an angle can also be achieved by straightedge and compass. Gauss proved the sufficiency of the following result.

Theorem 6.29 *A regular n-gon is constructible by straightedge and compass if and only if $n = 2^s p_1 \cdots p_t$, where $s \geq 0$ and $p_1, \ldots, p_t$ are distinct Fermat primes.*

The necessary direction of the theorem is due to P. LAURENT WANTZEL (1814–1848) [90]. See [60] for a discussion of Wantzel's work – an example of a mathematical proof of impossibility – and its reception.

6.3 Filling in: Burnside's $p^\alpha q^\beta$ Theorem*

Let us return to representation theory, equipped with some Galois theory, to prove an important result for finite groups. Recall that the values of the characters of a finite group are sums of roots of unity (Theorem 4.23). Let $E_n \subset \mathbb{C}$ denote the splitting field of $x^n - 1$. The nth roots of unity each satisfy some cyclotomic polynomial $\Phi_d(x)$ and hence are algebraic integers. The sum of algebraic integers is an algebraic integer, so the values $\chi_\rho(g)$ of the characters of representations are algebraic integers.

Lemma 6.30 *Suppose $\rho: G \to \mathrm{Gl}(W)$ is an irreducible representation and $[g]$ is the conjugacy class in G of an element g. Then $\dfrac{\#[g]\chi_\rho(g)}{\chi_\rho(e)}$ is an algebraic integer.*

Proof Consider the linear mapping $\rho([g]) = \sum_{h \in [g]} \rho(h): W \to W$. For any $l \in G$,

$$\rho(l) \circ \rho\big([g]\big) \circ \rho(l)^{-1} = \sum_{h \in [g]} \rho(l) \circ \rho(h) \circ \rho\big(l^{-1}\big) = \sum_{h \in [g]} \rho\big(lhl^{-1}\big)$$

$$= \sum_{h' \in [g]} \rho\big(h'\big) = \rho\big([g]\big).$$

Therefore $\rho([g]): W \to W$ is a G-equivariant linear transformation. Because W is irreducible, Schur's Lemma implies $\rho([g]) = \lambda \,\mathrm{Id}$ for some $\lambda \in \mathbb{C}$. Taking the trace of $\rho([g])$, we get $\#[g]\chi_\rho(g) = \lambda\chi_\rho(e)$. As shown in the proof of Proposition 4.36 and in Exercises 4.12–14 of Chapter 4, λ is an algebraic integer equal to $\dfrac{\#[g]\chi_\rho(g)}{\chi_\rho(e)}$. The fact that λ is an algebraic integer follows from the observation that the product set $[g_i][g_j] = \{gh \mid g \in [g_i], h \in [g_j]\}$ is closed under conjugation. $\qquad\square$

Theorem 6.31 *Suppose $\rho: G \to \mathrm{Gl}(W)$ is a nontrivial irreducible representation and $[g]$ is the conjugacy class in G of an element $g \neq e$. If $\gcd(\#[g], \chi_\rho(e)) = 1$, then either $\chi_\rho(g) = 0$, or G contains a nontrivial normal subgroup.*

Proof Suppose $\gcd(\#[g], \chi_\rho(e)) = 1$. There are integers u, v such that $u(\#[g]) + v\chi_\rho(e) = 1$. If we divide by $\chi_\rho(e)$ and multiply by $\chi_\rho(g)$, we

get the relation $u\dfrac{\#[g]\chi_\rho(g)}{\chi_\rho(e)} + v\chi_\rho(g) = \dfrac{\chi_\rho(g)}{\chi_\rho(e)}$. By Lemma 6.30, $\dfrac{\#[g]\chi_\rho(g)}{\chi_\rho(e)}$
is an algebraic integer as is $\chi_\rho(g)$. Integral linear combinations of algebraic
integers are also algebraic integers, and so $\dfrac{\chi_\rho(g)}{\chi_\rho(e)}$ is an algebraic integer. The
value of $\chi_\rho(g)$ is a sum of $\chi_\rho(e) = n$ many nth roots of unity. As we showed
in the proof of Theorem 4.39, $|\chi_\rho(g)| \le \chi_\rho(e)$ with equality if and only if
$\chi_\rho(g) = n\omega$ for an nth root ω of unity. When $\left|\dfrac{\chi_\rho(g)}{\chi_\rho(e)}\right| < 1$, we can involve the
Galois group of E_n over $\mathbb{Q}$. For any $\sigma \in \mathcal{G}al(E_n/\mathbb{Q})$, $\sigma(\chi_\rho(g))$ is also a sum of
nth roots of unity, and if the sum $\chi_\rho(g)$ does not have all equal summands, then
the same is true for $\sigma(\chi_\rho(g))$. Thus, if $\left|\dfrac{\chi_\rho(g)}{\chi_\rho(e)}\right| < 1$, then $\left|\dfrac{\sigma(\chi_\rho(g))}{\chi_\rho(e)}\right| < 1$.
Form the product

$$\left|\prod_{\sigma \in \mathcal{G}al(E_n/\mathbb{Q})} \frac{\sigma(\chi_\rho(g))}{\chi_\rho(e)}\right| = \prod_{\sigma \in \mathcal{G}al(E_n/\mathbb{Q})} \left|\frac{\sigma(\chi_\rho(g))}{\chi_\rho(e)}\right| < 1.$$

The product of the quotients $\sigma(\chi_\rho(g))/\chi_\rho(e)$ is fixed by every $\tau \in \mathcal{G}al(E_n/\mathbb{Q})$,
and so it is a rational number. Because $\sigma \in \mathcal{G}al(E_n/\mathbb{Q})$ takes nth roots of
unity to nth roots of unity, the product is also an algebraic integer. Since a
rational number that is an algebraic integer is an integer (Proposition 4.22),
the inequality implies that $\prod_{\sigma \in \mathcal{G}al(E_n/\mathbb{Q})} \dfrac{\sigma(\chi_\rho(g))}{\chi_\rho(e)} = 0$. It follows that some
$\sigma(\chi_\rho(g)) = 0$, and because σ is an automorphism of E_n, $\chi_\rho(g) = 0$.

If $|\chi_\rho(g)| = \chi_\rho(e)$, then $\rho(g) = \omega\,\mathrm{Id}$ for some nth root ω of unity. Since
multiples of the identity commute with all linear transformations, $\rho(gh) = \rho(g)\rho(h) = \rho(h)\rho(g) = \rho(hg)$ for all $h \in G$. Either $g \in Z(G)$ with $g \ne e$
and $Z(G)$ is a nontrivial normal subgroup of G, or there is an $h \in G$ such
that $gh \ne hg$. Then $\rho(ghg^{-1}h^{-1}) = \rho(e) = \mathrm{Id}$, and $ghg^{-1}h^{-1}$ is in $\ker \rho$.
Because ρ is nontrivial, $\ker \rho \ne G$. If some $ghg^{-1}h^{-1} \ne e$, then $\ker \rho \ne \{e\}$,
and $\ker \rho$ is a nontrivial normal subgroup of G. $\qquad\square$

This theorem and certain conditions on a group allow us to prove important
results of Burnside [9].

Theorem 6.32 *Suppose p is a prime number and G is a group which contains
a conjugacy class with cardinality p^r with $r \ge 1$. Then G is not a simple
group.*

Proof Let $g \in G$ with $\#[g] = p^r, r \ge 1$. Let $\mathrm{Irr}(G) = \{\chi_1, \chi_2, \ldots, \chi_t\}$
denote the set of irreducible characters of G, where χ_1 denotes the trivial character. In the character table for G the values in the column under $[e]$ and under

[g] satisfy, by the Second Orthogonality Theorem (Chapter 4),

$$\sum_{j=1}^{t} \chi_j(e)\overline{\chi_j(g)} = 1 + \sum_{j=2}^{t} \chi_j(e)\overline{\chi_j(g)} = 0.$$

This implies $\sum_{j=2}^{t} \frac{\chi_j(e)}{p}\overline{\chi_j(g)} = \frac{-1}{p}$. Because $-1/p$ is not an algebraic integer, there must be an index l such that $\chi_l(e)\overline{\chi_l(g)}/p$ is not an algebraic integer. It follows that $\overline{\chi_l(g)} \neq 0$. Because $\overline{\chi_l(g)}$ is an algebraic integer, $\chi_l(e)/p$ is not, and so p does not divide $\chi_l(e)$. Therefore $\gcd(\#[g], \chi_l(e)) = 1$, and by Theorem 6.31 G contains a nontrivial normal subgroup and is not a simple group. □

Burnside's $p^\alpha q^\beta$ Theorem *If p and q are distinct primes, G is a finite group, and there are integers α and β such that $\#G = p^\alpha q^\beta$ and $\alpha + \beta \geq 2$, then G is not a simple group. In fact, G is a solvable group.*

Proof If $\alpha = 0$, then G is a q-group of order q^β with $\beta \geq 2$. Then $Z(G) \neq \{e\}$ by Proposition 2.16, and G is solvable and not simple. The same argument holds when $\beta = 0$ and $\alpha \geq 2$.

Suppose $\alpha > 0$ and $\beta > 0$. By the Sylow Theorems there is a Sylow p-subgroup P of order p^α in G. Then $Z(P) \neq \{e\}$; let $g \neq e$ be an element of $Z(P)$. Since g commutes with everything in P, P is a subgroup of $C_G(g)$. It follows that $\#[g] = [G : C_G(g)] = q^c$ for some c. If $q^c = 1$, then $G = C_G(g)$ and $g \in Z(G)$. Therefore G has a nontrivial center and is not simple. If $\#[g] = q^c > 1$, then Theorem 6.32 establishes that G is not simple.

Recall that G is solvable if it has a composition series all of whose sub-quotients are of the form $\mathbb{Z}/p_i\mathbb{Z}$ for primes p_i. We proceed by induction on $\alpha + \beta$. If $\alpha + \beta = 1$, then G is one of $\mathbb{Z}/p\mathbb{Z}$ or $\mathbb{Z}/q\mathbb{Z}$, which are both solvable. Suppose $\alpha + \beta \geq 2$. We have shown that G is not simple, so let N be a maximal normal subgroup of G. The groups N and G/N have orders of the form $p^\gamma q^\delta$ with $\gamma + \delta < \alpha + \beta$. By induction both N and G/N are solvable. By Proposition 1.12, G is also solvable. □

Get to know the classical formulas

1. The earliest form of the quadratic formula solves the following problem: Given the sum and product of two numbers, find the numbers. We used this formulation in the derivation of the successive quadratic extensions that Gauss used to construct the regular 17-gon. Let us derive

the quadratic formula from this idea: Suppose $f(x) = x^2 + bx + c = 0$. In the splitting field E_f, $f(x) = (x-r)(x-s)$. Deduce that $r+s = -b$ and $rs = c$. The additional ingredient is $(r-s)^2 = (r+s)^2 - 4rs$. Show that this leads to the quadratic formula and to the solution of the ancient problem.

2. Let us extend this idea to finding the roots of a cubic polynomial. Suppose $f(x) = x^3 + bx^2 + cx + d$. A simplifying step is a linear substitution: Let $y = x + (b/3)$. Show that the cubic now takes the form $g(y) = y^3 + By + C$ with B and C functions of $b, c,$ and d.

3. Here is a clever idea. Suppose $y = \sqrt[3]{u} + \sqrt[3]{v}$. Compute y^3 to deduce that

$$y^3 - 3\sqrt[3]{u}\sqrt[3]{v}\,y - (u+v) = 0,$$

and so $B = -3\sqrt[3]{u}\sqrt[3]{v}$ and $-C = u + v$.

4. Finally, by cubing the value for B we obtain the sum $u + v = -C$ and the product $uv = -\dfrac{B^3}{27}$. Use the quadratic formula to obtain values for u and v in terms of B and C. Show that you have found a root of $g(y)$.

5. Based on the form of the solution to the cubic, show that the splitting field E_g of $g(y)$ can be constructed by adjoining a square root of an element w of k, the base field that contains B and C, the cubic roots of unity, followed by a cube root in $k(\sqrt{w})$. Thus we have established $E_g = k(\omega_3, \sqrt{w}, \sqrt[3]{v})$.

6.4 Solvability by Radicals

With cyclotomic extensions, we have analyzed completely the polynomial $x^n - 1$ and determined its Galois group when k is a field whose characteristic does not divide n. The Galois group is abelian in this case, and for $k = \mathbb{Q}$, it is isomorphic to $(\mathbb{Z}/n\mathbb{Z})^\times$, the group of multiplicative units in $\mathbb{Z}/n\mathbb{Z}$. With this structure to hand, Gauss was able to express $\cos(2\pi/17)$ through a series of square root extractions of elements in successive extensions of $\mathbb{Q}$. The classical formula for the solution of cubics is credited to GIROLAMO CARDANO (1501–1576), though he learned important cases from NICCOLO FONTANA "Tartaglia" (~1500–1557) and SCIPIONE DEL FERRO (1465–1526). (See [85] for a description of the history of this equation.) As sketched out in *Get to know the classical formulas*, the solution leads to a series of root extractions to get to the splitting field of a cubic polynomial.

Let us generalize that construction.

In a field the arithmetic operations of addition, subtraction, multiplication, and division are not sufficient to solve polynomial equations. Take $x^3 - 2 = 0$ as an example. As we saw in Chapter 5, to find all three solutions, we needed the irrational number $\sqrt[3]{2}$ and the root of unity $\omega = e^{2\pi i/3}$ to be adjoined to the rational field.

Definition 6.33 A field extension $k \subset K$ is a *simple radical extension* if $K = k(u)$, where u is a root of the polynomial $f(x) = x^n - a$, and $a \in k$. A field extension $k \subset E$ is a *radical extension* if there is a tower of field extensions

$$k = \mathbb{F}_0 \subset \mathbb{F}_1 \subset \cdots \subset \mathbb{F}_{t-1} \subset \mathbb{F}_t = E,$$

where for each $1 \leq i \leq t$, $\mathbb{F}_i$ is a simple radical extension of $\mathbb{F}_{i-1}$. A polynomial equation $p(x) = 0$ for $p(x) \in k[x]$ is *solvable by radicals* if the splitting field E_p of $p(x)$ over k is a subfield of a radical extension E of k.

Gauss' tower for the construction of $\cos(2\pi/17)$ is a radical extension

$$\mathbb{Q} \subset \mathbb{Q}(\sqrt{17}) \subset \mathbb{Q}(\sqrt{34 - 2\sqrt{17}})$$
$$\subset \mathbb{Q}(\sqrt{17 + 3\sqrt{17} - \sqrt{34 - 2\sqrt{17}} - 2\sqrt{34 + 2\sqrt{17}}}).$$

If follows that the polynomial equations associated with α, α_1, and η_4 are solvable by radicals.

Each simple radical extension in the tower may be analyzed in the following manner: Suppose k is a field and $a \in k$. The polynomial $x^n - a$ has $\sqrt[n]{a}$ as a root. However, as we saw in the case of $x^3 - 2$ in Chapter 5, the extension $k(\sqrt[n]{a})$ is not necessarily a splitting field for $x^n - a$ unless k contains the nth roots of unity. Suppose $\omega \in k$ is a primitive nth root of unity. Then $\omega^i \sqrt[n]{a}$ is also a root of $x^n - a$ for $0 \leq i < n$; $(\omega^i \sqrt[n]{a})^n = (\omega^n)^i a = a$. These roots are distinct: if $\omega^i \sqrt[n]{a} = \omega^j \sqrt[n]{a}$, then $\omega^i = \omega^j$ and $i \equiv j \pmod{n}$. With the assumption $\omega \in k$, the extension $k(\sqrt[n]{a})$ is a Galois extension because $x^n - a$ is separable and splits completely over $k(\sqrt[n]{a})$.

Lemma 6.34 *Suppose k is a field whose characteristic does not divide n and suppose k contains the nth roots of unity. If $a \in k$, then $k(\sqrt[n]{a})$ is a Galois extension of k with Galois group isomorphic to a subgroup of $\mathbb{Z}/n\mathbb{Z}$.*

Proof Suppose $\omega \in k$ is a primitive nth root of unity. The complete set of roots of $x^n - a$ is given by $\{\sqrt[n]{a}, \omega\sqrt[n]{a}, \ldots, \omega^{n-1}\sqrt[n]{a}\}$. Let $E = k(\sqrt[n]{a})$. If

$\sigma \in \mathcal{G}al(E/k)$, then $\sigma(\sqrt[n]{a}) = \omega^i \sqrt[n]{a}$ for some $0 \leq i < n$. Consider the mapping $\phi \colon \mathcal{G}al(E/k) \to \mathbb{Z}/n\mathbb{Z}$ given by $\phi(\sigma) = i$. If $\tau(\sqrt[n]{a}) = \omega^j \sqrt[n]{a}$, then $\sigma\tau(\sqrt[n]{a}) = \sigma(\omega^j \sqrt[n]{a}) = \sigma(\omega^j)\sigma(\sqrt[n]{a}) = \omega^j \omega^i \sqrt[n]{a} = \omega^{i+j} \sqrt[n]{a}$ and $\phi(\sigma\tau) = i + j = \phi(\sigma) + \phi(\tau)$. Thus ϕ is a homomorphism. Since $\sigma \in \mathcal{G}al(E/k)$ is determined by its value on $\sqrt[n]{a}$, if $\sigma(\sqrt[n]{a}) = \sqrt[n]{a}$, then $\phi(\sigma) = 0$ and $\sigma = \text{Id}$. Thus ϕ is injective, and $\mathcal{G}al(k(\sqrt[n]{a})/k)$ is isomorphic to a subgroup of $\mathbb{Z}/n\mathbb{Z}$. Subgroups of cyclic groups are cyclic, so we find that $\mathcal{G}al(k(\sqrt[n]{a})/k)$ is a cyclic group with order that divides n.   □

Suppose $n = p_1^{e_1} p_2^{e_2} \cdots p_m^{e_m}$. Then extracting an nth root can be decomposed into e_1 times taking a p_1th root, then e_2 times taking a p_2th root, and continuing on to taking e_m times a p_mth root. Thus an nth root may be accomplished by a finite sequence of prime-order roots. For pth roots, Lemma 6.34 has a partial converse:

Lemma 6.35 *Suppose p is a prime that is not equal to the characteristic of k, and k contains the pth roots of unity. If E is an extension of k with a cyclic group $\mathcal{G}al(E/k)$ and $[E : k] = p$, then E is a Galois extension of k. Furthermore, there is an element $u \in E$ with $E = k(u)$ and $u^p \in k$.*

Proof (Following [16]) Suppose $\mathcal{G}al(E/k)$ is a cyclic group generated by σ. By Theorem 6.15, $[E : E^{\mathcal{G}al(E/k)}] = \#\mathcal{G}al(E/k)$, and $[E : E^{\mathcal{G}al(E/k)}]$ divides $[E : k] = p$. So $E^{\mathcal{G}al(E/k)} = k$, and E is a Galois extension of k. Also, σ has order p.

Let $v \in E$ but $v \notin k$. Consider $k(v) \subset E$. Since $p = [E : k] = [E : k(v)][k(v) : k]$, and $k \neq k(v)$, it must be that $k(v) = E$. However, we do not know if $v^p \in k$. Consider the collection of elements in E given by the *Lagrange resolvent*: For $i = 0, \ldots, p-1$ and a primitive pth root ω of unity,

$$(\omega^i, v) = v + \sigma(v)\omega^i + \sigma^2(v)\omega^{2i} + \cdots + \sigma^{p-1}(v)\omega^{(p-1)i}.$$

When we apply σ to (ω^i, v), because $\omega^{(p-1)i} = \omega^{-i}$ and $\sigma^p = \text{Id}$, we get

$$\sigma(\omega^i, v) = \sigma(v) + \sigma^2(v)\omega^i + \sigma^3(v)\omega^{2i} + \cdots + \sigma^p(v)\omega^{(p-1)i} = \omega^{-i}(\omega^i, v).$$

This leads to an element that is fixed by $\mathcal{G}al(E/k)$:

$$\sigma((\omega^i, v)^p) = (\sigma(\omega^i, v))^p = (\omega^{-i}(\omega^i, v))^p = (\omega^p)^{-i}(\omega^i, v)^p = (\omega^i, v)^p.$$

Thus $(\omega^i, v)^p$ is fixed by every σ^j, so it must lie in k.

It suffices to find that some $(\omega^i, v) \neq 0$ with (ω^i, v) in E but not in k. With $(\omega^0, v) = v + \sigma(v) + \cdots + \sigma^{p-1}(v)$, a computation shows that $\sigma(\omega^0, v) = (\omega^0, v)$, so $(\omega^0, v) \in k$. If $i > 0$ and $(\omega^i, v) \neq 0$, then $\sigma(\omega^i, v) = \omega^{-i}(\omega^i, v) \neq (\omega^i, v)$, so $(\omega^i, v) \in E$, but $(\omega^i, v) \notin k$. Suppose $(\omega^i, v) = 0$ for

$i = 1, \ldots, p - 1$. Then

$$\left(\omega^0, v\right) = \left(\omega^0, v\right) + \left(\omega^1, v\right) + \cdots + \left(\omega^{p-1}, v\right)$$
$$= pv + \left(1 + \omega + \cdots \omega^{p-1}\right)\sigma(v) + \left(1 + \omega^2 + \cdots + \omega^{2(p-1)}\right)\sigma^2(v)$$
$$\cdots + \left(1 + \omega^{p-1} + \cdots + \omega^{(p-1)(p-1)}\right)\sigma^{p-1}(v).$$

Because $x^p - 1 = (x-1)(1+x+x^2+\cdots+x^{p-1})$, each of the $\omega^i \neq 1$ is a root of $1+x+\cdots+x^{p-1}$, and so we deduce $(\omega^0, v) = pv$. Therefore $v = (\omega^0, v)/p$, an element of k. This contradicts our choice of v, and so $(\omega^i, v) \neq 0$ for some $1 \leq i \leq p - 1$. Taking $u = (\omega^i, v)$ proves the lemma. $\qquad \square$

Recall that a finite group G is *solvable* if G has a composition series

$$\{e\} \lhd G_1 \lhd G_2 \lhd \cdots \lhd G_{t-1} \lhd G_t = G$$

whose subquotients satisfy $G_i/G_{i-1} \cong \mathbb{Z}/p_i\mathbb{Z}$ with prime p_i for $i = 1, \ldots, t$. If a finite group is abelian, it is solvable, so we could weaken the solvability condition on G to having a normal series with abelian subquotients.

The adjunction of roots of unity leads to abelian Galois groups (Theorem 6.28). The next lemma shows how such adjunctions interact with solvability of a Galois group.

Lemma 6.36 *Suppose E is a finite Galois extension of a field k. If the characteristic of k does not divide n and ω is a primitive nth root of unity, then $\mathcal{G}al(E/k)$ is a solvable group if and only if $\mathcal{G}al(E(\omega)/k(\omega))$ is solvable.*

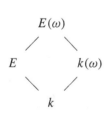

Proof Because E is a finite Galois extension of k, E is the splitting field of a polynomial $f(x) \in k[x]$. It follows that $E(\omega)$ is the splitting field of $f(x)(x^n - 1)$ and $E(\omega)$ is a Galois extension of k and of E. Similarly, $E(\omega)$ is a Galois extension of $k(\omega)$. By Proposition 6.20 and Corollary 6.21 we have $\mathcal{G}al(E(\omega)/E) \lhd \mathcal{G}al(E(\omega)/k)$ and $\mathcal{G}al(E(\omega)/k(\omega)) \lhd \mathcal{G}al(E(\omega)/k)$, along with two short exact sequences:

$$1 \to \mathcal{G}al\left(E(\omega)/E\right) \to \mathcal{G}al\left(E(\omega)/k\right) \to \mathcal{G}al(E/k) \to 1,$$
$$1 \to \mathcal{G}al\left(E(\omega)/k(\omega)\right) \to \mathcal{G}al\left(E(\omega)/k\right) \to \mathcal{G}al\left(k(\omega)/k\right) \to 1.$$

Both $\mathcal{G}al(E(\omega)/E)$ and $\mathcal{G}al(k(\omega)/k)$ are solvable abelian groups. If $\mathcal{G}al(E/k)$ is a solvable group, then, by Proposition 1.12, $\mathcal{G}al(E(\omega)/k)$ is a solvable group being the extension in the first short exact sequence. Also, the subgroup $\mathcal{G}al(E(\omega)/k(\omega))$ is solvable.

On the other hand, if we assume that $\mathcal{G}al(E(\omega)/k(\omega))$ is a solvable group, then the second short exact sequence implies $\mathcal{G}al(E(\omega)/k)$ is solvable. From the first short exact sequence, $Gal(E/k)$ is a quotient of a solvable group by a solvable group, and so it is also solvable.   □

Galois's Theorem *A polynomial equation $f(x) = 0$, where $f(x) \in k[x]$ is a separable polynomial, is solvable by radicals if and only if the Galois group $Gal(E_f/k)$ is a solvable group for the splitting field E_f of $f(x)$ over k.*

Proof Let us first assume that $G = Gal(E_f/k)$ is a solvable group. We have a composition series

$$\{e\} = G_0 \lhd G_1 \lhd G_2 \lhd \cdots \lhd G_{t-1} \lhd G_t = G$$

for which $G_i/G_{i-1} \cong \mathbb{Z}/p_i\mathbb{Z}$ for $i = 1, \ldots, t$. Thus $\#G$ equals the product of powers of the p_i, and by Lemma 6.36 we can assume that k contains the p_ith roots of unity for all i. We apply the fixed subfield mapping Φ of the Fundamental Theorem of Galois Theory to the composition series to obtain a tower of intermediate fields

$$E_f \supset L_1 \supset L_2 \supset \cdots \supset L_{t-1} \supset L_t = k,$$

where each $L_i = E_f^{G_i}$. Let us proceed by induction on n in $t - n$ starting at $t = 0$ and examine $E_f \supset L_{t-1} \supset L_t = k$. Because E_f is a Galois extension of k, and $G_{t-1} = Gal(E_f/L_{t-1})$ is a normal subgroup of G, we have that L_{t-1} is a Galois extension of k and $Gal(L_{t-1}/L_t) \cong G_t/G_{t-1} \cong \mathbb{Z}/p_{t-1}\mathbb{Z}$. Lemma 6.35 implies that $L_{t-1} = L_t(\gamma_t)$ with $\gamma_t^{p_{t-1}} \in L_t = k$, so $L_t \subset L_{t-1}$ is a simple radical extension.

By induction let us assume that $L_{t-n+1} \supset L_{t-n+2} \supset \cdots \supset L_{t-1} \supset L_t = k$ is a radical extension and E_f is Galois over L_{t-n+1}. Then the same argument applies to $E_f \supset L_{t-n} \supset L_{t-n+1}$, and so, continuing up the tower, we have proved that E_f is a radical extension of k and $f(x)$ is solvable by radicals.

In the other direction, suppose E_f is the splitting field over k of $f(x)$, a separable polynomial in $k[x]$, and $k \subset K$ is a radical extension with $E_f \subset K$:

$$k = K_0 \subset K_1 \subset \cdots \subset K_{t-1} \subset K_t = K.$$

Here each $K_i = K_{i-1}(u_i)$ and $u_i^{e_i} \in K_{i-1}$ for $i = 1, \ldots, t$. We can summarize these data by writing $K = k(u_1, \ldots, u_t)$ with $u_i^{e_i} \in k(u_1, \ldots, u_{i-1})$. By Lemma 6.36 we can assume that k contains the nth roots of unity for $n = \operatorname{lcm}(e_1, \ldots, e_t)$. Then each K_i is the splitting field of $x^{e_i} - u_i^{e_i}$ in $K_{i-1}[x]$, and $Gal(K_i/K_{i-1})$ is a cyclic group.

A possible drawback of the radical extension K of k is that it may not be a Galois extension. If it were, then the Fundamental Theorem of Galois Theory would obtain from the radical extension a normal series of subgroups of $Gal(K/k)$, showing $Gal(K/k)$ to be solvable (in this case, all subquotients are cyclic). We can overcome this problem by extending the radical extension by the normal closure of K over k. We form the splitting field M of the product of the distinct minimal polynomials $m_{u_i}(x)$ for $i = 1, \ldots, t$. Because u_i is a root of $x^{e_i} - u_i^{e_i}$, the minimal polynomial $m_{u_i}(x)$ divides this polynomial in $K_{i-1}[x]$, which contains $k[x]$. Since the roots of unity needed to completely split $x^{e_i} - u_i^{e_i}$ are available already in k, $m_{u_i}(x)$ is separable. Thus M is a Galois extension of k containing K. We now take the normal closure of K in M, which is the compositum of the conjugates of K by the Galois group $\mathcal{G}al(M/k)$ (Proposition 6.25). To present this field more concretely, recall that $K = k(u_1, \ldots, u_t)$. Then for $\sigma \in \mathcal{G}al(M/k)$, $\sigma K = k(\sigma(u_1), \ldots, \sigma(u_t))$. Notice that $(\sigma(u_1))^{e_1} = \sigma(u_1^{e_1})$ and since $u_1^{e_1} \in k$, $\sigma(u_1^{e_1}) = u_1^{e_1} \in k$, and we have the first step of a radical extension. For $\sigma(u_i^{e_i}) \in \sigma k(u_1, \ldots, u_{i-1})$, we have $\sigma k(u_1, \ldots, u_{i-1}) = k(\sigma(u_1), \ldots, \sigma(u_{i-1}))$, and so σK is a radical extension of k. The compositum of all σK takes the form

$$L = k\big(u_1, \ldots, u_t, \sigma_2(u_1), \ldots, \sigma_2(u_t), \sigma_3(u_1), \ldots\big),$$

which also gives a radical extension. So $k \subset K \subset L \subset M$ extends the radical extension K of k to a normal and separable extension L of k that is also radical. Since L is Galois over k, $L = M$. Thus we can substitute L for K as the radical extension of k containing E_f that is also a Galois extension.

Suppose the radical extension K over k takes the form

$$k = K_0 \subset K_1 \subset K_2 \subset \cdots \subset K_{t-1} \subset K_t = K.$$

By Lemma 6.34 we can adjoin to k and every field in the tower a primitive nth root of unity where $n = \text{lcm}(e_0, e_1, \ldots, e_{t-1})$. The new intermediate fields K_i' still satisfy $K_i' = K_{i-1}'(\gamma_i)$ and $\gamma_i^{e_i} \in K_{i-1}'$ for $i = 0, \ldots, t-1$. With the e_ith roots of unity in K_{i-1}', K_i' is the splitting field of $x^{e_i} - (\gamma_i)^{e_i}$ over K_{i-1}'. Also, if $k \subset k' = k(\omega)$, then $\mathcal{G}al(k'/k)$ is abelian.

By the Fundamental Theorem of Galois Theory, K' being Galois over k' implies K is Galois over K_i for all the intermediate fields K_i. According to Lemma 6.35, $\mathcal{G}al(K_{i-1}/K_i)$ is a cyclic group. Apply the Galois group mapping Γ of the Fundamental Theorem of Galois Theory to get a sequence of groups

$$\mathcal{G}al(K'/k') = G \supset G_1 \supset \cdots \supset G_{t-2} \supset G_{t-1} \supset G_t = \{\text{Id}\}$$

with $G_i = \mathcal{G}al(K'/K_i')$. Consider the portion of the tower given by $K_i' \subset K_{i+1}' \subset K'$. Since K' is Galois over k', it is Galois over K_i'. Since K_{i+1}' is a splitting field over K_i', K_{i+1}' is a normal extension of K_i', and we have

$$\mathcal{G}al\big(K_{i+1}'/K_i'\big) \cong \mathcal{G}al\big(K'/K_i'\big)/\mathcal{G}al\big(K'/K_{i+1}'\big) = G_i/G_{i+1}.$$

The series of subgroup inclusions becomes a normal series with abelian subquotients. By refining the series we can obtain simple abelian subquotients, and so $\mathcal{G}al(K'/k')$ is solvable. By Lemma 6.36, $\mathcal{G}al(K/k)$ is solvable. Finally, $\mathcal{G}al(E_f/k)$ is solvable because $\mathcal{G}al(E_f(\omega)/k')$ is solvable, where $E_f(\omega)$ is the splitting field of $f(x)(x^n - 1)$ over k'. $\square$

Get to know pth roots in fields of characteristic p

1. Suppose p is a prime integer and $\mathbb{F}$ is a field of characteristic p. Suppose further that E is a Galois extension of $\mathbb{F}$ of degree p. Then show that $\mathcal{G}al(E/\mathbb{F})$ is cyclic of order p. Show that there is some $v \in E$ such that $v + \sigma(v) + \sigma^2(v) + \cdots + \sigma^{p-1}(v) \neq 0$ where σ is a generator of $\mathcal{G}al(E/\mathbb{F})$.

2. Observe that the element $b = v + \sigma(v) + \cdots + \sigma^{p-1}(v)$ satisfies $\sigma(b) = b$ and deduce that $b \in \mathbb{F}$.

3. Define $\beta = \sigma(v) + 2\sigma^2(v) + \cdots + (p-1)\sigma^{p-1}(v)$. Show that $\sigma(\beta) = \beta - b$. Define $\alpha = -\beta/b$. Show that $\sigma(\alpha) = \alpha + 1$. Deduce that $E = \mathbb{F}(\alpha)$.

4. Define $a = \alpha^p - \alpha$. Show that $a \in \mathbb{F}$ and that α is a root of the polynomial $x^p - x - a \in \mathbb{F}[x]$. Show that $\alpha + 1$ is also a root of $x^p - x - a$. Conclude that E is the splitting field of $x^p - x - a \in \mathbb{F}[x]$ over $\mathbb{F}$.

5. Reverse the argument: Let $\mathbb{F}$ be a field of characteristic p, and let $x^p - x - a$ be an irreducible polynomial in $\mathbb{F}[x]$. Show further that $x^p - x - a$ is separable and that its splitting field E over $\mathbb{F}$ has degree p and $\mathcal{G}al(E/\mathbb{F})$ is cyclic of order p.

Galois' Theorem explains how there can be formulas for roots of quadratic, cubic, and quadric polynomials: Σ_n is solvable for $n = 2, 3$, and 4. However, we know that Σ_5 is not solvable: the composition series is $\{\text{Id}\} \lhd A_5 \lhd \Sigma_5$ because A_5 is a simple group. If a polynomial $f(x) \in \mathbb{Q}[x]$ has a Galois group $\mathcal{G}al(E_f/\mathbb{Q}) \cong \Sigma_5$ or A_5, then there cannot be a solution by radicals for the roots of $f(x)$. To find such a polynomial, we first prove the following:

Lemma 6.37 *If p is a prime and H is a subgroup of Σ_p that contains a transposition $\tau = (a, b)$ and an p-cycle $(a_1, \ldots, a_p)$, then $H = \Sigma_p$.*

Proof By relabeling we can take $a = 1$ and the p-cycle to begin at 1, $\alpha = (1, a_2', \ldots, a_p')$. Because α is a p-cycle, every power of α is a p-cycle. There is some power of α that begins $(1, b, \ldots)$, and so another relabeling allows us to assume that $\alpha = (1, 2, \ldots, p)$ and $\tau = (1, 2)$.

Conjugation with α gives $\alpha(1, 2)\alpha^{-1} = (2, 3)$. Conjugating repeatedly gives $(3, 4), \ldots, (p - 1, p), (p, 1)$. Next, we conjugate τ with $(2, 3)$ to give $(2, 3)(1, 2)(2, 3) = (1, 3)$, and then conjugating $(1, 3)$ with α gives $(2, 4)$. In this way, we get everything of the form $(a, a + 2)$. Continuing in this manner, we get $(a, a + k)$ for every a and k, but these transpositions generate all of Σ_p. $\square$

When $p = 5$, we can apply this lemma to Galois groups. Suppose $P(x) = x^5 + bx^4 + cx^3 + dx^2 + ex + f$ is an integral polynomial that is irreducible and has exactly two complex roots and three real roots. Let $\rho \colon \mathcal{G}al(E_P/\mathbb{Q}) \to \Sigma(\{\alpha_1, \ldots, \alpha_5\})$ denote the action of the Galois group on the roots of $P(x)$. Let $\hat{\rho} \colon \mathcal{G}al(E_P/\mathbb{Q}) \to \Sigma_5$ denote the action on $[5] = \{1, \ldots, 5\}$ induced by the isomorphism $\Sigma(\{\alpha_1, \ldots, \alpha_5\}) \to \Sigma_5$. By Proposition 6.9 $\mathcal{G}al(E_P/\mathbb{Q})$ is a transitive group acting on the five roots of $P(x)$.

Transitivity implies there is an orbit of the action ρ of cardinality five. By the Orbit-Stabilizer Theorem, five divides the order of $\mathcal{G}al(E_P/\mathbb{Q})$. Then Cauchy's Theorem implies there is an element of order five in $\hat{\rho}(\mathcal{G}al(E_P/\mathbb{Q}))$, which in turn implies that $\hat{\rho}(\mathcal{G}al(E_P/\mathbb{Q}))$ contains a 5-cycle. Since complex conjugation acts on the roots to exchange the complex roots while leaving the real roots fixed, there is a transposition in $\hat{\rho}(\mathcal{G}al(E_P/\mathbb{Q}))$. By Lemma 6.37, $\mathcal{G}al(E_P/\mathbb{Q}) \cong \Sigma_5$, which is not solvable. Hence $P(x)$ cannot be solved by radicals, that is, E_P is not a subfield of a radical extension of $\mathbb{Q}$.

For example, we can take $P(x) = x^5 - 6x + 3$. By Eisenstein's criterion $P(x)$ is irreducible. The derivative $P'(x) = 5x^4 - 6$ is zero at points $\pm\sqrt[4]{6/5}$, and we can easily see that $p(-\sqrt[4]{6/5}) > 0$ and $p(\sqrt[4]{6/5}) < 0$. It follows that $P(x)$ has exactly three real roots and two complex roots. Since $\mathcal{G}al(E_P/\mathbb{Q})$ is isomorphic to a transitive subgroup of Σ_5, it has to contain a 5-cycle. The complex roots are conjugates giving a transposition in $\mathcal{G}al(E_P/\mathbb{Q})$. The lemma implies $\mathcal{G}al(E_P/\mathbb{Q}) \cong \Sigma_5$. By a similar argument the reader can construct integral polynomials of degree a prime p greater than 5 with Galois group Σ_p.

6.5 Universal Examples

In the cases of cyclotomic polynomials and radical extensions, we knew enough about the roots of the polynomials involved to establish the properties of their Galois groups. The general case is more complicated. A useful approach is via a *universal example*: Suppose k is a field and $x_1, \ldots, x_n$ are variables. By Corollary 5.16 the ring $k[x_1, x_2, \ldots, x_n]$ of polynomials in the

x_i is a unique factorization domain, whose field of quotients is denoted by $k(x_1, \ldots, x_n)$.

The polynomial $F(x) = \prod_{i=1}^{n}(x - x_i)$ is in $k(x_1, \ldots, x_n)[x]$ and factors completely over $k(x_1, \ldots, x_n)$. The symmetric group Σ_n acts on the set $\{x_1, \ldots, x_n\}$ by $\rho(\sigma)(x_i) = x_{\sigma(i)}$, and this action extends to $k[x_1, \ldots, x_n]$ by $\rho(\sigma)(P(x_1, \ldots, x_n)) = P(x_{\sigma(1)}, \ldots, x_{\sigma(n)})$. When we view the indeterminates x_i as the roots of $F(x)$, Σ_n acts transitively on the set of roots, and Σ_n is the Galois group of $F(x)$ over the field $k(x_1, \ldots, x_n)^{\Sigma_n}$, which contains the coefficients of $F(x)$. Expanding $F(x)$ as a polynomial, we obtain

$$F(x) = x^n - E_1 x^{n-1} + \cdots + (-1)^k E_k x^{n-k} + \cdots + (-1)^n E_n,$$

where $E_i = E_i(x_1, \ldots, x_n)$ denote the *elementary symmetric polynomials*:

$$E_1 = x_1 + \cdots + x_n, \quad E_2 = \sum_{1 \le i < j \le n} x_i x_j, \ldots, \quad E_n = x_1 x_2 \cdots x_n.$$

Theorem 6.38 *The polynomials in $k[x_1, \ldots, x_n]$ that are fixed by the action of Σ_n, that is, the fixed set $k[x_1, \ldots, x_n]^{\Sigma_n}$, form the polynomial ring $k[E_1, E_2, \ldots, E_n]$. Furthermore, the rational functions fixed by the action of Σ_n are the rational functions in $E_1, \ldots, E_n$, that is, $k(x_1, \ldots, x_n)^{\Sigma_n} = k(E_1, \ldots, E_n)$.*

Proof A polynomial $P(x_1, \ldots, x_n)$ in $k[x_1, \ldots, x_n]$ is a sum of homogeneous polynomials, that is, polynomials $h(x_1, \ldots, x_n)$ that are sums of monomials $c x_1^{d_1} x_2^{d_2} \cdots x_n^{d_n}$ whose degree $d_1 + d_2 + \cdots + d_n$ is the same for all summands. The action of the symmetric group on monomials preserves their degree. We can focus on homogeneous polynomials in this proof without loss of generality.

The monomials in $k[x_1, \ldots, x_n]$ can be ordered according to *lexicographic order*: Fix $x_1 \succ x_2 \succ \cdots \succ x_n$, and we declare that $x_1^{e_1} \cdots x_n^{e_n} \succ x_1^{f_1} \cdots x_n^{f_n}$ if for some i, $e_1 = f_1, \ldots, e_{i-1} = f_{i-1}$ and $e_i > f_i$. Suppose $H(x_1, \ldots, x_n)$ is a homogeneous polynomial fixed by the action of Σ_n. Each monomial summand belongs to a Σ_n-orbit, $\{\rho(\sigma)(c x_1^{f_1} \cdots x_n^{f_n}) = c x_{\sigma(1)}^{f_1} \cdots x_{\sigma(n)}^{f_n} \mid \sigma \in \Sigma_n\}$, whose elements are summands of $H(x_1, \ldots, x_n)$. Lexicographically order each orbit and choose the *leading term* in the orbit, that is, the monomial appearing first in the ordering. If $c x_1^{f_1} \cdots x_n^{f_n}$ is a leading term, then it satisfies $f_1 \ge f_2 \ge \cdots \ge f_n$: if this ordering of the exponents were not the case, then a transposition would promote another monomial in the orbit to a higher lexicographic place. For example, consider the polynomial

$$H(x_1, x_2, x_3) = x_1 x_2^2 x_3^3 + x_2 x_1^2 x_3^3 + x_3 x_2^2 x_1^3 + x_1 x_3^2 x_2^3 + x_2 x_3^2 x_1^3 + x_3 x_1^2 x_2^3.$$

The ordering of the monomials is

$$x_1^3 x_2^2 x_3 \succ x_1^3 x_2 x_3^2 \succ x_1^2 x_2^3 x_3 \succ x_1^2 x_2 x_3^3 \succ x_1 x_2^3 x_3^2 \succ x_1 x_2^2 x_3^3.$$

For the leading term $c x_1^{f_1} \cdots x_n^{f_n}$, consider the monomial in the elementary symmetric polynomials given by

$$c E_1^{f_1 - f_2} E_2^{f_2 - f_3} \cdots E_{n-1}^{f_{n-1} - f_n} E_n^{f_n}.$$

Since $\deg E_i = i$, this monomial has the same degree as $H(x_1, \ldots, x_n)$. When we expand the monomial in E_i into its expression as a polynomial in $x_1, \ldots, x_n$, the leading term of the product is given by

$$c x_1^{f_1 - f_2} (x_1 x_2)^{f_2 - f_3} \cdots (x_1 x_2 \cdots x_{n-1})^{f_{n-1} - f_n} (x_1 x_2 \cdots x_n)^{f_n}$$

$$= c x_1^{f_1} x_2^{f_2} \cdots x_n^{f_n}.$$

Since $c E_1^{f_1 - f_2} E_2^{f_2 - f_3} \cdots E_{n-1}^{f_{n-1} - f_n} E_n^{f_n}$ is fixed by the Σ_n-action, its expansion as a polynomial in $x_1, \ldots, x_n$ is also fixed by Σ_n, and so its monomials lie in the orbit of the leading term. Since every $c x_{\sigma(1)}^{e_1} \cdots x_{\sigma(n)}^{e_n}$ occurs in the expansion, the monomials fill out the orbit of the leading term. In the example of $H(x_1, x_2, x_3)$, we begin with $E_1^{3-2} E_2^{2-1} E_3^1$.

Thus $H(x_1, \ldots, x_n) - c E_1^{f_1 - f_2} E_2^{f_2 - f_3} \cdots E_n^{f_n}$ is a symmetric polynomial without the leading term $c x_1^{f_1} x_2^{f_2} \cdots x_n^{f_n}$ and all its conjugates under Σ_n. Passing on to the next orbit, we can cancel it with a monomial in the E_i. After finitely many steps, we obtain

$$H(x_1, \ldots, x_n) - c E_1^{f_1 - f_2} E_2^{f_2 - f_3} \cdots E_n^{f_n}$$

$$- c' E_1^{g_1 - g_2} E_2^{g_2 - g_3} \cdots E_n^{g_n} - \cdots = 0,$$

and so $H(x_1, \ldots, x_n)$ can be written as a polynomial in the E_i. When carried out for $H(x_1, x_2, x_3)$, the first difference is $H(x_1, x_2, x_3) - E_1 E_2 E_3 = -3 E_3^2$. Since E_3 is already symmetric, we see that $H(x_1, x_2, x_3) = E_1 E_2 E_3 - 3 E_3^2$.

We already know that $k[E_1, \ldots, E_n] \subset k[x_1, \ldots, x_n]^{\Sigma_n}$. Therefore $k[x_1, \ldots, x_n]^{\Sigma_n} = k[E_1, \ldots, E_n]$.

For the field of rational functions, we know that $k(E_1, \ldots, E_n) \subset k(x_1, \ldots, x_n)^{\Sigma_n}$. Suppose $\dfrac{p(x_1, \ldots, x_n)}{q(x_1, \ldots, x_n)} = \dfrac{p}{q} \in k(x_1, \ldots, x_n)^{\Sigma_n}$. Denote $\rho(\sigma)(q)$ by q^σ and suppose $Q = \dfrac{1}{q} \prod_{\sigma \in \Sigma_n} q^\sigma$. Then $\dfrac{p}{q} = \dfrac{pQ}{qQ}$. By construction, $qQ \in k[x_1, \ldots, x_n]^{\Sigma_n} = k[E_1, \ldots, E_n]$. The polynomial $pQ = \dfrac{p}{q} qQ$ is also Σ_n-invariant and so in $k[E_1, \ldots, E_n]$. It follows that $\dfrac{p}{q} = \dfrac{pQ}{qQ} \in k(E_1, \ldots, E_n)$, and so $k(x_1, \ldots, x_n)^{\Sigma_n} = k(E_1, \ldots, E_n)$. $\qquad\square$

The representation of a symmetric polynomial as a polynomial in E_i can be shown to be unique: if $p(x_1, \ldots, x_n) \neq q(x_1, \ldots, x_n)$ but $p(E_1, \ldots, E_n) = q(E_1, \ldots, E_n)$, then the polynomial $f = p - q$ satisfies $f \neq 0$ and $f(E_1, \ldots, E_n) = 0$. We show by induction on n that this cannot happen. Suppose $n = 1$. Then $E_1 = x_1$ and $f(x_1) \neq 0$ implies $f(E_1) \neq 0$. So we suppose that no such relation among E_i holds for the elementary symmetric polynomials in variables $x_1, \ldots, x_{n-1}$. Suppose $f(x_1, \ldots, x_n) \neq 0$ and $f(E_1, \ldots, E_n) = 0$, where $E_i = E_i(x_1, \ldots, x_n)$. Because $k[x_1, \ldots, x_n] = k[x_1, \ldots, x_{n-1}][x_n]$, we can write

$$f(x_1, \ldots, x_n) = h_m(x_1, \ldots, x_{n-1})x_n^m + \cdots + h_1(x_1, \ldots, x_{n-1})x_n$$
$$+ h_0(x_1, \ldots, x_{n-1}).$$

Suppose that the degree m of f in x_n is the least degree for which $f(E_1, \ldots, E_n) = 0$. This implies that $h_0(x_1, \ldots, x_{n-1}) \neq 0$: If $h_0(x_1, \ldots, x_{n-1}) = 0$, then $f = x_n g(x_1, \ldots, x_n)$, and $g(E_1, \ldots, E_n) = 0$, a lower degree relation than f. Consider $f(E_1, \ldots, E_n)|_{x_n=0}$, that is, evaluate f on $(E_1, \ldots, E_n)$ and then let $x_n = 0$. Since $f(E_1, \ldots, E_n) = 0$, we get $h_0(E_1|_{x_n=0}, \ldots, E_{n-1}|_{x_n=0}) = 0$. However, this is a polynomial relation among E_i in the variables $x_1, \ldots, x_{n-1}$. By the induction hypothesis, $h_0 = 0$ is the only such polynomial, and we arrive at a contradiction.

This establishes the *Fundamental Theorem of Symmetric Polynomials*, which states that any polynomial $f(x_1, \ldots, x_n) \in k[x_1, \ldots, x_n]$ that is invariant under the action of the symmetric group on the indeterminates x_i is uniquely expressible as a polynomial in the elementary symmetric polynomials. Furthermore, the extension $k(E_1, \ldots, E_n) \subset k(x_1, \ldots, x_n)$ is a Galois extension being the splitting field of $F(x)$ with Galois group Σ_n.

The universal polynomial $F(x) = \prod_{i=1}^{n} x - x_i$ relates to an arbitrary monic and separable polynomial $f(x) \in k[x]$ when we factor $f(x)$ over its splitting field as $f(x) = (x - \alpha_1) \cdots (x - \alpha_n)$. If we evaluate $F(x) \in E_f[x_1, \ldots, x_n][x]$ by letting $x_i \mapsto \alpha_i$, then $\mathrm{ev}(F(x)) = f(x)$. This implies that $E_i(\alpha_1, \ldots, \alpha_n) = c_{n-i}$ where $f(x) = x^n + c_{n-1}x^{n-1} + \cdots + c_1 x + c_0$. Hence $E_i(\alpha_1, \ldots, \alpha_n) \in k$, and any polynomial in $E_f[x_1, \ldots, x_n][x]$ that is symmetric with respect to the Σ_n-action on $\{x_i\}$, when evaluated by $x_i \mapsto \alpha_i$, gives a polynomial in $k[x]$.

Get to know symmetric polynomials

1. Suppose $P(x_1, \ldots, x_n) \in k[x_1, \ldots, x_n]$ is a polynomial in $x_1, \ldots, x_n$ with coefficients in k. Suppose further that P is symmetric, that is,

$P(x_{\sigma(1)}, \ldots, x_{\sigma(n)}) = P(x_1, \ldots, x_n)$ for all $\sigma \in \Sigma_n$. If $f(x)$ is a separable polynomial with coefficients in k with roots $\alpha_1, \ldots, \alpha_n$ in E_f, the splitting field of $f(x)$ over k, then show that $P(\alpha_1, \ldots, \alpha_n)$ is in k.

2. Using the algorithm that leads to the proof of the Fundamental Theorem of Symmetric Polynomials, consider the polynomial in $\mathbb{Q}[x_1, x_2, x_3]$ given by

$$F(x_1, x_2, x_3) = x_1^3 x_2 + x_1^3 x_3 + x_2 x_3^3 + x_1 x_2^3 + x_2^3 x_3 + x_1 x_3^3 + 2x_1^2 x_2^2$$
$$+ 2x_2^2 x_3^2 + 2x_1^2 x_3^2 + 4x_1^2 x_2 x_3 + 4x_1 x_2^2 x_3 + 4x_1 x_2 x_3^2.$$

   Show that F is symmetric and express F as a polynomial in E_1, E_2, E_3.

3. ALBERT GIRARD (1595–1632) considered the polynomials $s_k(x_1, \ldots, x_n) = x_1^k + \cdots + x_n^k$ for x_i, the roots of a known polynomial. Since the coefficients of a polynomial are given in terms of elementary symmetric polynomials, Girard expressed the s_k, which are also clearly symmetric, in terms of E_i. ISAAC NEWTON (1643–1727) published a relation between the s_k and E_i in his 1707 *Arithmetica universalis* based on the work he did in 1665–1666. Newton's relations can be derived by letting $F(x) = (x - x_1) \cdots (x - x_n) = x^n - E_1 x^{n-1} + \cdots + (-1)^n E_n$ and equating two expressions for the derivative of $F(x)$:

$$f'(x) = nx^{n-1} - (n-1)E_1 x^{n-2} + \cdots + (-1)^{n-1} E_{n-1}$$
$$= \frac{F(x)}{x - x_1} + \cdots + \frac{F(x)}{x - x_n}.$$

   Derive this identity. Using long division of polynomials, compute $F(x)/(x - x_i)$ and sum to prove *Newton's identity*: $kE_k = s_1 E_{k-1} + s_2 E_{k-2} - \cdots + (-1)^{k-1} s_k$ for $1 \leq k < n$.

4. Newton's relations between the power sums s_k and the elementary symmetric functions can be used recursively to give formulas for the s_k in terms of E_i, and vice versa. Give the first few such explicit formulas. Deduce by induction that the subrings of $k[x_1, \ldots, x_n]$ given by $k[s_1, \ldots, s_n]$ and $k[E_1, \ldots, E_n]$ are the same.

5. The universal polynomial $F(x) = (x - x_1) \cdots (x - x_n)$ can be altered by changing the roots x_i to their squares x_i^2. Let $F_2(x) = (x - x_1^2) \cdots (x - x_n^2)$. Deduce that $F_2(x) \in k[E_1, \ldots, E_n]$. How are the coefficients of $F_2(x)$ related to the coefficients of $F(x)$? Suppose $g(x) = x^3 - 2x^2 + 8x - 6$? What is the polynomial whose roots are the squares of the roots of $g(x)$?

6.6 Discriminants and Resolvents

Suppose $f(x)$ is a polynomial with coefficients in k that is monic, separable, and irreducible. Then the Galois group $Gal(E_f/k)$ acts transitively on the set $\{\alpha_1, \dots, \alpha_n\}$ of roots of $f(x)$, which lie in the splitting field E_f of $f(x)$ over k. Indexing the roots allows us to identify $Gal(E_f/k)$ as a subgroup G_f of Σ_n via the faithful action $\rho: Gal(E_f/k) \to \Sigma(\{\alpha_1, \dots, \alpha_n\}) \cong \Sigma_n$, given by $\rho(\phi)(i) = j$ when $\phi(\alpha_i) = \alpha_j$. By Cauchy's Formula this action depends on the indexing up to conjugation. One way to get to know G_f is through its intersection with various subgroups of Σ_n.

Let Σ_n act on $\{x_1, \dots, x_n\}$ by $\rho(\sigma)(x_i) = x_{\sigma(i)}$ and extend this action to the polynomial ring $k[x_1, \dots, x_n]$. Associate to a polynomial $P(x_1, \dots, x_n)$ its stabilizer subgroup $St_P \subset \Sigma_n$ given by

$$St_P = \{\sigma \in \Sigma_n \mid \rho(\sigma)\big(P(x_1, \dots, x_n)\big) = P(x_{\sigma(1)}, \dots, x_{\sigma(n)})$$
$$= P(x_1, \dots, x_n)\}.$$

Certain specific and useful subgroups of Σ_n can be identified with stabilizers of carefully chosen P. The best-known example of such polynomials is given by

$$\Delta(x_1, \dots, x_n) = \prod_{1 \le i < j \le n} (x_i - x_j).$$

Every permutation $\sigma \in \Sigma_n$ is a product of transpositions (i, j). To see the effect of (i, j) on Δ, display Δ as the product of rows:

$$\Delta = (x_1 - x_2)(x_1 - x_3)(x_1 - x_4) \cdots (x_1 - x_n)$$
$$\times (x_2 - x_3)(x_2 - x_4) \cdots (x_2 - x_n)$$
$$\times (x_3 - x_4) \cdots (x_3 - x_n)$$
$$\ddots \quad \cdots \quad \vdots$$
$$\times (x_{n-1} - x_n).$$

The transposition (i, j) acting on Δ changes the sign of $(x_i - x_j)$, interchanges the terms $(x_i - x_k)$ with $(x_k - x_j)$ for $i < k < j$ – a partial row with part of a column – and finally interchanges the terms $(x_i - x_l)$ with $(x_j - x_l)$ for $j < l \le n$. The first interchange involves two sign changes, and the last interchange involves no sign change, so $\rho((i, j))(\Delta) = -\Delta$. It follows that even permutations preserve Δ and the stabilizer $St_\Delta = A_n$.

The square of Δ is called the *discriminant* and denoted by

$$D(x_1, \dots, x_n) = \big(\Delta(x_1, \dots, x_n)\big)^2.$$

By eliminating the sign, $D(x_1, \ldots, x_n)$ is fixed by all of Σ_n. The Fundamental Theorem of Symmetric Polynomials implies that $D(x_1, \ldots, x_n)$ can be expressed as a polynomial in the elementary symmetric polynomials. The simplest example in the case $n = 2$ is $D(x_1, x_2) = (x_1 - x_2)^2 = x_1^2 + x_2^2 - 2x_1 x_2 = (x_1 + x_2)^2 - 4x_1 x_2 = E_1^2 - 4E_2$. In the case of three variables, a straightforward computation gives

$$D(x_1, x_2, x_3) = E_1^2 E_2^2 + 18E_1 E_2 E_3 - 4E_2^3 - 4E_1^3 E_3 - 27E_3^2.$$

The computation is tedious if done directly, but there are simpler approaches outlined in the exercises.

To bring in the Galois group of a polynomial $f(x)$, consider the ring homomorphism

$$\mathrm{ev} \colon k[x_1, \ldots, x_n] \to E_f$$

given by $\mathrm{ev}(q(x_1, \ldots, x_n)) = q(\alpha_1, \ldots, \alpha_n)$.

Suppose that the characteristic of k is zero. Our monic, separable, and irreducible polynomial $f(x) \in k[x]$ has roots $\alpha_1, \alpha_2, \ldots, \alpha_n$ in E_f, the splitting field of $f(x)$ over k. Since

$$f(x) = x^n - c_{n-1}x^{n-1} + \cdots + (-1)^n c_0 = (x - \alpha_1) \cdots (x - \alpha_n)$$
$$= x^n - E_1(\alpha_1, \ldots, \alpha_n)x^{n-1} + \cdots + (-1)^n E_n(\alpha_1, \ldots, \alpha_n),$$

$c_{n-i} = E_i(\alpha_1, \ldots, \alpha_n)$. For example, when $f(x) = x^3 - bx^2 + cx - d$ and α_1, α_2, and α_3 are the roots of $f(x)$ in the splitting field E_f, we have

$$D(\alpha_1, \alpha_2, \alpha_3) = b^2 c^2 + 18bcd - 4c^3 - 4b^3 d - 27d^2 \in k.$$

In general, if $P(x_1, \ldots, x_n) \in k[x_1, \ldots, x_n]$, then the orbit of P under the Σ_n-action on $\{x_1, \ldots, x_n\}$ is determined by a transversal $\{\sigma_1 = \mathrm{Id}, \sigma_2, \ldots, \sigma_t\}$ for $\Sigma_n / St_P = \{St_P, \sigma_2 St_P, \ldots, \sigma_t St_P\}$. Define the *resolvent of P* with respect to $f(x)$ by

$$\mathrm{Res}(P, f)(x) = \prod_{i=1}^{t} \left(x - P(\alpha_{\sigma_i(1)}, \ldots, \alpha_{\sigma_i(n)}) \right)$$
$$= \prod_{i=1}^{t} \left(x - \mathrm{ev} \circ \rho(\sigma_i)\left(P(x_1, \ldots, x_n) \right) \right).$$

Because Σ_n acts on Σ_n / St_p by $\hat{\rho}(\tau)(\sigma St_P) = (\tau\sigma)St_P$, Σ_n acts on the roots of $\mathrm{Res}(P, f)(x)$ by $\rho(\tau)(P(\alpha_{\sigma_i(1)}, \ldots, \alpha_{\sigma_i(n)}) = P(\alpha_{\sigma_j(1)}, \ldots, \alpha_{\sigma_j(n)})$, where $(\tau\sigma_i)St_P = \sigma_j St_P$. It follows that $\mathrm{Res}(P, f)(x)$ is fixed by Σ_n and so is in $k[x]$.

Proposition 6.39 *If $f(x)$ is a monic separable polynomial over k with roots $\alpha_1, \ldots, \alpha_n$ in E_f, its splitting field, and $P(x_1, \ldots, x_n) \in k[x_1, \ldots, x_n]$, then*

$G_f = \hat{\rho}(\mathcal{Gal}(E_f/k)) \subset \Sigma_n$ *is conjugate to a subgroup of* St_P *whenever* Res$(P, f)(x)$ *has a root in* k. *Conversely, if* Res$(P, f)(x)$ *has a root of multiplicity one in* k, *then* G_f *is conjugate to a subgroup of* St_P.

Proof Suppose there is $\sigma \in \Sigma_n$ with $\sigma G_f \sigma^{-1} \subset St_P$. Then, for all $\tau \in G_f$, $\rho(\sigma\tau\sigma^{-1})(P) = P$. It follows that $\rho(\tau)(\rho(\sigma^{-1})(P)) = \rho(\sigma^{-1})(P)$, and so $P(x_{\sigma^{-1}(1)}, \ldots, x_{\sigma^{-1}(n)})$ is fixed by τ. Evaluation commutes with the action of the Galois group, and so

$$\text{ev}(P(x_{\sigma^{-1}(1)}, \ldots, x_{\sigma^{-1}(n)})) = P(\alpha_{\sigma^{-1}(1)}, \ldots, \alpha_{\sigma^{-1}(n)})$$

is fixed by $\mathcal{Gal}(E_f/k)$ and hence takes its value in k. This shows that $P(\alpha_{\sigma^{-1}(1)}, \ldots, \alpha_{\sigma^{-1}(n)}) \in k$ is a root of Res$(P, f)(x)$.

Conversely, suppose $\beta = P(\alpha_{\sigma_i(1)}, \ldots, \alpha_{\sigma_i(n)})$ is in k and β is a root of Res$(P, f)(x)$ of multiplicity one. By Lemma 2.10, $St_{\rho(\sigma_i)(P)} = \sigma_i St_P \sigma_i^{-1}$. For all $\tau \in \mathcal{Gal}(E_f/k)$, $\tau(\beta) = \beta$. This shows that the evaluation of $P(x_{\sigma_i(1)}, \ldots, x_{\sigma_i(n)})$ is fixed by τ. If τ does not fix $P(x_{\sigma_i(1)}, \ldots, x_{\sigma_i(n)})$, then we have $\rho(\tau)(P(x_{\sigma_i(1)}, \ldots, x_{\sigma_i(n)})) = P(x_{\sigma_j(1)}, \ldots, x_{\sigma_j(n)})$, and the evaluation map makes $P(\alpha_{\sigma_j(1)}, \ldots, \alpha_{\sigma_j(n)}) = \beta$ another root of Res$(P, f)(x)$. But then β is a multiple root of Res$(P, f)(x)$, which contradicts the assumption that β is a root of multiplicity one. Therefore $G_f \subset St_{\rho(\sigma_i)(P)}$. □

In the case of Δ, we have $St_\Delta = A_n$ and $(\text{ev}(\Delta))^2 = D(\alpha_1, \ldots, \alpha_n) \in k$. Then $\Sigma_n/A_n = \{A_n, (1, 2)A_n\}$ and

$$\text{Res}(\Delta, f)(x) = (x - \Delta)(x - \rho((1, 2))(\Delta)) = x^2 - D(\alpha_1, \ldots, \alpha_n) \in k[x].$$

If $D(\alpha_1, \ldots, \alpha_n)$ is a square in k, then $\Delta \in k$, and $x^2 - D = (x - \Delta)(x + \Delta)$. When $f(x)$ is separable, $\Delta \neq 0$, and so Δ is a root of multiplicity one of Res$(\Delta, f)(x)$. By Proposition 6.39, G_f is conjugate to a subgroup G'_f of A_n.

In general, by the Galois correspondence,

$$1 \text{ or } 2 = [G_f : G_f \cap A_n] = \left[E_f^{G_f \cap A_n} : E_f^{G_f}\right] = \left[E_f^{G_f \cap A_n} : k\right].$$

If $\Delta(\alpha_1, \ldots, \alpha_n) \in k$, then $[E_f^{G_f \cap A_n} : k] = 1$ and $G_f \cap A_n = G_f$, that is, $G_f \subset A_n$. If $[E_f^{G_f \cap A_n} : k] = 2$, then there is a field automorphism of E_f that sends ev(Δ) to $-\text{ev}(\Delta)$. In this case, $E_f^{G_f \cap A_n} = k(\sqrt{D})$ is the extension of k.

The Galois group for an irreducible separable cubic $f(x) = x^3 + bx^2 + cx + d$ is a transitive subgroup of Σ_3, that is, $G_f = \Sigma_3$ or $G_f = A_3 = C_3$. The discriminant in this case distinguishes between the cases: if $D(\alpha_1, \alpha_2, \alpha_3) = b^2c^2 + 18bcd - 4c^3 - 4b^3d - 27d^2$ is a square in k, then $G_f \cong C_3$; otherwise, $G_f \cong \Sigma_3$. For example, $f(x) = x^3 - 4x + 2$ has the discriminant $-4(-4)^3 - 27(-2)^2 = 148$, which is not a square in $\mathbb{Q}$, and $\mathcal{Gal}(E_f/\mathbb{Q}) \cong \Sigma_3$. On the other hand, $F(x) = x^3 + x^2 - 2x - 1$ has the discriminant 49, a square in $\mathbb{Q}$. It is easy to check that $F(x)$ is irreducible mod 3 and so is irreducible in $\mathbb{Q}[x]$. Hence $\mathcal{G}al(E_F/\mathbb{Q}) \cong C_3$.

Let $f(x) = x^4 + bx^3 + cx^2 + dx + e$ be a separable irreducible quadric over k. The discriminant of $f(x)$ takes the (rather daunting) form

$$D(f(x)) = -27e^2b^4 + 18eb^3cd - 4b^3d^3 - 4eb^2c^3 + b^2c^2d^2 + 144e^2b^2c$$
$$- 6eb^2d^2 - 80ebc^2d + 18bcd^3 - 192e^2bd + 16ec^4 - 4c^3d^2$$
$$- 128e^2c^2 + 144ecd^2 - 27d^4 + 256e^3.$$

The subgroups of Σ_4 that act transitively on $\{1, 2, 3, 4\}$ (Exercise 2.9) can be organized in the diagram

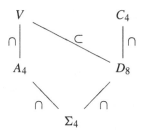

Consider the polynomial $P(x_1, x_2, x_3, x_4) = x_1x_3 + x_2x_4$ in $k[x_1, x_2, x_3, x_4]$. The subgroup St_P contains $(1, 3)$ and $(1, 2, 3, 4)$, and so it contains D_8 where we identify $f = (1, 3)$ and $r = (1, 2, 3, 4)$ acting on the vertices of a square. The set of cosets of D_8 in Σ_4 is $\Sigma_4/D_8 = \{D_8, (2, 3)D_8, (3, 4)D_8\}$, and $\rho((2, 3))(P) = x_1x_2 + x_3x_4$, $\rho((3, 4))(P) = x_1x_4 + x_2x_3$, which are distinct. Hence $St_P = D_8$.

Combining the conjugates of P into a polynomial, we get the symmetric polynomial

$$G(x) = \big(x - (x_1x_3 + x_2x_4)\big)\big(x - (x_1x_2 + x_3x_4)\big)\big(x - (x_1x_4 + x_2x_3)\big),$$

which is called the *cubic resolvent* or the *Ferrari resolvent* of a quadric. Since $G(x)$ is invariant under Σ_4, it is in $k[E_1, E_2, E_3, E_4]$. A calculation leads to the expression

$$G(x) = x^3 - E_2x^2 + (E_1E_3 - 4E_4)x + \big(4E_2E_4 - E_3^2 - E_1^2E_4\big).$$

When we evaluate $G(x)$ by assigning the roots α_i of $f(x)$ to the variables x_i, we get

$$\text{ev}\big(G(x)\big) = \text{Res}(P, f)(x)$$
$$= x^3 - cx^2 + (bd - 4e)x + \big(4ce - d^2 - b^2e\big) \in k[x].$$

The differences between the roots of $\text{Res}(P, f)(x)$ are generally, for $1 \leq i < j, k < l \leq 4$,

$$(\alpha_i\alpha_j + \alpha_k\alpha_l) - (\alpha_i\alpha_k + \alpha_j\alpha_l) = (\alpha_i - \alpha_l)(\alpha_j - \alpha_k),$$

from which it follows that $\Delta(\text{Res}(P, f)(x)) = \Delta(f)$. Therefore $\text{Res}(P, f)(x)$ is separable because $f(x)$ is. The behavior of the roots of $\text{Res}(P, f)(x)$ will determine $\mathcal{G}al(E_f/k)$.

Suppose $\text{Res}(P, f)(x)$ is irreducible. For all $\sigma \in \Sigma_n$, the value $\text{ev}(\rho(\sigma)(P))$ is in E_f, so E_f contains E_{Res}, the splitting field for $\text{Res}(P, f)(x)$. Since $\text{Res}(P, f)(x)$ is an irreducible cubic, $k[x]/(\text{Res}(P, f)(x))$ is isomorphic to a subfield of E_{Res}, and so three divides $[E_{\text{Res}} : k]$. Because $[E_f : k] = [E_f : E_{\text{Res}}][E_{\text{Res}} : k]$ and $[E_f : k] = \#\mathcal{G}al(E_f/k)$, we have that three divides $\#G_f$. The transitivity of the action of G_f means there is only one orbit of cardinality four. Hence four divides the order of $\mathcal{G}al(E_f/k)$, and 12 must divide $\#\mathcal{G}al(E_f/k)$. Therefore G_f must be isomorphic to one of A_4 or Σ_4. The discriminant of $f(x)$ distinguishes these cases.

Suppose $\text{Res}(P, f)(x)$ has a root in k. Being a cubic, $\text{Res}(P, f)(x)$ has either a single root in k or all three roots in k. If all three roots live in k, then $\mathcal{G}al(E_f/k)$ fixes $\text{ev}(P)$, $\text{ev}(\rho((2, 3))(P))$, and $\text{ev}(\rho((3, 4))(P))$. Proposition 6.39 applied to each root implies that G_f lies in every conjugate of D_8. But $V = D_8 \cap ((2, 3)D_8(2, 3)) \cap ((3, 4)D_8(3, 4))$. Thus $G_f \cong V$.

Suppose $\text{Res}(P, f)(x)$ has exactly one root in k. Since $\text{Res}(P, f)(x)$ is separable, this root has multiplicity one, so G_f is conjugate to a subgroup of D_8, that is, $G_f \cong D_8$ or $G_f \cong C_4$. We can renumber the roots so that the root in k is $r = \alpha_1\alpha_3 + \alpha_2\alpha_4$. This implies that $\text{Res}(P, f)(x)$ factors as

$$\text{Res}(P, f)(x) = (x - r)\big(x - (\alpha_1\alpha_2 + \alpha_3\alpha_4)\big)\big(x - (\alpha_1\alpha_4 + \alpha_2\alpha_3)\big)$$
$$= (x - r)\big(x^2 + sx + t\big),$$

where $x^2 + sx + t$ is irreducible in $k[x]$. The splitting field for $\text{Res}(P, f)(x)$ becomes $E_{\text{Res}} = k(\sqrt{s^2 - 4t})$, a quadratic extension of k. Following [50], we introduce the following polynomials over k:

$$q_1(x) = (x - \alpha_1\alpha_3)(x - \alpha_2\alpha_4) = x^2 - rx + e,$$
$$q_2(x) = \big(x - (\alpha_1 + \alpha_3)\big)\big(x - (\alpha_2 + \alpha_4)\big) = x^2 - bx + (c - r).$$

If $q_1(x)$ and $q_2(x)$ both factor over E_{Res}, then $\alpha_1 + \alpha_3$ and $\alpha_1\alpha_3$ are in E_{Res}. The quadratic polynomial $x^2 - (\alpha_1 + \alpha_3)x + \alpha_1\alpha_3 \in E_{\text{Res}}[x]$ has roots α_1 and α_3. Then $E_f = E_{\text{Res}}(\alpha_1, \alpha_2)$ is a quadratic extension, and $[E_f : E_{\text{Res}}] \leq 2$. However, $[E_{\text{Res}} : k] = 2$, so $[E_f : k] \leq 4$, which implies that $\mathcal{G}al(E_f/k) \cong$

C_4. If at least one of $q_1(x)$ or $q_2(x)$ does not factor over E_{Res}, then $D_8 \cong Gal(E_f/k)$. We put these ideas to use in the following exercises.

Get to know quartic polynomials

Fix an irreducible quartic polynomial $f(x) = x^4 + cx^2 + e \in k[x]$.

1. Compute the cubic resolvent for $f(x)$ and show that it factors as a linear function times a quadratic polynomial.
2. Show that if $e \neq 0$ is a square in k, then $Gal(E_f/k) \cong V$.
3. Suppose e is not a square in k, but $e(c^2 - 4e)$ is a square in k. Then $Gal(E_f/k) \cong C_4$.
4. Show that if neither e nor $e(c^2-4e)$ is a square in k, then $Gal(E_f/k) \cong D_8$.
5. For $k = \mathbb{Q}$ and $f(x) = x^4 + cx^2 + e \in \mathbb{Z}[x]$ as above, give values of c and e for each case discussed with irreducible $f(x)$.

6.7 Another Theorem of Dedekind

When a mathematical object has two presentations, it may be possible to play one off against the other to be able to get enough information about the object to know it explicitly. For the Galois group of a polynomial, such an alternate avatar is available through a resolvent introduced by Kronecker [55]. Let $f(x) \in k[x]$ by a monic, irreducible, and separable polynomial of degree n with splitting field E_f over k. Let $\alpha_1, \ldots, \alpha_n$ be a numbering of the roots of $f(x)$. Let $u_1, \ldots, u_n$ denote n indeterminates, and let $w = u_1\alpha_1 + \cdots u_n\alpha_n$, an element of $E_f[u_1, \ldots, u_n]$.

There are several actions associated with these sets:

$$\hat{\rho}: Gal(E_f/k) \to \Sigma(\{\alpha_1, \ldots, \alpha_n\}) \cong \Sigma_n,$$
$$\rho_\alpha: \Sigma_n \to \Sigma(\{\alpha_1, \ldots, \alpha_n\}),$$
$$\rho_u: \Sigma_n \to \Sigma(\{u_1, \ldots, u_n\}),$$

given by $\hat{\rho}(\phi)(i) = j$ if $\phi(\alpha_i) = \alpha_j$, $\rho_\alpha(\sigma)(\alpha_i) = \alpha_{\sigma(i)}$, and $\rho_u(\sigma)(u_j) = u_{\sigma(j)}$. The subgroup $\hat{\rho}(Gal(E_f/k))$ is denoted by $G_f \subset \Sigma_n$. Define the polynomial

$$F_u(x) = \prod_{\sigma \in \Sigma_n} \left(x - \rho_\alpha(\sigma)(w)\right) = \prod_{\sigma \in \Sigma_n} \left(x - (u_1\alpha_{\sigma(1)} + \cdots + u_n\alpha_{\sigma(n)})\right).$$

By construction $F_u(x)$ is an element of $E_f[u_1, \ldots, u_n][x]$, which is symmetric in the roots of $f(x)$. Therefore the coefficients of $F_u(x)$ are elementary symmetric polynomials in $\{\alpha_1, \ldots, \alpha_n\}$, which implies that $F_u(x) \in k[u_1, \ldots, u_n][x]$, a unique factorization domain.

Suppose $F_u(x) = F_1(x) \cdots F_s(x)$ with $F_i(x)$ irreducible in $k[u_1, \ldots, u_n][x]$. Number the factors in such a way that $x - w = x - (u_1\alpha_1 + \cdots + u_n\alpha_n)$ divides $F_1(x)$. Since $F_u(x)$ is invariant under the action ρ_α of Σ_n on $\{\alpha_1, \ldots, \alpha_n\}$, there is an action of Σ_n on the set of factors $\{F_1, \ldots, F_s\}$. Let us denote that action by $\rho' \colon \Sigma_n \to \Sigma(\{F_1, \ldots, F_s\})$, and denote the stabilizer of the action ρ' on $F_1(x)$ by

$$St_1 = \big\{\tau \in \Sigma_n \mid \rho'(\tau)\big(F_1(x)\big) = F_1(x)\big\}.$$

Theorem 6.40 *Let* $G_f = \hat{\rho}(\mathcal{G}al(E_f/k)) \subset \Sigma_n$. *Then* $G_f \cong St_1$.

Proof Let $G_u(x) = \prod_{\tau \in G_f}(x - (u_1\alpha_{\tau(1)} + \cdots + u_n\alpha_{\tau(n)}))$. This polynomial lies in $E_f[u_1, \ldots, u_n][x]$ and is invariant under the action of $\mathcal{G}al(E_f/k)$. It follows that $G_u(x) \in k[u_1, \ldots, u_n][x]$. Because $f(x)$ is separable, the α_i are distinct, and because the u_i are indeterminates, the roots of $G_u(x)$, $\{u_1\alpha_{\tau(1)} + \cdots + u_n\alpha_{\tau(n)} \mid \tau \in G_f\}$, are also distinct, and $G_u(x)$ is separable. Proposition 6.9 implies $G_u(x)$ is irreducible in $k[u_1, \ldots, u_n][x]$. However, $G_u(x)$ shares the root $w = u_1\alpha_1 + \cdots + u_n\alpha_n$ with $F_1(x)$, and so $G_u(x)$ and $F_1(x)$ share a factor in $E_f[u_1, \ldots, u_n][x]$. Since both are irreducible, $G_u(x) = F_1(x)$.

Suppose τ is in St_1; then the linear polynomial $\hat{\rho}(\tau)(x - (u_1\alpha_1 + \cdots + u_n\alpha_n)) = x - (u_1\alpha_{\tau(1)} + \cdots u_n\alpha_{\tau(n)})$ divides $F_1(x) = G_u(x)$. The factorization of $G_u(x)$ in $E_f[u_1, \ldots, u_n][x]$ implies that $\tau \in G_f$, and we have that $St_1 \subset G_f$.

If $\tau \notin St_1$, then $x - (u_1\alpha_{\tau(1)} + \cdots + u_n\alpha_{\tau(n)})$ does not divide $F_1(x) = G_u(x)$, and so $\tau \notin G_f$, and $G_f \subset St_1$. Hence $G_f = St_1$. $\qquad\square$

When $k = \mathbb{Q}$, the theorem gives us a toehold to reveal elements of the Galois group of the splitting field E_f over $\mathbb{Q}$ for a polynomial $f(x) \in \mathbb{Z}[x]$. We will assume that $f(x)$ is irreducible and monic. It follows that $f(x)$ is separable. The mapping $\mathbb{Z} \to \mathbb{F}_p$ given by reduction mod p is a ring homomorphism that induces a ring homomorphism $r_p \colon \mathbb{Z}[x] \to \mathbb{F}_p[x]$ that reduces each coefficient of an integral polynomial to its mod p residue. Let $r_p(f(x)) = \bar{f}(x) \in \mathbb{F}_p[x]$. Choose p so that it does not divide the discriminant $D(f)$. Then $\bar{f}(x)$ is also separable. However, it is possible that $\bar{f}(x)$ is no longer irreducible. Say $\bar{f}(x) = \bar{f}_1(x) \cdots \bar{f}_t(x)$ in $\mathbb{F}_p[x]$, and $d_i = \deg \bar{f}_i(x)$. It follows from Proposition 6.9 that $\mathcal{G}al(E_{\bar{f}}/\mathbb{F}_p)$ permutes the roots of $\bar{f}(x)$ in t blocks, each one of size d_i and consisting of the roots of $\bar{f}_i(x)$. This corresponds to the orbit structure of the set of roots of $\bar{f}(x)$.

The splitting field $E_{\bar{f}}$ is isomorphic to $\mathbb{F}_q$ for some $q = p^l$, and so the Galois group $\mathcal{G}al(E_{\bar{f}}/\mathbb{F}_p)$ is a cyclic group generated by the Frobenius automorphism. If β_i is a root of $\bar{f}_i(x)$, then the orbit of β_i, $\{\beta_i, \beta_i^p, \ldots, \beta_i^{p^{d_i-1}}\}$, is the set of roots of $\bar{f}_i(x)$. As a permutation of the roots, the Frobenius automorphism determines a d_i-cycle. Hence there is an element of $\hat{\rho}(\mathcal{G}al(E_{\bar{f}}/\mathbb{F}_p))$ representable as a product of disjoint cycles

$$(\cdots d_1 \cdots)(\cdots d_2 \cdots)\cdots(\cdots d_t \cdots).$$

An example of this decomposition is suggested by cyclotomic extensions. For example, consider $\Phi_8(x) = x^4 + 1$. This is an irreducible separable polynomial, which means that $\mathcal{G}al(E_{\Phi_8}/\mathbb{Q})$ is a transitive group on the four roots. We know the roots, and we have identified the Galois group as isomorphic to $(\mathbb{Z}/8\mathbb{Z})^\times$. Taking a page out of Galois (p. 114 of [65]), we can represent the elements of the Galois group as permutations of the roots in a table: Number the roots as one through four, and an element of $\mathcal{G}al(E_{\Phi_8}/\mathbb{Q})$ takes the first row of roots to their images on a subsequent row. The associated permutation is shown on the right.

1	2	3	4	
ω	ω^3	ω^5	ω^7	$(\,)$
ω^3	ω	ω^7	ω^5	$(1,2)\,(3,4)$
ω^5	ω^7	ω	ω^3	$(1,3)\,(2,4)$
ω^7	ω^5	ω^3	ω	$(1,4)\,(2,3)$

The group is isomorphic to the Klein *Viergruppe* as a subgroup of Σ_4.

Consider the integral polynomial $\Phi_8(x)$ over $\mathbb{F}_p$. The discriminant of $\Phi_8(x)$ is $256 = 2^8$, and so the mod p reduction of $\Phi_8(x)$ is separable for all odd primes. Since $x^4 + 1 = x^4 - (-1)$, when -1 is a quadratic residue mod p, we can factor $x^4 + 1$ as $(x^2 - a)(x^2 + a)$ with $a^2 \equiv -1 \pmod{p}$. If a is also a quadratic nonresidue, then $x^2 - a$ and $x^2 + a$ are irreducible over $\mathbb{F}_p$. The irreducible factors of a polynomial correspond to orbits of the permutation representation on the roots, and so these factorizations of $x^4 + 1$ correspond to permutations like the ones we find in the table. For example,

$$
\begin{aligned}
x^4 + 1 &\equiv (x^2 - 2)(x^2 + 2) \pmod 5 \\
&\equiv (x^2 - 5)(x^2 + 5) \pmod{13} \\
&\equiv (x^2 - 12)(x^2 + 12) \pmod{29}.
\end{aligned}
$$

It is a number-theoretic fact [38] that -1 is a quadratic residue when the prime p is congruent to one mod 4, and the square root of -1 is not a square in the cases shown, and many more. When p is congruent to three mod 4, $x^4 + 1$ is irreducible, since a root in $\mathbb{F}_p$ determines a square congruent to -1. The only other factorization possible is as a product of quadratic polynomials.

In an 1882 letter to Frobenius [29], Dedekind proved the following remarkable theorem.

Another Theorem of Dedekind *Suppose $f(x)$ is an integral polynomial that is monic and irreducible. Suppose p is a prime that does not divide the discriminant of $f(x)$. Let $\bar{f}(x)$ be the reduction mod p of $f(x)$ in $\mathbb{F}_p[x]$. Suppose $\bar{f}(x) \equiv \bar{f}_1(x) \cdots \bar{f}_t(x) \pmod{p}$ is the factorization of $\bar{f}(x)$ in $\mathbb{F}_p[x]$ into a product of irreducible polynomials. Let $d_i = \deg \bar{f}_i(x)$. Then there is at least one permutation in $\mathcal{G}al(E_f/\mathbb{Q})$ that is a product of disjoint cycles of lengths $d_1, d_2, \ldots, d_t$.*

Proof (Following [94] and [16].) Reduction mod p, $r_p \colon \mathbb{Z} \to \mathbb{F}_p$ extends to a ring homomorphism

$$r_p \colon \mathbb{Z}[u_1, \ldots, u_n][x] \to \mathbb{F}_p[u_1, \ldots, u_n][x],$$

where u_i are indeterminates.

Let G_f denote the image of the Galois group $\hat{\rho}(\mathcal{G}al(E_f/\mathbb{Q})) \subset \Sigma_n$, and let $\overline{G}$ denote the analogous group $\hat{\rho}(\mathcal{G}al(E_{\bar{f}}/\mathbb{F}_p)) \subset \Sigma_n$. By Proposition 6.9 we know that there is a permutation that is a product of disjoint cycles of lengths $d_1, \ldots, d_t$ in $\overline{G}$ by the factorization of $\bar{f}(x)$. The theorem will follow if we show that $\overline{G} \subset G_f$.

By Theorem 6.40 the Galois group $\mathcal{G}al(E_f/\mathbb{Q})$ is isomorphic to the stabilizer of a factor of

$$F_u(x) = \prod_{\sigma \in \Sigma_n} x - \rho_\alpha(\sigma)(w),$$

where $w = u_1\alpha_1 + \cdots + u_n\alpha_n$ with $\rho_\alpha(\sigma)(w) = u_1\alpha_{\sigma(1)} + \cdots + u_n\alpha_{\sigma(n)}$ for any $\sigma \in \Sigma_n$. The polynomial $F_u(x)$ is fixed by all $\sigma \in \hat{\rho}(\mathcal{G}al(E_f/\mathbb{Q}))$, and so $F_u(x)$ is in $\mathbb{Q}[u_1, \ldots, u_n][x]$. Since every coefficient is symmetric in α_i, $F_u(x) \in \mathbb{Z}[u_1, \ldots, u_n][x]$. Let $F_u(x) = F_1(x) \cdots F_s(x) \in \mathbb{Z}[u_1, \ldots, u_n][x]$ with $x - w$ a factor of $F_1(x)$ and all $F_i(x)$ irreducible. Then $\mathcal{G}al(E_f/\mathbb{Q}) \cong St_1 = \{\sigma \in \Sigma_n \mid \rho(\sigma)(F_1(x)) = F_1(x)\}$.

Since $f(x) \in \mathbb{Z}[x]$, its discriminant is an integer. Because the prime p does not divide the discriminant $D(f)$, we know, after reduction mod p, that $\bar{f}(x)$ is a separable polynomial in $\mathbb{F}_p[x]$. Let $\{\beta_1, \ldots, \beta_n\}$ denote the set of roots of $\bar{f}(x)$ and form the analogue of $F_u(x)$ for $\bar{f}(x)$:

$$\bar{F}_u(x) = \prod_{\sigma \in \Sigma_n} x - \rho_\beta(\sigma)(\bar{w}),$$

where $\bar{w} = u_1\beta_1 + \cdots + u_n\beta_n$ and $\rho_\beta(\sigma)(\bar{w}) = u_1\beta_{\sigma(1)} + \cdots + u_n\beta_{\sigma(n)}$. The argument showing that $F_u(x) \in \mathbb{Z}[u_1, \ldots, u_n][x]$ applies to $\bar{F}_u(x)$, and $\bar{F}_u(x) \in \mathbb{F}_p[u_1, \ldots, u_n][x]$.

When we write $F_u(x) = x^{n!} + c_1 x^{n!-1} + \cdots + c_0$, the coefficients are in $\mathbb{Z}[u_1, \ldots, u_n]$, and each c_i is a polynomial in $u_1, \ldots, u_n$ and the coefficients of $f(x)$. The same is true for $\bar{F}_u(x)$, and each coefficient will be represented by the same polynomial in $u_1, \ldots, u_n$ and the coefficients of $\bar{f}(x)$, which are the mod p reductions of the coefficients of $f(x)$. Therefore $r_p(F_u(x)) = \bar{F}_u(x)$.

Factoring $F_u(x)$ in $\mathbb{Z}[u_1, \ldots, u_n][x]$ into its irreducible factors $F_u(x) = F_1(x) \cdots F_s(x)$ gives a factorization in $\mathbb{F}_p[u_1, \ldots, u_n][x]$, $\bar{F}_u(x) = \bar{F}_1(x) \cdots \bar{F}_s(x)$. However, these $\bar{F}_i(x)$ need not be irreducible. Suppose $\bar{F}_u(x) = g_1(x) \cdots g_l(x)$ in $\mathbb{F}_p[u_1, \ldots, u_n][x]$ with all $g_j(x)$ irreducible. Suppose that $x - \bar{w}$ divides $g_1(x)$ and so $g_1(x)$ divides $\bar{F}_1(x)$. Then the stabilizer of $g_1(x)$, $st_1 = \{\tau \in \Sigma_n \mid \rho(\tau)(g_1(x)) = g_1(x)\}$, is isomorphic to $\mathcal{G}al(E_{\bar{f}}/\mathbb{F}_p)$. To prove the theorem, we need to show that st_1 is a subgroup of St_1.

Suppose $\tau \in st_1 \subset \Sigma_n$ but $\tau \notin St_1$. Then $\rho(\tau)(F_1(x)) \neq F_1(x)$; let's say $\rho(\tau)(F_1(x)) = F_i(x)$. We know that $g_1(x)$ divides $\bar{F}_1(x)$; then $\rho(\tau)(g_1(x))$ divides $\rho(\tau)(\bar{F}_1(x))$. Since reduction mod p commutes with the action of Σ_n, $\rho(\tau)(g_1(x)) = g_1(x)$ and $\rho(\tau)(\bar{F}_1(x)) = \bar{F}_i(x)$ for some $i \neq 1$. Thus $g_1(x)$ divides both $\bar{F}_1(x)$ and $\bar{F}_i(x)$. It follows that $(g_1(x))^2$ divides $\bar{F}_u(x)$, and so $\bar{F}_u(x)$ has repeated roots when considered in $E_{\bar{f}}[u_1, \ldots, u_n]$, a contradiction to the separability of $\bar{F}_u(x)$. Thus $st_1 \subset St_1$, and the theorem is proved. $\square$

There are two naive questions that we can ask at this stage:

(1) Given an integral polynomial $f(x)$, what is the Galois group $\mathcal{G}al(E_f/\mathbb{Q})$? Is $\mathcal{G}al(E_f/\mathbb{Q})$ solvable?

(2) Given a finite group G, is it the Galois group of a Galois extension of $\mathbb{Q}$? Or more generally, of a Galois extension of a given field k?

The second question is known as the *Inverse Galois Problem*, posed in this generality by EMMY NOETHER (1882–1935) [66]. In the case of k, an algebraically closed field, the answer is known. In the case that the group is abelian, problem (2) is answered by the *Kronecker–Weber Theorem*, which states that *every finite Galois extension of $\mathbb{Q}$ with an abelian Galois group is a subfield of a cyclotomic extension*. It follows that algebraic integers with abelian Galois groups are rational linear combinations of roots of unity. A proof of this result is beyond the scope of this book (see [91]).

Let us consider the naive questions for the case of quintic polynomials. There are some computational challenges. For example, the discriminant of a quintic polynomial has 59 terms. This fact alone should convince the reader that using computers for mathematical discoveries is a good thing!

The transitive subgroups of Σ_5 are given by C_5, D_{10}, F_{20}, A_5, and Σ_5. Recall that D_{10} is the dihedral group of 10 elements (the symmetries of a regular pentagon in the plane) and F_{20} is the Frobenius group of order $4 \cdot 5$ consisting of the affine transformations, $x \mapsto ax + b$ with $a \neq 0$ of $\mathbb{F}_5$ to itself.

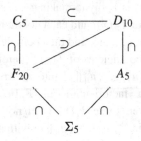

The subgroups C_5, D_{10}, and F_{20} are solvable; A_5 is simple, and Σ_5 is a nontrivial extension by $\mathbb{Z}/2\mathbb{Z}$ of A_5.

For a given quintic polynomial $f(x) \in \mathbb{Z}[x]$, a first pass toward determining the Galois group is to compute the discriminant (using *Mathematica* or *Sage*), from which we can determine if the Galois group is isomorphic to a subgroup of A_5, that is, one of C_5, D_{10}, or A_5, or if it is one of F_{20} or Σ_5. Finding a polynomial that realizes each group is possible from what we have established so far.

The cyclic group C_5 is abelian. The Kronecker–Weber Theorem tells us that a Galois extension of $\mathbb{Q}$, with C_5 as a Galois group, will be a subfield of a cyclotomic extension. The first case where these conditions hold is the cyclotomic extension for $\Phi_{11}(x) = x^{10} + \cdots + x + 1$. The splitting field $E_{\Phi_{11}}$ is $\mathbb{Q}(\omega)$ where ω is a primitive 11th root of unity. We know the Galois group $\mathcal{G}al(E_{\Phi_{11}}/\mathbb{Q})$ to be $(\mathbb{Z}/11\mathbb{Z})^{\times}$, the units in $\mathbb{Z}/11\mathbb{Z}$. This group is cyclic, generated by a primitive root mod 11. Conveniently, 2 is a primitive root, so $\mathcal{G}al(E_{\Phi_{11}}/\mathbb{Q}) \cong \langle 2 \rangle$, that is, the squaring map on the powers of ω generates the Galois group. Following the ideas of Gauss in *Disquisitones Arithmetica*, let $\alpha_1 = \omega + \omega^{-1} = \omega + \omega^{10} \in E_{\Phi_{11}}$. This element is fixed by $\sigma^5 \in \mathcal{G}al(E_{\Phi_{11}}/\mathbb{Q})$, where $\sigma(\omega) = \omega^2$, because $32 \equiv -1 \pmod{11}$.

The conjugates of α_1 are

$$\sigma(\alpha_1) = \alpha_2 = \omega^2 + \omega^{-2}, \quad \sigma^2(\alpha_1) = \alpha_3 = \omega^4 + \omega^{-4},$$

$$\sigma^3(\alpha_1) = \alpha_4 = \omega^8 + \omega^{-8}, \quad \sigma^4(\alpha_1) = \alpha_5 = \omega^5 + \omega^{-5}.$$

The fixed field of $\langle \sigma^5 \rangle$ contains the α_i and since σ^5 has order 2 in the Galois group, the extension $\mathbb{Q}(\alpha_1, \ldots, \alpha_5) = E_{\Phi_{11}}^{\langle \sigma^5 \rangle}$ with $[E_{\Phi_{11}} : E_{\Phi_{11}}^{\langle \sigma^5 \rangle}] = 2$. Since $E_{\Phi_{11}}^{\langle \sigma^5 \rangle}$ is Galois over $\mathbb{Q}$, we get a Galois extension of $\mathbb{Q}$ with Galois group $C_{10}/C_2 \cong C_5$.

The polynomial that generates the extension may be computed to be

$$f(x) = (x - \alpha_1)(x - \alpha_2)(x - \alpha_3)(x - \alpha_4)(x - \alpha_5)$$
$$= x^5 + x^4 - 4x^3 - 3x^2 + 3x + 1.$$

By Proposition 6.9, $f(x)$ is irreducible, and $\mathcal{G}al(E_f/\mathbb{Q}) \cong C_5$.

We turn next to F_{20}. The following result should be compared with the statement of Lemma 6.34.

Proposition 6.41 *For an integer a that is not a fifth power, the polynomial $f(x) = x^5 - a$ has a Galois group $\mathcal{G}al(E_f/\mathbb{Q}) \cong F_{20}$.*

Proof By assumption $\sqrt[5]{a}$ is not in $\mathbb{Q}$, but $\sqrt[5]{a} \in \mathbb{R}$. It follows that $f(x)$ is irreducible in $\mathbb{Q}[x]$. Let $k = \mathbb{Q}(\sqrt[5]{a}) \subset \mathbb{R}$. Then $[k : \mathbb{Q}] = 5$, and k is not the splitting field of $f(x)$. Adjoin a fifth root of unity ω_5 to k to obtain $E_f = \mathbb{Q}(\sqrt[5]{a}, \omega_5)$. Then $\#\mathcal{G}al(E_f/\mathbb{Q}) = [E_f : \mathbb{Q}] = [E_f : k][k : \mathbb{Q}] = 4 \cdot 5 = 20$. Since $f(x)$ is irreducible, $\mathcal{G}al(E_f/\mathbb{Q})$ acts transitively on the roots of $f(x)$, and so must be F_{20}. □

To apply Dedekind's Theorem using the mod p reduction of an integral polynomial, it is helpful to have a list of presentations of the transitive subgroups of Σ_5 with the appearing types of the elements and their frequencies.

type	C_5	D_{10}	F_{20}	A_5	Σ_5
1,1,1,1,1	1	1	1	1	1
2,1,1,1					10
3,1,1				20	20
2,2,1		5	5	15	15
4,1			10		30
3,2					20
5	4	4	4	24	24

Notice that the type $(3, 2)$ only appears in Σ_5. Working modulo a prime, an irreducible cubic times an irreducible quadratic can be used to determine an irreducible integral quintic whose mod p reduction is the product and whose Galois group is Σ_5. For example, working mod 5, the quadratic $x^2 + 2$ and the cubic $x^3 + 2x + 1$ are both irreducible (neither has a root in $\mathbb{F}_5$). Their product is $x^5 + 4x^3 + x^2 + 4x + 2$. If we add $5x^2$, then we get $f(x) = x^5 + 4x^3 + 6x^2 + 4x + 2$, which is irreducible in $\mathbb{Q}[x]$ by the Eisenstein criterion. The discriminant of $f(x)$ is 553,552 which is not divisible by 5. By Dedekind's Theorem, $\hat{\rho}(\mathcal{G}al(E_f/\mathbb{Q})) \subset \Sigma_5$ contains a permutation of type $(3, 2) = (1^0, 2^1, 3^1, 4^0, 5^0)$, which implies that $\mathcal{G}al(E_f/\mathbb{Q}) \cong \Sigma_5$.

When the discriminant of a polynomial is a square, the table shows that a permutation of type $(1^2, 2^0, 3^1, 4^0, 5^0)$ distinguishes the Galois group as A_5.

In [79] the polynomial $f(x) = x^5 + 20x + 16$ is given as having $\mathcal{G}al(E_f/\mathbb{Q}) \cong A_5$. The discriminant is given by $2^{16}5^6$, a square, and so the Galois group is a subgroup of A_5. Experimenting with small primes that are not 2 or 5, we find

$$x^5 + 20x + 16 \equiv (x+2)(x+3)(x^3 + 2x^2 + 5x + 5) \pmod{7}.$$

The appearance of this factorization into irreducibles mod 7 implies the appearance of a permutation of type 3,1,1, and that forces $\mathcal{G}al(E_f/\mathbb{Q}) \cong A_5$.

The case of a quintic $f(x)$ with Galois group isomorphic to D_{10} requires deeper methods. From the table of types we can expect that factorizations of $f(x)$ modulo primes occur in only three ways: five linear factors, one linear and two quadratic factors, or one quintic irreducible. Dedekind's Theorem tells us that any appearance of such a factorization signals the existence of permutations of the roots of the same type. However, what if you have to wait a long time to witness such a factorization? Suppose for some big prime your polynomial factors into a permutation of type 3,1,1. With the aid of computational devices, a theorem of Frobenius [29] allows us to decide. For the theorem to make sense, we need the notion of the *density* of a collection of primes.

Definition 6.42 Suppose S is a subset of $\mathcal{P}$, the set of primes in $\mathbb{Z}$. The *density* of S, if it exists, is the limit

$$\lim_{x \to \infty} \frac{\#\{p \in S \mid p \leq x\}}{\#\{p \in \mathcal{P} \mid p \leq x\}}.$$

A factorization modulo p of a polynomial $f(x) \equiv u_1(x) \cdots u_t(x)$ with $\deg u_i(x) = d_i$ determines a partition of $n = \deg f(x)$ because $d_1 + d_2 + \cdots + d_t = n$ assuming that $f(x)$ is separable and p does not divide $D(f)$. If we arrange the partition by $d_1 \leq d_2 \leq \cdots \leq d_t$ and denote by e_i the number of times $i = d_j$ for $1 \leq j \leq t$, then $d_1 + \cdots + d_t = e_1 + 2e_2 + \cdots + ne_n = n$. We can associate the type $(1^{e_1}, 2^{e_2}, \ldots, n^{e_n})$ with this partition. Dedekind's Theorem makes the connection between mod p factorizations of $f(x)$ and the Galois group by showing that a factorization of type $(1^{e_1}, 2^{e_2}, \ldots, n^{e_n})$ means a permutation of the roots of $f(x)$ in $\mathcal{G}al(E_f/\mathbb{Q})$ of the same type. The *Frobenius Density Theorem* states the following:

Theorem 6.43 *The density of the set of primes p for which $f(x)$ has a factorization leading to the type $(1^{e_1}, 2^{e_2}, \ldots, n^{e_n})$ exists and is equal to*

$$\frac{1}{\#G} \#\{\sigma \in \hat{\rho}(\mathcal{G}al(E_f/\mathbb{Q})) \mid type\ of\ \sigma = (1^{e_1}, 2^{e_2}, \ldots, n^{e_n})\}.$$

A proof of this theorem is beyond the scope of this book. A proof of this result and related results follows from the more general *Chebotarev Density Theorem* [13], a proof of which may be found in [27].

The polynomial $f(x) = x^5 - 5x + 12$ found in [79] has the discriminant $2^{12}5^6$, a square, and hence its Galois group is one of A_5, D_{10}, or C_5. Because the polynomial has at most three real roots (check the derivative), a complex conjugate pair of roots implies that 2 divides $\#\mathcal{G}al(E_f/\mathbb{Q})$, and so C_5 can be eliminated. Theorem 6.43 implies that if $\mathcal{G}al(E_f/\mathbb{Q}) \cong D_{10}$, then the mod p factorizations of $f(x)$, for N many primes p, should garner approximately $N/10$ many 1^5 factorizations (that is, $f(x)$ has all its roots in $\mathbb{F}_p$), $N/2$ many $(1, 2^2)$ factorizations, and $2N/5$ many instances of $f(x)$ as irreducible over $\mathbb{F}_p$. Using an online resource (planetcalc.com/8332/), I tested the first fifty primes, excluding 2 and 5, getting three factorizations of type 1^5 (6%), 29 factorizations of type $(1, 2^2)$ (59%), and 18 factorizations of type 5^1 (36%). These are reasonably close to the expected (10%, 50%, 40%) for D_{10}, and the nonappearance of any $(1^2, 3)$ factorizations excludes A_5.

To be more exacting, we can apply a theorem of HEINRICH WEBER (1842–1913). The *Weber resolvent* [92] is based on the polynomial

$$P(x_1, \ldots, x_5) = x_1x_2 + x_2x_3 + x_3x_4 + x_4x_5 + x_5x_1$$
$$- (x_1x_3 + x_3x_5 + x_5x_2 + x_2x_4 + x_4x_1).$$

This polynomial is fixed by the action of D_{10} as a subgroup of Σ_5, and its square is fixed by the action of F_{20}. There are six cosets in A_5/D_{10} (and in Σ_5/F_{20}) represented by the transversal $\{(\ ), (1, 2, 3), (1, 3, 2), (1, 2, 5), (1, 5, 2), (1, 3, 4)\}$. The associated resolvent $\mathrm{Res}(P, f)$ has degree six.

Theorem 6.44 *Let $f(x) = x^5 + ax + b \in \mathbb{Z}[x]$ be an irreducible quintic with $a \neq 0$. Suppose further that $a = \dfrac{5^5\lambda\mu^4}{(\lambda - 1)^4(\lambda^2 - 6\lambda + 25)}$ and $b = a\mu$ for some $\lambda, \mu \in \mathbb{Q}$ with $\lambda \neq 1$, $\mu \neq 0$. Then $\mathcal{G}al(E_f/\mathbb{Q}) \cong D_{10}$ if and only if $D(f)$ is a square; otherwise, $\mathcal{G}al(E_f/\mathbb{Q}) \cong F_{20}$.*

See [50] for details.

6.8 The Primitive Element Theorem

The indeterminates in the Kronecker resolvent $u_1, \ldots, u_n$ land us in the ring $k[u_1, \ldots, u_n][x]$. Evaluating the u_i with integer values leads to more familiar field extensions. The separability of the resolvent can be retained after evaluation.

Proposition 6.45 *There are integers $m_1, m_2, \ldots, m_n$ for which the linear combinations $m_1\alpha_{\sigma(1)} + \cdots + m_n\alpha_{\sigma(n)}$ are all distinct as σ varies over Σ_n.*

Proof Let $W \subset E_f$ denote the span over the rational numbers of the set of roots $\{\alpha_1, \ldots, \alpha_n\}$. There is a permutation representation $\rho \colon \Sigma_n \to \mathrm{Gl}(W)$ given by

$$\rho(\sigma)(m_1\alpha_1 + \cdots + m_n\alpha_n) = m_1\alpha_{\sigma(1)} + \cdots + m_n\alpha_{\sigma(n)}.$$

Given two distinct permutations σ and τ in Σ_n, define the subspace of W given by $U_{\sigma,\tau} = \ker(\rho(\sigma) - \rho(\tau))$. Notice that $U_{\sigma,\tau} \neq W$ for $\sigma \neq \tau$: If $U_{\sigma,\tau}$ were all of W, then we would have $\sigma(\alpha_i) = \tau(\alpha_i)$ for all i and $\sigma = \tau$.

Let us prove the useful fact that a vector space over an infinite field is not a union of finitely many proper subspaces. Denote the set of $U_{\sigma,\tau}$ by $\{U_1, U_2, \ldots, U_\ell\}$, where $\ell = \binom{n!}{2}$ in some order and suppose that $\bigcup_{j=1}^{\ell} U_j = W$. Let $w_1 \in U_1$ and $w_2 \in W - U_1$. Consider the set of all expressions of the form $w_1 + cw_2$ with $c \in \mathbb{Q}$. There are infinitely many such vectors scattered among U_2 to U_ℓ; however, none of them, except $w_1 + 0w_2$, lies in U_1. If there were such a vector, subtracting w_1 would imply $w_2 \in U_1$. It follows from the Pigeonhole Principle that there are two distinct elements $w_1 + c_1w_2$ and $w_1 + c_2w_2$ in some U_j. Since U_j are subspaces of W, the value of the difference $w_1 + c_1w_2 - (w_1 + c_2w_2) = (c_1 - c_2)w_2$ implies that $w_2 \in U_j$, and it follows then that $w_1 \in U_j$. By varying w_1 we have shown that $U_1 \subset \bigcup_{j=2}^{\ell} U_j$, and so $W = \bigcup_{j=2}^{\ell} U_j$. Repeat the argument for U_2 to obtain $W = \bigcup_{j=3}^{\ell} U_j$. Continuing in this fashion, we deduce $W = U_\ell$, a contradiction. Hence $W \neq \bigcup_{j=1}^{\ell} U_j$, and there is some rational linear combination $r_1\alpha_1 + \cdots + r_n\alpha_n = w$ that is not in any $\ker(\rho(\sigma) - \rho(\tau))$. By clearing denominators we get an integral linear combination $m_1\alpha_1 + \cdots + m_n\alpha_n = v$ for which the set $\{\rho(\sigma)(v) \mid \sigma \in \Sigma_n\}$ has cardinality $n!$. $\qquad\square$

Denote $\rho(\sigma)(v) = m_1\alpha_{\sigma(1)} + \cdots + m_n\alpha_{\sigma(n)}$ by v^σ. Consider the polynomial

$$F(x) = \prod_{\sigma \in \Sigma_n} \left(x - v^\sigma\right) \in E_f[x].$$

For any $\tau \in \Sigma_n$, $\rho(\tau)(F(x)) = \rho(\tau)(\prod_{\sigma \in \Sigma_n} x - v^\sigma) = \prod_{\sigma \in \Sigma_n} x - v^{\tau\sigma} = F(x)$, and so $F(x)$ is a polynomial in $\mathbb{Q}[x]$.

Because $f(x) \in \mathbb{Z}[x]$ is a monic irreducible polynomial, the coefficients of $F(x)$ are rational numbers and algebraic integers. By Proposition 4.22, $F(x) \in \mathbb{Z}[x]$.

In $\mathbb{Q}[x]$, we can factor $F(x) = u_1(x)u_2(x)\cdots u_s(x)$ into a product of irreducible polynomials. Because $F(x)$ is a monic integral polynomial, by Gauss' Lemma we can take $u_i(x) \in \mathbb{Z}[x]$ and monic. Suppose that $u_1(v) = 0$. Then $m_v(x)$ divides $u_1(x)$, and because both are irreducible and monic, $u_1(x) = m_v(x)$.

Let $St_v = \{\tau \in \Sigma_n \mid \rho(\tau)(m_v(x)) = m_v(x)\}$. For any $\tau \in St_v$, the linear combination v^τ is a root of $m_v(x)$ by Proposition 6.3. If $\sigma \in \Sigma_n$ takes $x - v$ to $x - v^\sigma$, a divisor of $u_i(x)$, then $x - v^\sigma$ is not a divisor of $m_v(x)$: if it were, then $u_i(x)$ and $u_1(x) = m_v(x)$ would share a divisor and $F(x)$ would fail to be separable, a contradiction to our choice of the integers m_j. It follows that $m_v(x) = \prod_{\tau \in St_v} x - v^\tau$. Thus $\deg m_v(x) = \#St_v = [\mathbb{Q}[v] : \mathbb{Q}] = [\mathbb{Q}(v) : \mathbb{Q}]$.

Because the automorphisms in $\mathcal{G}al(E_f/\mathbb{Q})$ permute the roots of $f(x)$ while fixing $\mathbb{Q}$, for $\phi \in \mathcal{G}al(E_f/\mathbb{Q})$, the permutation representation gives $\hat{\rho}(\phi) \in \Sigma_n$ and $\hat{\rho}(\phi)(m_v(x)) = m_v(x)$, and so $\hat{\rho}(\phi) \in St_v$. Thus $\hat{\rho}(\mathcal{G}al(E_f/\mathbb{Q}))$ is a subgroup of St_v. Also, $\mathbb{Q}(v) \subset E_f$. Our goal is to show that $E_f = \mathbb{Q}(v)$ and that $\mathcal{G}al(E_f/\mathbb{Q}) \cong St_v$.

The field $\mathbb{Q}(v) \cong \mathbb{Q}[x]/(m_v(x))$ is a simple extension. We employ a construction due to Lagrange to show that $v^\tau \in \mathbb{Q}(v)$ for all $\tau \in St_v$. For each $\tau \in St_v$, consider the polynomial

$$G_\tau(x) = \sum_{\sigma \in \Sigma_n} v^{\sigma\tau} \frac{F(x)}{x - v^\sigma}.$$

For $\zeta \in \Sigma_n$, $\rho(\zeta)(G_\tau(x)) = \sum_{\sigma \in \Sigma_n} v^{(\zeta\sigma)\tau} \frac{F(x)}{x - v^{(\zeta\sigma)}} = G_\tau(x)$. Hence $G_\tau(x)$ is a rational polynomial. The evaluation mapping $\mathrm{ev}_v \colon \mathbb{Q}[x] \to \mathbb{Q}[v]$ takes $G_\tau(x)$ to $G_\tau(v)$. When we evaluate $G_\tau(x)$ on v, because v is a root of $F(x)$, all summands vanish except when $\sigma = \mathrm{Id}$ where $G_\tau(v) = v^\tau \prod_{\mathrm{Id} \neq \sigma \in \Sigma_n} v - v^\sigma$. It follows that $G_\tau(v)$ is in $\mathbb{Q}[v]$.

Next, consider the derivative $F'(x)$. Using the same observation with which we related the sums of powers to the elementary symmetric polynomials, we can express $F'(x) = \sum_{\sigma \in \Sigma_n} \frac{F(x)}{x - v^\sigma}$. Since $F(x) \in \mathbb{Z}[x]$, so also is $F'(x) \in \mathbb{Z}[x]$, and we have $F'(v) = \prod_{\mathrm{Id} \neq \sigma \in \Sigma_n} v - v^\sigma$, and this product is in $\mathbb{Q}[v]$. Since $\mathbb{Q}[v] \cong \mathbb{Q}[x]/(m_v(x))$ is a field, the value $G_\tau(v)/F'(v) = v^\tau$ lies in $\mathbb{Q}[v]$ for all $\tau \in \Sigma_n$. Therefore $\mathbb{Q}[v]$ is the splitting field of $m_v(x)$ over $\mathbb{Q}$. The Galois group $\mathcal{G}al(\mathbb{Q}(v)/\mathbb{Q})$ contains St_v and satisfies

$$\#\mathcal{G}al(\mathbb{Q}(v)/\mathbb{Q}) = [\mathbb{Q}(v) : \mathbb{Q}] = \#St_v,$$

and therefore $\mathcal{G}al(\mathbb{Q}(v)/\mathbb{Q}) \cong St_v$.

The method of Lagrange can be extended further. Define the polynomial $g_j(x) = \sum_{\sigma \in \Sigma_n} \alpha_{\sigma(j)} \frac{F(x)}{x - v^\sigma}$. For $\tau \in \Sigma_n$, we have that $\rho(\tau)(g_j(x)) = \sum_{\sigma \in \Sigma_n} \alpha_{(\tau\sigma)(j)} \frac{F(x)}{x - v^{\tau\sigma}} = g_j(x)$. Therefore $g_j(x) \in \mathbb{Q}[x]$ and $g_j(v) \in \mathbb{Q}[v]$. But $g_j(v) = \alpha_j \prod_{\mathrm{Id} \neq \sigma \in \Sigma_n} v - v^\sigma$. We have shown that $F'(v) \in \mathbb{Q}[v]$, so $\alpha_j = g_j(v)/F'(v) \in \mathbb{Q}[v]$. It follows that $E_f \subset \mathbb{Q}[v]$, and hence $E_f = \mathbb{Q}[v]$.

Finally, we have $Gal(E_f/\mathbb{Q}) \cong Gal(\mathbb{Q}(v)/\mathbb{Q}) \cong St_v$. We have proved the following result.

Theorem 6.46 *Let $f(x) \in \mathbb{Z}[x]$ be a monic, separable, and irreducible polynomial with roots $\alpha_1, \ldots, \alpha_n$ in E_f, the splitting field of $f(x)$ over $\mathbb{Q}$. For a choice of integers $m_1, \ldots, m_n$ for which the conjugates of $v = m_1\alpha_1 + \cdots + m_n\alpha_n$ under the action of Σ_n on $\{\alpha_1, \ldots, \alpha_n\}$ are all distinct, we have $E_f = \mathbb{Q}(v) \cong \mathbb{Q}[x]/(m_v(x))$ and $Gal(E_f/\mathbb{Q}) \cong St_v = \{\tau \in \Sigma_n \mid \rho(\tau)(m_v(x)) = m_v(x)\}$.*

The element $v = m_1\alpha_1 + \cdots + m_n\alpha_n$ is called a *primitive element* for the extension E_f over $\mathbb{Q}$. More generally, any finite extension of a field of characteristic zero has a primitive element, as outlined in the exercises.

6.9 The Normal Basis Theorem

A finite extension K of k has the Galois group $Gal(K/k)$ acting linearly on K as a vector space over k, that is, if $u \in k$ and $\alpha \in K$, then $\sigma(u\alpha) = u\sigma(\alpha)$ for all $\sigma \in Gal(K/k)$. These facts combine to give a representation $\rho: Gal(K/k) \to Gl(K/k)$, where $Gl(K/k)$ is the group of k-linear isomorphisms $K \to K$. Every field automorphism of K that fixes k is such a mapping, and the representation is given by inclusion. In Chapter 4, we characterized representations as sums of irreducible representations. Which representation is determined by ρ?

Suppose that K is a Galois extension of k. We know that $[K : k] = \dim_k K = \#Gal(K/k)$. This condition is shared by the regular representation of a finite group G. The vector space of the regular representation is $kG = \{\sum_{g \in G} a_g g \mid a_g \in k\}$ with a basis given by the elements of G. In the context of fields, we introduce the following idea.

Definition 6.47 A *normal basis* for an extension K of a field k is given by the set $\{u, \sigma_2(u), \ldots, \sigma_n(u)\}$ of conjugates of a particular element $u \in K$, which form a basis for K as a vector space over k.

Given any basis $\{\beta_1, \ldots, \beta_n\}$ for K over k, we can form the matrix of conjugates $M = (\sigma_i(\beta_j))$.

Proposition 6.48 *For a Galois extension K of a field k, a set of elements $\{\beta_1, \ldots, \beta_n\} \subset K$ is a basis for K over k if and only if $\det(\sigma_i(\beta_j)) \neq 0$.*

Proof The set $\{\beta_1, \ldots, \beta_n\}$ is linearly dependent if and only if there are elements $c_1, \ldots, c_n \in k$, not all zero, such that $c_1\beta_1 + \cdots + c_n\beta_n = 0$. This holds if and only if there is a nonzero vector in $k^{\times n}$ satisfying

$$\begin{bmatrix} \beta_1 & \beta_2 & \cdots & \beta_n \\ \sigma_2(\beta_1) & \sigma_2(\beta_2) & \cdots & \sigma_2(\beta_n) \\ \vdots & \vdots & \cdots & \vdots \\ \sigma_n(\beta_1) & \sigma_n(\beta_2) & \cdots & \sigma_n(\beta_n) \end{bmatrix} \begin{bmatrix} c_1 \\ c_2 \\ \vdots \\ c_n \end{bmatrix} = \begin{bmatrix} 0 \\ 0 \\ \vdots \\ 0 \end{bmatrix},$$

but this holds if and only if $\det(\sigma_i(\beta_j)) = 0$. $\qquad \square$

Corollary 6.49 *For an element* $\alpha \in K$, $\{\alpha, \sigma_2(\alpha), \ldots, \sigma_n(\alpha)\}$ *is a normal basis for K over k if and only if the matrix* $(\sigma_i \sigma_j(\alpha))$ *has nonzero determinant.*

With this condition for a normal basis, we can prove the following result.

The Normal Basis Theorem *Suppose k is a field containing infinitely many elements. Let K be a Galois extension of a field k with* $\mathcal{G}al(K/k) = \{\mathrm{Id} = \sigma_1, \sigma_2, , \ldots, \sigma_n\}$. *Then there is an element* $\alpha \in K$ *for which the set* $\{\alpha, \sigma_2(\alpha), \ldots, \sigma_n(\alpha)\}$ *is a normal basis for K over k.*

Proof Let $K = E_f$ denote a Galois extension of k where K is the splitting field of $f(x) \in k[x]$, a monic, separable, and irreducible polynomial. By the Primitive Element Theorem there is an element $v \in K$ with $K = k(v)$. Let $m_v(x)$ denote the minimal polynomial of v in $k[x]$. Over K, $m_v(x) = (x - v)(x - v_2) \cdots (x - v_n)$ where $v_i = \sigma_i(v)$ for $\sigma_i \in \mathcal{G}al(K/k)$. The derivative of $m_v(x)$ is given by

$$m'_v(x) = \sum_{i=1}^{n}(x - v_1) \cdots \widehat{(x - v_i)} \cdots (x - v_n) = \sum_{i=1}^{n} \frac{m_v(x)}{x - v_i},$$

where $v_1 = v$, and $\widehat{(x - v_i)}$ means "omit this factor." Therefore $m'_v(v_i) = \prod_{j \neq i}(v_i - v_j)$.

Let $g_i(x) = \dfrac{m_v(x)}{(x - v_i)m'_v(v_i)}$. For $\sigma_j \in \mathcal{G}al(K/k)$,

$$\sigma_j\big(g_1(x)\big) = \sigma_j\left(\frac{m_v(x)}{(x - v)m'_v(v)}\right) = \frac{m_v(x)}{(x - \sigma_j(v))m'_v(\sigma_j(v))}$$
$$= \frac{m_v(x)}{(x - v_j)m'_v(v_j)} = g_j(x),$$

and $\mathcal{G}al(K/k)$ permutes the set $\{g_j(x)\}$. The polynomials $g_1(x), \ldots, g_n(x)$ each have degree $n - 1$. Each satisfies the relation $g_i(v_j) = 0$ if $i \neq j$ and $g_i(v_i) = 1$. Sum the $g_i(x)$ and subtract one giving $g_1(x) + g_2(x) + \cdots + g_n(x) - 1$, a polynomial of degree $n - 1$ or less, which evaluates to zero for

$x = v_1, v_2, \ldots, v_n$. This is only possible if $g_1(x) + g_2(x) = \cdots + g_n(x) - 1 = 0$.

For $i \neq j$, another relation among the $g_i(x)$ holds, $g_i(x)g_j(x) \equiv 0 \pmod{m_v(x)}$:

$$g_i(x)g_j(x) = \frac{m_v(x)}{(x - v_i)m'_v(v_i)} \frac{m_v(x)}{(x - v_j)m'_v(v_j)}$$

$$= \frac{(x - v)^2(x - v_1)^2 \cdots (x - v_n)^2}{(x - v_i)(x - v_j)m'_v(v_i)m'_v(v_j)} \equiv 0 \pmod{m_v(x)}.$$

This relation holds in $k[x]/(m_v(x))$, which is isomorphic to K. To exploit this isomorphism, we work in $k[x]/(m_v(x))$. Multiplying $g_1(x) + g_2(x) + \cdots + g_n(x) - 1 = 0$ by $g_i(x)$, we obtain $g_i(x)g_i(x) \equiv g_i(x) \pmod{m_v(x)}$. This gives us two relations among the $g_i(x)$:

$$\text{for all } i, g_i(x)g_i(x) \equiv g_i(x) \pmod{m_v(x)}, \quad (\diamondsuit)$$

$$\text{for } i \neq j, g_i(x)g_j(x) \equiv 0 \pmod{m_v(x)}. \quad (\clubsuit).$$

Consider the matrix with entries in $k[x]/(m_v(x))$ that is given by $A(x) = (\sigma_i\sigma_j(g_1(x)))$. We can write $\sigma_i\sigma_j = \sigma_{k(i,j)}$, and so $\sigma_i\sigma_j(g_1(x)) = g_{k(i,j)}(x)$. The isomorphism between $k[x]/(m_v(x))$ and K means that if $A(x)$ is invertible in $k[x]/(m_v(x))$, then there is an element in K with $(\sigma_i\sigma_j(\alpha))$ invertible, giving us a normal basis. To proceed in $k[x]/(m_v(x))$, let us compute $A(x)A(x)^T$, where $A(x)^T$ is the transpose of $A(x)$. The (i, j)th entry of $A(x)A(x)^T$ is given by

$$\left(A(x)A(x)^T\right)_{(i,j)} = g_{k(i,1)}(x)g_{k(j,1)}(x) + g_{k(i,2)}(x)g_{k(j,2)}(x)$$

$$+ \cdots + g_{k(i,n)}(x)g_{k(j,n)}(x).$$

Such a product in $k[x]/(m_v(x))$ can be analyzed by applying relations $(\diamondsuit)$ and $(\clubsuit)$. For $i \neq j$, $g_{k(i,l)}(x) = g_{k(j,l)}(x)$ if and only if $\sigma_i\sigma_l(g_1(x)) = \sigma_j\sigma_l(g_1(x))$ if and only if $\sigma_i(g_l(x)) = \sigma_j(g_l(x))$. This means that the polynomials share the same roots, which only happens if $\sigma_i = \sigma_j$. Hence every off diagonal entry in $A(x)A(x)^T$ is congruent to zero modulo $m_v(x)$. For the diagonal entries, we have

$$g^2_{k(i,1)}(x) + \cdots + g^2_{k(i,n)}(x) \equiv g_{k(i,1)}(x) + \cdots + g_{k(i,n)}(x) \pmod{m_v(x)}$$

$$= g_1(x) + \cdots + g_n(x) \equiv 1 \pmod{m_v(x)}.$$

Thus $A(x)A(x)^T \equiv \text{Id} \pmod{m_v(x)}$, and so $\det A(x) \equiv \pm 1 \pmod{m_v(x)}$.

As a polynomial, $\det A(x)$ has only finitely many roots in k, and so there is $u \in k$ with $A(u) = (\sigma_i\sigma_j(g_1(u)))$ invertible. If we let $\alpha = g_1(u)$, then $\{\alpha, \sigma_2(\alpha), \ldots, \sigma_n(\alpha)\}$ is a normal basis. $\qquad\square$

The existence of a normal basis for a Galois extension K of a finite field k reduces to a problem in linear algebra that we will not discuss here (see [44] for more detail). Suppose p is a prime and $q = p^m$ for some $m \geq 1$. The field $\mathbb{F}_{q^n}$ is a Galois extension of $\mathbb{F}_q$ given as the splitting field of $x^{q^n} - x$. The Galois group $\mathcal{G}al(\mathbb{F}_{q^n}/\mathbb{F}_q)$ is a cyclic group generated by the Frobenius automorphism $\text{Fr}(a) = a^q$. As a linear transformation over $\mathbb{F}_q$, $\text{Fr}: \mathbb{F}_{q^n} \to \mathbb{F}_{q^n}$ satisfies the relation $\text{Fr}^n = \text{Id}$, that is, Fr satisfies the polynomial $x^n - 1$ in $\text{Hom}_{\mathbb{F}_q}(\mathbb{F}_{q^n}, \mathbb{F}_{q^n})$. If $Q(x) = a_0 + a_1 x + \cdots + a_{n-1} x^{n-1}$ is any other polynomial of degree less than n, then $Q(\text{Fr}) \neq 0$ as a field automorphism by the Dedekind Independence Theorem. Therefore $x^n - 1$ is both the minimal polynomial of Fr and its characteristic polynomial. This fact implies that Fr has a *cyclic vector*, that is, an element α of $\mathbb{F}_{q^n}$ with $\{\alpha, \text{Fr}(\alpha), \text{Fr}^2(\alpha), \ldots, \text{Fr}^{n-1}(\alpha)\}$ a basis for $\mathbb{F}_{q^n}$ over $\mathbb{F}_q$. But this is a normal basis. See [35], Chapter XIII for details.

The Normal Basis Theorem was proved in 1931 by Emmy Noether in the context of algebraic number theory [68]. The more general result [21] is proved by MAX DEURING (1907–1984), a student of Noether. Using a normal basis, for $G = \mathcal{G}al(K/k)$, there is a linear mapping $\Gamma: kG \to K$ over the field k, given by extending $\Gamma(\phi) = \phi(\alpha)$. Under the assumption that K is a finite Galois extension of k, Γ is a G-invariant isomorphism of representations of $\mathcal{G}al(K/k)$, and so we conclude:

Theorem 6.50 *The inclusion* $\rho: \mathcal{G}al(K/k) \to \text{Gl}(K/k)$, *where* $\text{Gl}(K/k)$ *is the group of k-linear automorphisms of K, is a representation of* $\mathcal{G}al(K/k)$ *isomorphic to the regular representation* $\mathcal{G}al(K/k)$.

Exercises

6.1   Let us derive the Ferrari solution of a quartic following Descartes [85]. After eliminating the cubic summand, suppose $x^4 + ax^2 + bx + c = (x^2 + px + q)(x^2 + rx + s)$. The vanishing of the cubic term implies that $r = -p$, so $x^4 + ax^2 + bx + c = (x^2 + px + q)(x^2 - px + s)$. Multiplying the quadratics gives the relations

$$q + s - p^2 = a, \quad p(s - q) = b, \quad sq = c.$$

These imply $s - q = b/p$ and $s + q = a + p^2$. Adding and subtracting these equations yield $2s = p^2 + a + b/p$ and $2q = p^2 + a - b/p$. Multiply these together to obtain

$$p^6 + 2ap^4 + \left(a^2 - 4c\right)p^2 - b^2 = 0,$$

a cubic in p^2. Solving for p^2 via the Cardano Formula, we then obtain p, q, and s, and we can solve the original quartic from the quadratic formula. Fill in the details.

6.2 Suppose K is an extension of k with $\mathcal{G}al(K/k) \cong C_2$. Show that K is a normal extension of k. If the characteristic of k is not two, then show that $K = k(\alpha)$ for which $\alpha^2 \in k$. What goes wrong when the characteristic is two?

6.3 Suppose that $Q = \{\pm 1, \pm i, \pm j, \pm k\}$, the unit quaternion group of order 8, is isomorphic to $\mathcal{G}al(K/k)$. Work out the Hasse diagram for Q and use it to describe the lattice of subfields of K that contain k.

6.4 Let $f(x) = (x^2 - 3)(x^3 - 5)$. Determine the splitting field E_f of $f(x)$ over $\mathbb{Q}$. What is the Galois group of E_f over $\mathbb{Q}$?

6.5 Field extensions can be tricky. For example, Galois extensions are not transitive. Consider the tower $\mathbb{Q} \subset \mathbb{Q}(\sqrt{2}) \subset \mathbb{Q}(\sqrt[4]{2})$. Show that $\mathbb{Q}(\sqrt{2}) \supset \mathbb{Q}$ and $\mathbb{Q}(\sqrt[4]{2}) \supset \mathbb{Q}(\sqrt{2})$ are both Galois extensions but $\mathbb{Q}(\sqrt[4]{2}) \supset \mathbb{Q}$ is not a Galois extension.

6.6 Suppose K is a Galois extension of k and $\mathcal{G}al(K/k) = \{\sigma_1, \ldots, \sigma_n\}$. For $a \in K$, define the *trace* $\mathrm{tr}(a)$ and the *norm* $N(a)$ by

$$\mathrm{tr}(a) = \sigma_1(a) + \cdots + \sigma_n(a), \quad N(a) = \sigma_1(a)\sigma_2(a)\cdots\sigma_n(a).$$

Prove that $\mathrm{tr}(a)$ and $N(a)$ are in k, $\mathrm{tr}(a+b) = \mathrm{tr}(a)+\mathrm{tr}(b)$, and $N(ab) = N(a)N(b)$. What values do the trace and norm take on elements of k? If $\mathcal{G}al(K/k) \cong C_n = \langle \sigma \rangle$, then show that if $a = b - \sigma(b)$, then $\mathrm{tr}(a) = 0$. If $a = b/\sigma(b)$ for some nonzero $b \in K$, then $N(a) = 1$. Hilbert's Theorem 90 [70] shows the converse of these conditions.

6.7 Using Gauss' method for constructing the 17-gon, work out the simpler case of a regular pentagon. Prove that the regular 15-gon is constructible using straightedge and compass.

6.8 Prove that a quartic $f(x) = x^4 + bx^3 + cx^2 + dx + e \in \mathbb{Q}[x]$ has the discriminant $D(f) < 0$ if and only if $f(x)$ has exactly two real roots.

6.9 Prove the theorem of Steinitz [81] that a finite extension K of a field k has a finite number of intermediate fields if and only if K is a simple extension of k, that is, there is an element $\alpha \in K$ with $K = k(\alpha)$. This is a version of the *Primitive Element Theorem*. Suppose k is infinite (the finite case is Corollary 5.21) and suppose $k(\alpha)$ is the subfield of K with maximal $[k(\alpha) : k]$. If $k(\alpha) \neq K$, then there is $\beta \in K$ not in $k(\alpha)$. The fields $k(\alpha + u\beta)$ for $u \in k$ cannot be infinite in number by assumption. Thus some $k(\alpha + u_1\beta) = k(\alpha + u_2\beta)$. Show that this implies $\beta \in k(\alpha)$, so no such β exists, and $K = k(\alpha)$. Conversely, if $K = k(\alpha)$, then the minimal polynomial $m_\alpha(x)$ of α over k is in $k[x]$. If L is an intermediate field, then let $n_\alpha(x)$ be the minimal polynomial of α over L. Show that $n_\alpha(x)$ determines L uniquely. Show that there are only finitely many possible choices of $n_\alpha(x)$.

6.10 What polynomial is the cyclotomic polynomial $\Phi_{12}(x) \in \mathbb{Z}[x]$? We know that the cyclotomic polynomials are irreducible. Show that the mod p reduction of $\Phi_{12}(x)$ is reducible for all primes p.

6.11 Consider the following quintics and try to determine their Galois groups:

$$x^5 + 11x + 44, \qquad x^5 + 15x^3 + 81,$$
$$x^5 - 14x + 7, \qquad x^5 - 7x^2 + 7.$$

(You might want to use some online computational site to obtain the discriminants.)

6.12 Consider the matrix

$$V = \begin{bmatrix} 1 & x_1 & x_1^2 & \cdots & x_1^{n-1} \\ 1 & x_2 & x_2^2 & \cdots & x_2^{n-1} \\ 1 & x_3 & x_3^2 & \cdots & x_3^{n-1} \\ \vdots & \vdots & \vdots & \cdots & \vdots \\ 1 & x_n & x_n^2 & \cdots & x_n^{n-1} \end{bmatrix}.$$

We want to show that $\det(V) = \Delta(U(x))$, where $U(x) = \prod_{i=1}^n (x - x_i)$. First observe that $\deg \det(V) = \binom{n}{2}$ as a polynomial in $\mathbb{Z}[x_1, \ldots, x_n]$. Show that this is the degree of $\Delta(U(x)) = \prod_{1 \le i < j \le n}(x_i - x_j)$. In $\mathbb{Z}[x_1, \ldots, \widehat{x_i}, \ldots, x_n][x_i]$, notice that if $x_i = x_j$ for some $j \ne i$, then $\det(V)$ is zero. Therefore $(x_i - x_j)$ divides $\det(V)$ for all $i < j$, and $\Delta(U(x))$ divides $\det(V)$. It follows that $\det(V) = c\Delta(U(x))$ with a constant c. To see the value of c, notice that the product of the diagonal entries of V is $x_2 x_3^2 \cdots x_n^{n-1}$. The final summand in $\Delta(U(x))$, the product of the x_j, is exactly $(-1)^{\binom{n}{2}} x_2 x_3^2 \cdots x_n^{n-1}$, and so $\det(V) = \pm\Delta(U(x))$.

6.13 Suppose $f(x) \in k[x]$ is a monic, separable, and irreducible polynomial. Let $f(x) = (x - \alpha)(x - \alpha_2) \cdots (x - \alpha_n)$ over E_f, the splitting field of $f(x)$ over k. Show that the discriminant of $f(x)$ satisfies

$$D(\alpha, \alpha_2, \ldots, \alpha_n) = (-1)^{\binom{n}{2}} \prod_{\sigma \in \mathcal{G}al(E_f/k)} \sigma\big(f'(\alpha)\big).$$

Use this fact to compute the discriminant of $x^3 - 2$ and polynomials like it.

7
Epilogue

In writing this book, I have had to make many choices that allowed me to focus the story I wanted to tell about the actions of groups. Every choice leaves behind a wealth of missed opportunity. This epilogue is a guide to other experiences you might seek.

7.1 History

The narrative I have tried to construct in this book is not at all the historical narrative. The main ideas of simple groups, representations of groups, and Galois theory have their origins in the work of Cauchy and Lagrange on permutation groups and polynomial equations, and in the work of Galois, who introduced the term *le groupe* to describe the collections of permutations of roots associated with a polynomial. The story of how Galois' theory of groups reached the mathematical community is as extraordinary as the story of his short life (see [25, 65, 85] and [87]). Stories of the development of the mathematics of Galois can be found in [16, 52, 97] and [89].

The origins of representation theory are described in the works of THOMAS HAWKINS (1938–) [39–42], in the book [18] of CHARLES CURTIS (1926–), and in [14]. A curious role is played by a question of Dedekind concerning the *group determinant*: If $g \in G$ is a finite group, then let x_g denote a variable indexed by g. The matrix $(x_{gh^{-1}})$ can be constructed from the group table of G, and its determinant is a polynomial in the variables x_g, which factors into irreducible polynomials

$$\det(x_{gh^{-1}}) = \prod_{i=1}^{s} P_i(x_e, x_{g_2}, \ldots, x_{g_n}).$$

This determinant plays a role in the Normal Basis Theorem. Dedekind had observed to Frobenius in 1896 that the irreducible polynomials P_i had degree 1 in the case of an abelian group, but for a nonabelian group, polynomials of higher degree appeared. Dedekind further noticed that if one worked over a certain noncommutative ring, the factors would be linear. Frobenius had seen similar decompositions of determinants in his study of theta functions. With his interest in finite groups, Frobenius developed the general properties of the group determinant, answering a question of Dedekind of whether the number of linear factors of $\det((x_{gh^{-1}}))$ equaled the order of G/G', the abelianization of G [29]. His nonabelian generalization identified properties of a finite group that are now clarified through the lens of representation theory.

7.2 Other Approaches

I have taken the viewpoint of linear algebra in the development of representation theory pioneered by Schur. A fruitful, alternative viewpoint is based on the fact that the vector space $\mathbb{F}G$, for a field $\mathbb{F}$ and a finite group G, is also a ring. The multiplication on $\mathbb{F}G$ is given by $(\sum_{g\in G} a_g g)(\sum_{h\in G} b_h h) = \sum_{g,h\in G} a_g b_h gh$. A ring A that is also a vector space over a field $\mathbb{F}$ is called an *algebra* over $\mathbb{F}$ if the vector space structure intertwines with the multiplication $((\alpha v)(\beta w) = (\alpha\beta)(vw)$ for $\alpha, \beta \in \mathbb{F}$ and $v, w \in A)$. Thus $\mathbb{F}G$ is an algebra over $\mathbb{F}$.

A representation $\rho \colon G \to \mathrm{Gl}(W)$ with a vector space W over $\mathbb{F}$ determines a *module* over the algebra $\mathbb{F}G$. A (left) module M over a ring R is like a vector space over a field. The module M is an abelian group. The multiplication by a scalar is replaced with a left multiplication by the ring on the module, $\mu \colon R \times M \to M$, satisfying the familiar axioms of a vector space.

The $\mathbb{F}G$-module structure on W is defined by

$$\mu\left(\sum_{g\in G} a_g g, v\right) = \sum_{g\in G} a_g \rho(g)(v).$$

On the other hand, if W is a $\mathbb{F}G$-module, then, because $\mathbb{F} = \mathbb{F}e$ is a subring of $\mathbb{F}G$, W is a vector space over $\mathbb{F}$ as follows: the multiplicative identity in $\mathbb{F}G$ is $1e$, so $\mu(1e, v) = v$. For $\alpha \in \mathbb{F}$, $\mu(\alpha e, v) = (\alpha e) \cdot v$ determines the vector space structure. For $g \in G$, let $\rho(g) \colon W \to W$ be given by $\rho(g)(w) = \mu(g, w) = g \cdot w$. Then

$$\rho(g)(\alpha w_1 + \beta w_2) = g \cdot (\alpha w_1 + \beta w_2) = \big(g(\alpha e)\big) \cdot w_1 + \big(g(\beta e)\big) \cdot w_2$$
$$= (\alpha g) \cdot w_1 + (\beta g) \cdot w_2 = \alpha\rho(g)(w_1) + \beta\rho(g)(w_2),$$

and $\rho(g)$ is a linear transformation on W. Because $\rho(g^{-1}) = \rho(g)^{-1}$, $\rho(g)$ is invertible and so in $\mathrm{Gl}(W)$. Finally, $\rho(gh)(w) = (gh) \cdot w = g \cdot (h \cdot w) = \rho(g)(\rho(h)(w))$ as in the multiplicative axiom for a vector space $r(sw) = (rs)w$. Thus an $\mathbb{F}G$-module gives a representation of G on W.

The fundamental concepts for modules such as submodules, product modules, etc. translate into representation theory notions [67]. For example, an $\mathbb{F}G$-submodule V of W is a G-invariant subspace of W. Irreducible (or *simple*) modules have only themselves and the zero submodule as submodules. Given two submodules V_1 and V_2 of an $\mathbb{F}G$-module W, let $V_1 + V_2 = \{u + v \in W \mid u \in V_1, v \in V_2\}$. Then $V_1 + V_2$ is also a submodule of W. Maschke's Theorem for a finite group G and a field whose characteristic does not divide the order of G translates to the fact that, for any submodule V of W, there is another submodule V^0 such that $W = V + V^0$.

One of the features of module theory over a ring R that is different from vector space theory is the existence of modules that are smaller than R. In particular, an ideal $I \subset R$ is also a module over R. The canonical examples of algebras over a field $\mathbb{F}$ are the matrix algebras $M_n(\mathbb{F})$ of $n \times n$-matrices with entries in $\mathbb{F}$. In a celebrated paper of 1908 [93], Wedderburn showed that *semisimple* algebras (that is, a sum of submodules that are each simple) over a field $\mathbb{F}$ are isomorphic to sums of matrix algebras $M_{n_i}(D_i)$ where D_i is a division algebra over $\mathbb{F}$. A *division algebra* is an algebra for which every nonzero element has a multiplicative inverse (a commutative division ring is a field). The matrix algebras are simple algebras. For an algebraically closed field, the simple algebras are $M_n(\mathbb{F})$. From such ingredients characters can be defined, and the main features of representation theory can be proved using methods of ring theory and the theory of algebras.

The interested reader seeking a new land of rich ideas can pursue this formulation of representation theory and its power to unify various constructions in [19, 49, 73] and [80].

One remarkable development is the case of fields $\mathbb{F}$ of characteristic $p > 0$. When p does not divide the order of G, then the proof of Maschke's Theorem carries over for representations of G on V, a vector space over $\mathbb{F}$. When p does divide the order of G, new ideas are needed. Such developments were dubbed *modular representation theory* by LEONARD DICKSON (1874–1954) [22]. The first important question to settle is the number of irreducible representations of a finite group over an algebraically closed field $\overline{\mathbb{F}}$ of characteristic $p > 0$. Dickson determined the number in the case of a finite abelian group. RICHARD BRAUER (1901–1977) took up this question in 1935 [8] to prove that the number of equivalence classes of irreducible representations over $\overline{\mathbb{F}}$ is the number of conjugacy classes of the group containing only elements of order prime to p. The methods were based on the study of algebras over $\overline{\mathbb{F}}$, an alternative notion of characters, and what is called the theory of *blocks*. Brauer utilized these new insights to prove that a finite simple group of order pqr^n, where p, q, and r are primes, and $n > 0$, must have order 60 or 168. Brauer and his collaborators refined the theory of blocks in the 1940s and used it to characterize certain families of finite simple groups. See [18] for the history of these developments and [19] for an exposition of these ideas.

7.3 The Dodecahedron

What about the dodecahedron and Galois theory? It is not this polyhedron but its dual, the icosahedron, that takes center stage. In 1858, Charles Hermite [43] proved that adjoining the values of Jacobi theta functions to the rational numbers, along with roots, allowed the determination of the roots of a polynomial of degree five. The algebra associated with the study of elliptic functions, theta function, and modular forms was well developed in the latter half of the nineteenth century. In a remarkable display of technical prowess [53], FELIX KLEIN (1849–1925) wove together geometric, analytic, and algebraic ideas to reframe the solution of Hermite. The icosahedron has the same rotation group as the dodecahedron, $\mathbf{Doc}^+ \cong A_5$. Being the smallest nonsolvable group, Klein sought to understand the Galois theory of A_5 through the geometry of an icosahedron. The regular icosahedron has a circumscribed sphere, and central projection to the sphere gives a spherical icosahedron that is invariant under the rotation group of the icosahedron acting on the sphere. The sphere takes many guises: it is the complex projective line $\mathbb{P}^1$ and the Riemann sphere $\widehat{\mathbb{C}} = \mathbb{C} \cup \{\infty\}$. The Möbius transformations $z \mapsto \dfrac{az+b}{cz+d}$ determine an embedding of the symmetry group $\mathbf{Doc}^+$ into $\mathrm{PSl}_2(\mathbb{C})$. As the complex projective line, the vertices of the icosahedron are seen to satisfy a homogeneous polynomial of degree 12 given by $f(z_1, z_2) = z_1 z_2 (z_1^{10} + 11 z_1^5 z_2^5 - z_2^{10})$.

A quintic equation can be transformed into $y^5 + 5ay^2 + 5by + c = 0$ by the use of Tschirnhaus transformations. If a quintic has roots $z_1, \ldots, z_5$, then the point $[z_1, z_2, z_3, z_4, z_5]$ in four-dimensional complex projective space lies in the set $F = \{[w_1, w_2, w_3, w_4, w_5] \mid \sum w_i = 0 = \sum w_i^2\}$, a quadric surface in $\mathbb{C}P^4$. This surface is doubly ruled, and A_5 acts on it by permuting coordinates. The quotient by this action on one ruling yields a complex projective line. Thus a solution is a point in the quotient and is determined by certain invariant polynomials. Recovering the roots from this formulation takes one through some hypergeometric functions. The interested reader will find thorough discussions of these ideas in [64, 77, 86], and of course, in [53].

7.4 Compact Lie Groups

The representation theory of finite groups has a generalization to certain infinite groups, in particular, to $\mathrm{Gl}_n(\mathbb{F})$ and its subgroups, where $\mathbb{F} = \mathbb{R}$ or $\mathbb{C}$. In this setting the fact that $\mathrm{Gl}_n(\mathbb{F})$ is a differentiable manifold adds extra structure that can be required of a representation. If $G \subset \mathrm{Gl}_n(\mathbb{F})$ is a closed compact subgroup that is also a smooth manifold for which matrix multiplication and the taking of inverses are smooth, then we call G a (matrix) *Lie group*. The best known example is the *special orthogonal group* SO(3) $\subset \mathrm{Gl}_3(\mathbb{R})$ of 3×3 matrices with real entries and determinant one, satisfying $A^{-1} = A^t$, that is, the inverse of A is its transpose. This is the group of rotations of $\mathbb{R}^3$, a group of linear isometries of space. When restricted to the unit sphere in $\mathbb{R}^3$, SO(3) is the subgroup of rotations of the unit sphere that, with reflections across planes, play the role of congruences in spherical geometry. The property of compactness of SO(3) equips the group with some properties of a finite set. Taking care to substitute appropriate constructions (integrals for sums, Haar measure for averaging, etc.) the representation theory of finite groups can be generalized to a representation theory of compact Lie groups. Some statements fail to generalize, as might be expected. For example, the number of irreducible representations of a Lie group may be infinite.

Lie groups also come equipped with a *Lie algebra* that has associated representations as an algebra over $\mathbb{F}$. If $\mathbb{F} = \mathbb{C}$, then we can speak of *unitary representations* of $G \subset \mathrm{Gl}_n(\mathbb{C})$, which are continuous homomorphisms of G into $U(W)$, where W is a Hilbert space, and $U(W)$ is the group of unitary operators on W (that is, $A^{-1} = \overline{A}^t$). In this case the Lie algebra $\mathfrak{g}$ has an associated representation $\mathfrak{g} \to \mathfrak{gl}(W)$. There is an imperfect correspondence between representations of G and representations of $\mathfrak{g}$ that depends upon the fundamental group of G. However, representations of $\mathfrak{g}$ are easier to determine. In the case of SO(3), irreducible representations correspond to spherical harmonics, that is, homogeneous polynomials in three real variables with complex coefficients that are harmonic ($\nabla^2 p = 0$). But spherical harmonics play a role in describing the energy levels of an electron in a spherically symmetric electrostatic field. This connection can be generalized: a Lie group of symmetries has a representation theory whose irreducible representations may be interpreted in the formalism of quantum mechanics. Expositions of these ideas can be found in [7, 54, 78, 96] and [37].

7.5 Great Reads

I learned much of the content of this book from some wonderful sources. The classic books on algebra of van der Waerden [88] and Jacobson [48] are rich with detail and scope, and newer texts by Artin [4], Rotman [71], and Dummit and Foote [24] provide broad perspectives on lots of algebra. For group theory, I learned a lot from books by Robinson [69], Isaacs [46], and Serre [75]. The paper by Shapiro [76] hooked me on group actions while I was a graduate student. It is still worth a look. For representation theory, I relied on Curtis and Reiner [19], James and Liebeck [49], Serre [73], and Isaacs [45]. In an account of Galois theory, a reader wants the history and the mathematics. The book of Cox [16] is an incredible achievement, meeting the highest expectations. I learned a lot about some of the deeper parts of Galois theory from the books of Gaal [30] and Swallow [82]. Course notes of Andy Baker [5, 6] on representation theory and Galois theory were most helpful. Notes by Milne on group theory and Galois theory [62, 63] broadened my horizons considerably. I hope you will find your way to some of these books.

References

[1] George E. Andrews. *The theory of partitions*, volume 2 of *Encyclopedia of Mathematics and Its Applications*. Addison-Wesley Publishing Co., Reading, MA–London–Amsterdam, 1976.

[2] Tom M. Apostol. *Introduction to analytic number theory*. Undergraduate Texts in Mathematics. Springer-Verlag, New York–Heidelberg, 1976.

[3] Emil Artin. *Galois theory*. Notre Dame Mathematical Lectures, No. 2. University of Notre Dame Press, South Bend, IN, 1959. Edited and supplemented with a section on applications by Arthur N. Milgram, Second edition, with additions and revisions, Fifth reprinting.

[4] Michael Artin. *Algebra*. Prentice Hall, Inc., Englewood Cliffs, NJ, 1991.

[5] A. Baker. *Representations of finite groups*. AMS Open Math Notes, version one, 2/07/2019, www.ams.org/open-math-notes/files/course-material/OMN-202003-110820-1-Course notes-v1.pdf, 2019.

[6] A. Baker. *An introduction to Galois theory*. AMS Open Math Notes, version one, 8/03/2020, www.ams.org/open-math-notes/files/course-material/OMN-202003-110819-1-Course notes-v1.pdf, 2020.

[7] A. Baker. *Matrix groups: An introduction to Lie group theory*. Springer Undergraduate Mathematics Series, Springer-Verlag, London, 2002.

[8] Richard Brauer. *Über die Darstellung von Gruppen in Galoisschen Feldern*. Hermann & Cie., Paris, 1935.

[9] W. Burnside. On groups of order $p^{\alpha} q^{\beta}$. *Proc. Lond. Math. Soc. (2)*, **1**:388–392, 1904.

[10] W. Burnside. *Theory of groups of finite order*. Cambridge University Press, London, UK, second edition, 1911. First edition, 1897.

[11] A. L. Cauchy. *Mémoire sur les arrangements que l'on peut former avec des lettres données, et sur les permutations ou substitutions à l'aide desquelles on passe d'un arrangement à un autre*. Paris, 1844. In: *Exercises d'analyse et de physique matheématique*, tome III.

[12] A. Cayley. On the theory of groups as depending on the symbolic equation $\theta^n = 1$. *Philos. Mag.*, **7**:40–47, 1854.

[13] N. G. Chebotarev. Die Bestimmung der Dichtigkeit einer Menge von Primzahlen, welche zu einer gegebenen Substitutionsklasse gehören. *Math. Ann.*, **95**:191–228, 1925.

[14] Keith Conrad. The origin of representation theory. *Expo. Math.*, **41**:361–392, 1998.

[15] J. H. Conway, R. T. Curtis, S. P. Norton, R. A. Parker, and R. A. Wilson. *Atlas of finite groups. Maximal subgroups and ordinary characters for simple groups. With comput. assist. from J. G. Thackray.* Clarendon Press, Oxford, XXXIII, 1985.

[16] David A. Cox. *Galois theory*. Pure and Applied Mathematics. John Wiley & Sons, Inc., Hoboken, NJ, second edition, 2012.

[17] H. S. M. Coxeter. *Regular polytopes*. Dover Publications, Inc., New York, third edition, 1973. First edition, 1949, Mehuen, London; second edition, 1963, Collier-MacMillan, London.

[18] Charles W. Curtis. *Pioneers of representation theory: Frobenius, Burnside, Schur, and Brauer*, volume 15 of *History of Mathematics*. American Mathematical Society, Providence, RI; London Mathematical Society, London, 1999.

[19] Charles W. Curtis and Irving Reiner. *Methods of representation theory. Vol. I.* Wiley Classics Library. John Wiley & Sons, Inc., New York, 1990. With applications to finite groups and orders, Reprint of the 1981 original, A Wiley-Interscience Publication.

[20] R. Dedekind. Beweis für die Irreductibilität der Kreistheilungs-Gleichungen. *J. Reine Angew. Math.*, **54**:27–30, 1857.

[21] Max Deuring. Galoissche Theorie und Darstellungstheorie. *Math. Ann.*, **107**:140–144, 1932.

[22] Leonard Eugene Dickson. *Algebras and their arithmetics*. Univ. of Chicago Press, Chicago, 1923.

[23] Underwood Dudley. *Elementary number theory*. W. H. Freeman and Co., San Francisco, CA, second edition, 1978. Reprinted by Dover Publications, Mineola, NY, 2008.

[24] David S. Dummit and Richard M. Foote. *Abstract algebra*. John Wiley & Sons, Inc., Hoboken, NJ, third edition, 2004.

[25] Harold M. Edwards. *Galois theory*, volume 101 of *Graduate Texts in Mathematics*. Springer-Verlag, New York, 1984.

[26] G. Eisenstein. Über die Irreductibilität und einige andere Eigenschaften der Gleichung, von welcher die Theilung der ganzen Lemniscate abhängt. I. *J. Reine Angew. Math.*, **39**:160–179, 1850.

[27] Michael D. Fried and Moshe Jarden. *Field arithmetic*, volume 11 of *Ergebnisse der Mathematik und ihrer Grenzgebiete. 3. Folge. A Series of Modern Surveys in Mathematics*. Springer-Verlag, Berlin, second edition, 2005.

[28] P. G. Frobenius. Neuer Beweis des Sylow'schen Satzes. *J. Reine Angew. Math.*, **100**:179–181, 1886.

[29] F. G. Frobenius. Über Beziehungen zwischen den Primidealen eines algebraischen Körpers und den Substitutionen seiner Gruppe. *S'ber. Preuß. Akad. Berlin*, pages 689–703, 1896. In Gesammelte Werke, volume II, 719–733, Springer, Berlin, 1968.

[30] Lisl Gaal. *Classical Galois theory with examples*. Markham Publishing Co., Chicago, IL, 1971.

[31] Joseph A. Gallian. *Contemporary abstract algebra*. Chapman and Hall/CRC, 10th edition, 2020.

[32] C.-F. Gauss. *Demonstratio nova theorematis omnen functionem algebraicam rationalem intgram unius variabilis in factores reales primi vel secundi gradus resolvi posse*. Apud C.G. Fleckeisen, Helmstadii, 1799, also in Werke, Bd. III, Georg Olms, Hildesheim, 1981, 1–30.

[33] C.-F. Gauss. *Disquisitiones Arithmeticae*. Königliche Gesselschaft der Wissenschaften Göttingen, 1863. In Werke, 1801, Band 1.

[34] J. J. Gray. A commentary on Gauss's mathematical diary, 1796–1814, with an English translation. *Expo. Math.*, **2**:97–130, 1984.

[35] Werner Greub. *Linear algebra*, volume 23 of *Graduate Texts in Mathematics*. Springer-Verlag, New York–Berlin, fourth edition, 1975.

[36] Robert L. Griess. A construction of F_1 as automorphisms of a 196,883-dimensional algebra. *Proc. Natl. Acad. Sci. USA*, **78**:689–691, 1981.

[37] Brian C. Hall. *Quantum theory for mathematicians*, volume 267 of *Graduate Texts in Mathematics*. Springer, New York, 2013.

[38] G. H. Hardy and E. M. Wright. *An introduction to the theory of numbers*. Oxford University Press, Oxford, sixth edition, 2008. Revised by D. R. Heath-Brown and J. H. Silverman, With a foreword by Andrew Wiles.

[39] Thomas Hawkins. The origins of the theory of group characters. *Arch. Hist. Exact Sci.*, **7**(2):142–170, 1971.

[40] Thomas Hawkins. Hypercomplex numbers, Lie groups, and the creation of group representation theory. *Arch. Hist. Exact Sci.*, **8**:243–287, 1972.

[41] Thomas Hawkins. New light on Frobenius' creation of the theory of group characters. *Arch. Hist. Exact Sci.*, **12**:217–243, 1974.

[42] Thomas Hawkins. *The mathematics of Frobenius in context: a journey through 18th to 20th century mathematics*. Sources and Studies in the History of Mathematics and Physical Sciences. Springer, New York, 2013.

[43] C. Hermite. Sur la résoultion de l'équation du cinquième degré. *Compt. Rend.*, **46**:508–515, 1858.

[44] Kenneth Hoffman and Ray Kunze. *Linear algebra*. Prentice-Hall, Inc., Englewood Cliffs, NJ, second edition, 1971.

[45] I. Martin Isaacs. *Character theory of finite groups*. Dover Publications, Inc., New York, 1994. Corrected reprint of the 1976 original: Academic Press, New York.

[46] I. Martin Isaacs. *Finite group theory*, volume 92 of *Graduate Studies in Mathematics*. American Mathematical Society, Providence, RI, 2008.

[47] K. Iwasawa. Über die Einfachheit der speziellen projektiven Gruppen. *Proc. Imp. Acad. Japan*, **17**:57–59, 1941.

[48] Nathan Jacobson. *Basic algebra. I*. W. H. Freeman and Company, New York, second edition, 1985.

[49] Gordon James and Martin Liebeck. *Representations and characters of groups*. Cambridge University Press, New York, second edition, 2001.

[50] Christian U. Jensen, Arne Ledet, and Noriko Yui. *Generic polynomials: constructive aspects of the inverse Galois problem*, volume 45 of *Mathematical Sciences Research Institute Publications*. Cambridge University Press, Cambridge, 2002.

[51] Arthur Jones, Sidney A. Morris, and Kenneth R. Pearson. *Abstract algebra and famous impossibilities*. Universitext. Springer-Verlag, New York, 1991.

[52] B. Melvin Kiernan. The development of Galois theory from Lagrange to Artin. *Arch. Hist. Exact Sci.*, **8**(1–2):40–154, 1971.

[53] Felix Klein. *Lectures on the icosahedron and the solution of equations of the fifth degree*. Dover Publications, Inc., New York, NY, revised edition, 1956. Translated into English by George Gavin Morrice.

[54] Yvette Kosmann-Schwarzbach. *Groups and symmetries: from finite groups to Lie groups*. Universitext. Springer, New York, 2010. Translated from the 2006 French 2nd edition by Stephanie Frank Singer.

[55] L. Kronecker. Grundzüge einer arithmetischen Theorie der algebraischen Grössen. (Festschrift zu Herrn Ernst Eduard Kummers fünfzigjährigem Doctor-Jubiläum, 10 September 1881). *J. Reine Angew. Math.*, **92**:1–122, 1882.

[56] D. C. Lay. *Linear algebra and its applications*. Pearson, London, UK, fourth edition, 2011.

[57] P. G. Lejeune-Dirichlet. *Vorlesungen über Zahlentheorie. Hrsg. und mit Zusätzen versehen von R. Dedekind. 4. umgearb. u. verm. Aufl.* F. Vieweg u. Sohn, Braunschweig, 1894.

[58] Rudolf Lidl and Harald Niederreiter. *Finite fields*, volume 20 of *Encyclopedia of Mathematics and Its Applications*. Cambridge University Press, Cambridge, second edition, 1997. With a foreword by P. M. Cohn.

[59] E. Lucas. Sur les congruences des nombres eulériens et les coefficients différentiels des functions trigonométriques suivant un module premier. *Bull. Soc. Math. France*, **6**:49–54, 1878.

[60] Jesper Lützen. Why was Wantzel overlooked for a century? The changing importance of an impossibility result. *Hist. Math.*, **36**:374–394, 2009.

[61] James H. McKay. Another proof of Cauchy's group theorem. *Amer. Math. Monthly*, **66**:119, 1959.

[62] J. S. Milne. *Fields and Galois theory*. Version 4.61, April 2020, www.jmilne.org/math/CourseNotes/ft.html, 2020.

[63] J. S. Milne. *Group theory*. Notes from March 20, 2020, www.jmilne.org/math/CourseNotes/GT.pdf, 2020.

[64] Oliver Nash. On Klein's icosahedral solution of the quintic. *Expo. Math.*, **32**(2):99–120, 2014.

[65] Peter M. Neumann. *The mathematical writings of Évariste Galois*. Heritage of European Mathematics. European Mathematical Society (EMS), Zürich, 2011.

[66] Emmy Noether. Gleichungen mit vorgeschriebener Gruppe. *Math. Ann.*, **78**:221–229, 1917.

[67] Emmy Noether. Hyperkomplexe Größen und Darstellungstheorie. *Math. Z.*, **30**:641–692, 1929.

[68] Emmy Noether. Normalbasis bei Körpern ohne höhere Verzweigung. *J. Reine Angew. Math.*, **167**:147–152, 1932.

[69] Derek J. S. Robinson. *A course in the theory of groups*, volume 80 of *Graduate Texts in Mathematics*. Springer-Verlag, New York, second edition, 1996.

[70] Steven Roman. *Field theory*, volume 158 of *Graduate Texts in Mathematics*. Springer, New York, second edition, 2006.

[71] Joseph J. Rotman. *Advanced modern algebra*, volume 114 of *Graduate Studies in Mathematics*. American Mathematical Society, Providence, RI, second edition, 2010. First edition Prentice Hall Inc., NJ, 2002.

[72] I. Schur. Neue Begründung der Theorie der Gruppencharaktere. *Berl. S'ber.*, **1904**:406–432, 1904.

[73] Jean-Pierre Serre. *Linear representations of finite groups*, volume 42 of *Graduate Texts in Mathematics*. Springer-Verlag, New York–Heidelberg, 1977. Translated from the second French edition by Leonard L. Scott.

[74] Jean-Pierre Serre. *Topics in Galois theory*, volume 1 of *Research Notes in Mathematics*. A K Peters, Ltd., Wellesley, MA, second edition, 2008. With notes by Henri Darmon.

[75] Jean-Pierre Serre. *Finite groups: an introduction*, volume 10 of *Surveys of Modern Mathematics*. International Press, Somerville, MA; Higher Education Press, Beijing, 2016. With assistance in translation provided by Garving K. Luli and Pin Yu.

[76] Louis W. Shapiro. Finite groups acting on sets with applications. *Math. Mag.*, **46**:136–147, 1973.

[77] Jerry Shurman. *Geometry of the quintic*. A Wiley-Interscience Publication, John Wiley & Sons, Inc., New York, 1997.

[78] Stephanie Frank Singer. *Linearity, symmetry, and prediction in the hydrogen atom*. Undergraduate Texts in Mathematics. Springer, New York, 2005.

[79] Leonard Soicher and John McKay. Computing Galois groups over the rationals. *J. Number Theory*, **20**:273–281, 1985.

[80] Benjamin Steinberg. *Representation theory of finite groups: an introductory approach*. Universitext. Springer, New York, 2012.

[81] Ernst Steinitz. Algebraische theorie der körpern. *J. Crelle*, **137**:167–309, 1910.

[82] John Swallow. *Exploratory Galois theory*. Cambridge University Press, Cambridge, 2004.

[83] L. Sylow. Théorèmes sur les groupes de substitutions. *Math. Ann.*, **5**:584–594, 1872.

[84] A. D. Thomas and G. V. Wood. *Group tables*, volume 2 of *Shiva Mathematics Series*. Shiva Publishing Ltd., Nantwich; distributed by Birkhäuser Boston, Inc., Cambridge, MA, 1980.

[85] J.-P. Tignol. *Galois' theory of algebraic equations*. World Scientific Publishing Co., Singapore, 2001.

[86] Gabor Toth. *Glimpses of algebra and geometry*. Undergraduate Texts in Mathematics. Springer-Verlag, New York, second edition, 2002. Readings in Mathematics.

[87] Laura Toti Rigatelli. *Evariste Galois 1811–1832*, volume 11 of *Vita Math.* Birkhäuser, Basel, 1996. Transl. from the Italian by John Denton.

[88] B. L. van der Waerden. *Modern algebra. Vol. I*. Frederick Ungar Publishing Co., New York, 1949. Translated from the 2nd revised German edition by Fred Blum, with revisions and additions by the author.

[89] B. L. van der Waerden. Die Galois-Theorie von Heinrich Weber bis Emil Artin. *Arch. Hist. Exact Sci.*, **9**(3):240–248, 1972.

[90] P. L. Wantzel. Recherches sur les moyens de reconnaître si un problème de géométrie peut se résoudre avec la règle et le compas. *J. Math. Pures Appl.*, pages 366–372, 1837.

[91] Lawrence C. Washington. *Introduction to cyclotomic fields*, volume 83 of *Graduate Texts in Mathematics*. Springer-Verlag, New York, second edition, 1997.

[92] Heinrich Weber. *Lehrbuch der Algebra. Zweite Auflage. Dritter Band: Elliptische Funktionen und algebraische Zahlen.* F. Vieweg & Sohn, Braunschweig, 1898.

[93] J. H. M. Wedderburn. On hypercomplex numbers. *Proc. Lond. Math. Soc. (2)*, **6**:77–118, 1908.

[94] S. H. Weintraub. *Galois theory*. Springer-Verlag, New York, NY, second edition, 2009.

[95] S. H. Weintraub. Several proofs of the irreducibility of the cyclotomic polynomials. *Amer. Math. Monthly*, **120**:537–545, 2013.

[96] Hermann Weyl. *The theory of groups and quantum mechanics*. Dover Publications, Inc., New York, 1950. Translated from the second (revised) German edition by H. P. Robertson, Reprint of the 1931 English translation.

[97] Hans Wussing. *The genesis of the abstract group concept: a contribution to the history of the origin of abstract group theory.* MIT Press, Cambridge, MA, 1984. Translated from the German by Abe Shenitzer and Hardy Grant.

Index

227

Printed in the United States
by Baker & Taylor Publisher Services